Regulating Chemical Risks

Johan Eriksson · Michael Gilek
Christina Rudén
Editors

Regulating Chemical Risks

European and Global Challenges

Springer

Editors
Johan Eriksson
Department of Political Science
School of Social Sciences
Södertörn University
Alfred Nobels allé 7
14189 Flemingsberg
Sweden
johan.eriksson@sh.se

Michael Gilek
School of Life Sciences and Centre
for Baltic and East European Studies
Södertörn University
Alfred Nobels allé 7
14189 Flemingsberg
Sweden
michael.gilek@sh.se

Christina Rudén
Royal Institute of Technology (KTH)
Teknikringen 78B
SE 100 44, Stockholm
Sweden
cr@abe.kth.se

ISBN 978-90-481-9427-8 e-ISBN 978-90-481-9428-5
DOI 10.1007/978-90-481-9428-5
Springer Dordrecht Heidelberg London New York

Library of Congress Control Number: 2010933855

Printed on acid-free paper

Springer is part of Springer Science+Business Media (www.springer.com)

Preface

This volume presents research on current trends in chemical regulations – a fast-growing, complex, and increasingly internationalized field. The book grew out from a multidisciplinary research project entitled 'Regulating Chemical Risks in the Baltic Sea Area: Science, Politics, and the Media', led by Michael Gilek at Södertörn University, Sweden. This research project involved scholars and experts from natural as well as social sciences, based at Södertörn University, Swedish Royal Institute of Technology (KTH), Karolinska Institutet, and Umeå University. The project group organized a multidisciplinary research conference on chemical risk regulations, held in Stockholm, August 15–17, 2007. Most of the contributions published in this book were, in draft form, first presented at this conference. The conference, like the ensuing edited volume, expanded the geographical focus beyond the Baltic Sea area to include wider European, and to some extent also global trends. Many thanks to all project colleagues and conference participants!

We are very grateful for the generous financial support received from The Foundation for Baltic and East European Studies (Östersjöstiftelsen), The Swedish Research Council Formas, and from Södertörn University. Without this support the present book would not have been possible.

Special thanks to all of our fellow contributors, all of whom have submitted topical papers based on high-quality research. Many thanks also to Tobias Evers, who assisted us with technical editing. Finally, we are grateful for the professionalism shown by our editors at Springer.

Södertörn University, Huddinge, Sweden — Johan Eriksson
Södertörn University, Huddinge, Sweden — Michael Gilek
Swedish Royal Institute of Technology, Stockholm, Sweden — Christina Rudén

Contents

Contributors

Anderson Alison
University of Plymouth, School of Applied Psychosocial Sciences,
Faculty of Health, Drake Circus, Plymouth, Devon, PL4 8AA, UK
aanderson@plymouth.ac.uk

Andonova Liliana B
Graduate Institute of International and Development Studies,
Case postale 136 - 1211, Genève 21 - Suisse
Liliana.Andonova@graduateinstitute.ch

Bal Roland
Institute of Health Policy and Management Erasmus University,
P.O. Box 1738, 3000 DR Rotterdam, The Netherlands
r.bal@bmg.eur.nl

Bengtsson Bengt-Erik
Department of Applied Environmental Science, Stockholm University, SE 106 91,
Stockholm, Sweden
e-mail: bengt-erik.bengtsson@itm.su.se

Bengtsson Gunnar
Luruddsvägen 32, Ekero, SE-17854, Sweden
gunnarbengt@telia.com

Breitholz Magnus
Department of Applied Environmental Science, Stockholm University,
SE 106 81, Stockholm, Sweden
magnus.breitholtz@itm.su.se

Bucht Bengt
Swedish Chemicals Agency, P.O. Box 2, SE-172 13, Sundbyberg, Sweden
bengt.bucht@kemi.se

Castaño Argelia
Centro Nacional de Sanidad Ambiental Instituto de Salud Carlos III,
Ctra. Majadahonda – Pozuelo, 28220, Majadahonda Madrid, Spain
castano@isciii.se

Dahl Ulrika
Department of Applied Environmental Science, Stockholm University,
SE 106 81, Stockholm, Sweden
ulrika.dahl@itm.su.se

Eklund Britta
Department of Applied Environmental Science, Stockholm University,
SE 106 81, Stockholm, Sweden
britta.eklund@itm.su.se

Eriksson Johan
Södertörn University, SE 141 89, Huddinge, Sweden
johan.eriksson@sh.se

Forbes Valery
Roskilde University, Universitetsvej 1, P.O. Box 260, DK 4000,
Roskilde, Denmark
vforbes@ruc.dk

Gilek Michael
Södertörn University, SE 141 89, Huddinge, Sweden
michael.gilek@sh.se

Greim Helmut
Technische Universität München, Institut für Toxikologie und Umwelthygiene,
85350, Freising-Weihensteph, Munich
helmut.greim@lrz.tu-muenchen.de

Halffman Willem
Department of Philosophy and Science Studies, Radboud University Nijmegen,
P.O. Box 9010, 6500 GL Nijmegen, The Netherlands
w.halffman@gmail.com

Hansson Sven Ove
KTH, Swedish Royal Institute of Technology, Skolan för arkitektur och
samhällsbyggnad, Avdelningen för filosofi, Drottning Kristinas väg 30,
SE 100 44, Stockholm, Sweden
soh@abe.kth.se

Heyvart Veerle
Department of Law, London School of Economics (LSE),
Houghton Street, London, WC2A 2AE, United Kingdom
v.heyvaert@lse.ac.uk

Johansson Vicki
School of Public Administration, University of Gothenburg, P.O. Box 100,
SE 405 30, Göteborg, Sweden
vicki.johansson@spa.gu.se

Karlsson Jenny
Department of Applied Environmental Science, Stockholm University,
SE 106 81, Stockholm, Sweden
jenny.karlsson@itm.su.se

Karlsson Mikael
School of Life Sciences, Södertörn University, SE 141 89, Huddinge, Sweden
mikael.karlsson@naturskyddsforeningen.se

Klinke Andreas
Eawag, Innovation Research in Utility Sectors,
Überlandstrasse 133, P.O. Box 611, 8600,
Dübendorf, Switzerland
andreas.klinke@eawag.ch

Lundström Elin
Department of Applied Environmental Science,
Stockholm University, SE 106 81, Stockholm, Sweden
elin.lundstrom@itm.sue.se

Ögren Kenneth
Umeå University, Institutionen för kultur- och medievetenskaper,
SE 901 87, Umeå, Sweden
kenneth.ogren@kultmed.umu.se

Pärt Peter
Institute for Environment and Sustainability,
European Commission Joint Research Center, Via E. Fermi 2749, I-21027,
Ispra (VA), Italy
peter.part@jrc.ec. europa. eu

Renn Ortwin
Universität Stuttgart, Institut für Sozialwissenschaften,
Abteilung für Technik- und Umweltsoziologie, Seidenstr. 36,
D-70174, Stuttgart, Germany
sekretariat.renn@sowi.uni-stuttgart.de

Reuter Marta
Department of Political Science, Stockholm University,
SE 106 91, Stockholm, Sweden
marta.reuter@statsvet.su.se

Rudén Christina
Royal Institute of Technology (KTH)
Teknikringen 78B, SE 100 44, Stockholm, Sweden
cr@abe.kth.se

Schenk Linda
Södertörn University, SE 141 89, Huddinge, Sweden
linda.schenk@sh.se

Sjölander Annika Egan
Umeå University, Institutionen för kultur- och medievetenskaper, SE 901 87, Umeå, Sweden
annika.egansjolander@kultmed.umu.se

Wolanik Katarzyna Boström
Umeå University, Institutionen för kultur- och medievetenskaper, SE 901 87, Umeå, Sweden
katarzyna@kultmed.umu.se

Chapter 1
Introduction

Michael Gilek, Johan Eriksson, and Christina Rudén

In this introductory chapter we wish to acquaint the reader briefly with the background, scope and structure of the book as well as to provide some pointers for how best to read and use this book. We also highlight some important issues and challenges connected with chemical risk regulation that are addressed by our contributors.

This book deals with some of the most significant developments in risk regulation witnessed in history. Furthermore, it addresses this multifaceted topic by adopting an unprecedented multidisciplinary approach and by focussing on both the multi-level (i.e. national, European and international) and multi-actor (e.g. scientific experts, decision-makers, inspectors, journalists, etc.) interactions that nurture these developments. Naturally, a complex field such as this, encompassing topics spanning from toxicological and ecological assessments of risks, to risk communication to global trends in chemical safety and general challenges for risk governance, cannot be comprehensively covered in a single book. We believe, however, that the following 19 chapters – written by acknowledged scientists, scholars and practitioners – provide the reader with a broad, up-to-date and multidisciplinary analysis of chemical risk regulation as well as a useful overview of significant processes and actors shaping chemical risk regulation. Several chapters contribute in-depth discussions of challenges posed by, and improvements required for, developing the efficiency, acceptability, sustainability and transparency of the assessment, communication and management of chemical risks. A number of the chapters focus on the development of (and the challenges posed by) the REACH[1] legislation in the European Union. In order to achieve the overall aim of establishing a general

[1]Registration, Evaluation, Authorisation and Restriction of Chemicals.

M. Gilek (✉) and J. Eriksson
Södertörn University, SE 141 89, Huddinge, Sweden
e-mail: michael.gilek@sh.se; johan.eriksson@sh.se

C. Rudén
Royal Institute of Technology (KTH),
Teknikringen 78B, SE 100 44, Stockholm, Sweden
e-mail: cr@abe.kth.se

J. Eriksson et al. (eds.), *Regulating Chemical Risks: European and Global Challenges*,
DOI 10.1007/978-90-481-9428-5_1, © Springer Science+Business Media B.V. 2010

understanding of how various processes interact in chemical risk regulation several chapters go beyond REACH (and the European Union) by covering other important topics, such as workplace safety, chemicals in products, international perspectives, risk governance, risk assessment, risk communication in the news media and the role of influential actors, including journalists, expert committees and other bureaucrats such as inspectors.

Risk regulation is a multidisciplinary field of study rapidly growing both in size and significance, which reflects a mounting concern for health and environmental risks in modern society and measures taken to prevent them. Nowhere is this trend more visible than within the domain of risks connected with chemical substances. One example of recent developments that emphasize the great need for analysing the nature, causes and consequences of chemical risk regulation is the REACH legislation, which is unprecedented in scope and complexity. Moreover, the 'reach' of REACH extends far beyond the member states of the European Union, not only to states seeking membership or whose chemical industry depends on access to the European market (Chapter 15 by Andonova) but also to global risk regulation (Chapter 13 by Heyvart). Consequently, several international developments, such as the Strategic Approach to International Chemicals Management (SAICM) and the Globally Harmonized System of Classification and Labelling of Chemicals (GHS), are currently reforming the global regulation of chemical risks (Chapter 12 by G. Bengtsson). Moreover, one chapter (Chapter 14 by Karlsson) provides a comparative analysis of regulatory policies in the U.S. and the European Union. The development towards international chemical risk regulation is fuelled by globalization of chemical production and trade and increasing production and demand of chemicals in developing countries with less stringent chemical regulation as well as by a growing awareness of the potential of many chemicals released into the environment to disseminate across vast geographical areas and constitute environmental and health risks far away from their source of production and use. Although internationalization can be seen as a necessary and positive development in chemical risk regulation, several problems and challenges remain.

The chemical industry today is a diverse and growing global business producing a multitude of substances and chemical products, many of which are used in other products such as pesticides, pharmaceuticals, solvents, paints, plastics, toys and electrical equipment. Although it is clear that there are substantial benefits connected with chemical use in terms of, for example, better products and more efficient processes, there are also obvious risks connected with the often hazardous and toxic properties of the chemicals used. Progress in the regulation and handling of these risks has been made in several areas connected with, for example, workplace and food safety as well as protection of the environment and public health. However, several problems and challenges remain to be sufficiently addressed. Consequently, modern societies are in urgent need of the development and implementation of a more sustainable regulation of chemical risks.

According to Klinke and Renn (Chapter 2), the design and development of efficient and sustainable regulation of risks of this kind need to be based on appropriate and legally prescribed procedures, sound scientific advice and widely acceptable trade-offs between benefits and risks. Based on the multidisciplinary analyses presented in this book, all of these prerequisites for good risk governance are in need of further improvement in relation to current chemical risk regulation. First of all, international chemical risk regulation has become complex, with around 100 international agreements, programmes and initiatives (see Chapter 12 by G. Bengtsson). Increasing complexity of regulatory arrangements implies high implementation costs, which is problematic, especially in developing countries (Chapter 16 by Bucht).

Secondly, complex international agreements take time to develop (e.g. as can be seen in the case of GHS). The time scale of development and implementation of international risk reduction measures – which often involves several decades – is often similar to the time scale of significant risk development following changing international production and use of chemicals. Furthermore, there are tens of thousands of industrial chemicals produced and used globally (e.g. 143,000 substances have been pre-registered within REACH). The resources required to generate data for so many substances comprise substantial barriers to the implementation of well-informed chemical risk management (Chapter 5 by Hansson and Rudén). More efficient approaches for describing and dealing with uncertainties and lack of data are also required in both risk assessment and risk management (Chapter 10 by Rudén and Gilek), as are approaches for increasing the ecological relevance of testing strategies (Chapter 6 by Breitholtz et al.) and methods for reducing the use of animals in chemical testing (Chapter 7 by B-E. Bengtsson et al.). Finally, one conclusion is that chemical risk regulation is to a large extent an expert driven endeavour, in which much of the knowledge concerning the severity and probability of potential negative effects is generated through natural science-based risk assessments and experts committees (Chapter 17 by Eriksson et al.). To counteract this lack of multidisciplinary perspective, chemical regulation would benefit from more consideration and assessment of concerns and actor interactions from a social science perspective than is the case today (see Chapters 3, 4 and 18 for discussions of the role of journalists and inspectors).

The book is organized in two main parts: Part I: Chemical Risks: Assessment and Communication and Part II: Chemicals Regulation: Politics, Policy and Management. Although every chapter can be read independently, we recommend the committed reader to read the chapters in each of the main parts in the order they appear. The chapters have been ordered in such a way that earlier chapters within each of the main parts are intended to facilitate the reading of later chapters in their respective parts of the book.

Part I begins by presenting a model for risk governance as well as discussing the contemporary and future challenges of risk governance (Chapter 2 by Klinke and Renn). Anderson (Chapter 3) proceeds by discussing how the media has framed chemical risks in particular cases, such as oil spills, and argues for the need for a more complex understanding of risk reporting reflecting both the nature of the 'risk' and the structure and workings of the media. Chemical risk communication is then further scrutinized by Egan Sjölander, Wolanik Boström and Ögren (Chapter 4) in a study of how chemical risks are framed by journalists in Swedish and Polish newspapers.

The remaining chapters in Part I focus on various specific challenges connected with the assessment of chemical risks. In Chapter 5 Hansson and Rudén highlight and discuss the major challenges posed by the substantial lack of toxicological data on chemicals at present. In the following chapter Breitholtz, Dahl and Forbes (Chapter 6) discuss how to increase the value and ecological relevance of standard toxicity tests by modelling the influence of ecological and environmental factors. In Chapter 7, Bengtsson, Castaño Calvo and Pärt also discuss toxicity testing in REACH but from a quite different perspective, namely how the 3R approach (i.e. a strategy to reduce the number of animals used in experiments) can be applied in ecotoxicity testing of chemicals. In Chapter 8 Greim introduces the fundamentals of chemical risk assessment from a toxicological perspective. Schenk continues in Chapter 9 on the theme of human health risks by presenting observations of differences and trends in occupational health limits among European countries, indicating significant differences in risk assessment and risk characterisation among expert organisations and national authorities. In the last two chapters of Part I Gilek and Rudén (Chapter 10) first use a case study of the risk assessment of three brominated flame retardants (i.e. penta-, octa- and decabromodiphenyl ethers) to discuss and problematize how scientific uncertainties are described and dealt with in the risk assessment of persistent and bioaccumulating chemicals. Eklund and Karlsson (Chapter 11) then use a case study of antifouling paints to highlight the need for assessing and evaluating not only the risks of single chemical substances but also those of complex chemical products.

Part II opens with a comprehensive overview of current global trends in chemical risk regulation (Chapter 12 by G. Bengtsson) and continues by exploring REACH in a global governance perspective (Chapter 13 by Heyvaert), followed by a comparative analysis of chemical risk regulation in the U.S. and in the European Union (Chapter 14 by Karlsson) and in Central and Eastern Europe (Chapter 15 by Andonova) as well as general requirements for countries that wish to increase their capacity for chemicals control (Chapter 16 by Bucht). Chapter 17 (by Eriksson, Karlsson and Reuter) discusses interactions between experts and decision-makers through a case study of SCHER – one of the scientific committees of the European Union. Ending Part II, Vicki Johansson (Chapter 18) studies the importance of the implementation of chemical risk regulation and, in particular, the vital role of inspectors.

In a concluding chapter (Chapter 19), a number of synthesizing observations are made, particularly with regard to the scenarios Bal and Halffman discussed in their 1998 book *The Politics of Chemical Risk: Possible Regulatory Futures* (Kluwer Academic) – an anthology that in many ways is a precursor to the present book. Trends of globalization regarding both risk development and policies for dealing with them were foreseen in the 1998 volume and are corroborated in the present book. Moreover, Halffman and Bal conclude that what they some 10 years ago called the 'International Expert Scenario' is the one that has come closest to a realization within Europe. They also make critical comments regarding the increasing complexity in chemical regulations, which is problematic in terms of transparency, and ultimately for the democratic legitimacy of regulation.

From a general risk governance perspective (as discussed in detail in Chapter 2 by Klinke and Renn), one conclusion of the book is that sustainable governance of chemical risk rests on three components: (i) knowledge in terms of scientific assessments, etc.; (ii) legally prescribed procedures; and (iii) social values. The structural arrangements of chemical risk regulation, thus, need to be developed in order to address these components and to facilitate their interaction. For example, risk appraisals need to be broadened not only to include natural science based risk assessments of what is known but also to incorporate assessments drawing on social science expertise. Approaches for improving the assessment and managing of uncertainties need to be developed. To avoid overlaps, duplication and unnecessary regulatory complexity, further coordination of international chemical regulation through, for example, the UN system and the OECD are required.

Furthermore, it is clear that the REACH legislation will be of paramount importance for the future governance of chemicals in Europe and internationally. Several issues connected with REACH and its future efficiency in promoting sustainable management of chemicals are raised and discussed in this book, yet they require further attention by both scientists and decision-makers. Above all, the substantial lack of toxicological data at present for most of the chemicals being used needs to be more properly addressed than is the case today by, for example, increasing data requirements for low volume chemicals (1–10 tonnes) and by including requirements and regulations for chemicals found in various products.

As a final word we argue that, in order to guarantee that chemical risk assessments and risk management measures effectively serve their purposes, it is important to reach a thorough understanding of the factors and processes that affect chemical risk regulation in terms of efficiency, acceptability, sustainability and transparency. In this respect we believe that this book provides an important contribution to the scientific understanding of chemical risk regulation by offering a coherent, comprehensive and updated multidisciplinary analysis. It is our hope that this book will spark and encourage future (multidisciplinary) research on the regulation of chemical risks and ultimately contribute to improved chemical risk regulation and management in Europe and elsewhere in the world.

Part I
Chemical Risk Assessment and Risk Communication

Chapter 2
Risk Governance: Contemporary and Future Challenges

Andreas Klinke and Ortwin Renn

Abstract The chapter will develop a general concept for integrative risk governance emphasizing procedural and structural mechanisms as well as precaution-oriented considerations. Key terms used in this chapter refer to seriousness, complexity, scientific uncertainty and socio-political ambiguity; the application of precaution in risk handling; the handling of risk issues that are subject to strongly divergent cultural attitudes, political perspectives or economic interests; the quest for more openness and transparency during the entire risk handling process; and the design of effective means and institutional arrangements for stakeholder and public involvement. The integrative concept refers to a set of procedural elements, which first of all embraces the classic components of risk analysis: pre-assessment, appraisal, and management. A further phase, comprising the characterization and evaluation of risk, is placed between the appraisal and management phase. The risk process also includes risk communication as a component that is a necessary complement to all risk phases. The chapter will first introduce the key challenges: seriousness, complexity, uncertainty and ambiguity. Section 2.3 is devoted to the explanation of the IRGC risk governance model and its components: pre-assessment, appraisal, characterization and evaluation, management and communication. The main lessons from using the risk governance model are summarized in Sect. 2.4.

Keywords Complexity • Governance • REACH • Risk • Uncertainty

A. Klinke (✉)
Eawag, Innovation Research in Utility Sectors, Überlandstrasse 133, P.O. Box 611 8600, Dübendorf, Switzerland
e-mail: andreas.klinke@eawag.ch

O. Renn
Universität Stuttgart, Institut für Sozialwissenschaften, Abteilung für Technik- und Umweltsoziologie, Seidenstr. 36, D-70174, Stuttgart, Germany
e-mail: sekretariat.renn@sowi.uni-stuttgart.de

J. Eriksson et al. (eds.), *Regulating Chemical Risks: European and Global Challenges*,
DOI 10.1007/978-90-481-9428-5_2,

2.1 Challenges Posed by Seriousness, Complexity, Uncertainty, and Ambiguity

2.1.1 Seriousness

Seriousness particularly refers to the inherent hazard potential of a risk agent to cause certainly and unambiguously significant harm to the environment or to human health (irrespective of exposure, dose–response relationships or intake quantity). This potential depends on special characteristics of the risk agent under investigation (Mueller-Herold et al. 2005). The new legislative framework on the regulation of chemicals of the European Union provides, for example, a number of specific exposure-based hazard criteria such as carcinogenicity, mutagenicity, toxicity for reproduction, and ecotoxicity, which are generally applicable to chemical threats and other areas as food safety.[1] In analogy to REACH, the Globally Harmonised System of Classification and Labelling of Chemicals (GHS) of the United Nations includes similar references to specific chemical hazards that could cause serious damage if released into the environment. Moreover, serious non-fatal health threats caused by endocrine disruptors, neurotoxins, immunotoxins or sensitising agents fall in this category. Additional criteria such as ubiquity, persistency, bio-accumulation, etc., help to qualify or even quantify the degree of hazard that is associated with a risk agent. Hazards may never materialize over time, if exposure is low or intake below the thresholds of causing any harm. These hazard characteristics may, however, be an excellent guide for setting up an early warning system, if effects are still unknown or ignorance about potential impacts prevails. The so-called 'grasshopper effect', i.e. repeated evaporation and condensation, causes an enhanced persistence and substance accumulation of persistent pollutants in the polar regions where they damage the Eskimo population and the ecosystem there. In other risk areas where there exist robust applicable data, seriousness may be formulated in terms of risk-based thresholds, such as concentrations for certain less hazardous toxicants.

2.1.2 Complexity

Complexity refers to the difficulty of identifying and quantifying causal links between a multitude of potential candidates and specific adverse effects (Renn and Walker 2007; Renn 2008). A crucial in this respect concerns the applicability of probabilistic risk assessment techniques. If the chain of events between a cause and an effect follows a linear relationship (as for example in car accidents, or in an

[1]See the EU-Regulation No 1907/2006 concerning the Registration, Evaluation, Authorisation and Restriction of Chemicals (REACH) and establishing a European Chemicals Agency.

overdose of pharmaceutical products), simple statistical models are sufficient to calculate the probabilities of harm. Such simple relationships may still be associated with high uncertainty, for example, if only few data are available or the effect is stochastic by its own nature. Sophisticated models of probabilistic inferences are required if the relationship between cause and effects becomes more complex. The nature of this difficulty may be traced back to interactive effects among these candidates (synergisms and antagonisms, positive and negative feedback loops), long delay periods between cause and effect, inter-individual variation, intervening variables, and others. It is precisely these complexities that make sophisticated scientific investigations necessary since the dose–effect relationship is neither obvious nor directly observable. Nonlinear response functions may also result from feedback loops that constitute a complex web of intervening variables. Complexity requires therefore sensitivity to non-linear transitions as well as to scale (on different levels). It also needs to take into account a multitude of exposure pathways and the composite effects of other agents that are present in the exposure situation. Examples of highly complex risk include sophisticated chemical facilities, synergistic effects of potentially toxic substances, failure risk of large interconnected infrastructures and risks of critical loads to sensitive ecosystems.

2.1.3 Scientific Uncertainty

Uncertainty is different from complexity, but most often results from an incomplete or inadequate reduction of complexity in modelling cause–effect chains. Whether the world is inherently uncertain is a philosophical question that is not pursued here. It is essential to acknowledge in the context of risk assessment that human knowledge is always incomplete and selective, and, thus, contingent upon uncertain assumptions, assertions and predictions (Functowicz and Ravetz 1992; Laudan 1996; Renn 2008). It is obvious that the modelled probability distributions within a numerical relational system can only represent an approximation of the empirical relational system that helps elucidate and predict uncertain events. It therefore seems prudent to include additional aspects of uncertainty (van Asselt 2000; van der Sluijs et al. 2003). Although there is no consensus in the literature on the best means of disaggregating uncertainties, the following categories appear to be an appropriate means of distinguishing between the key components of uncertainty:

- *Variability* refers to different vulnerability of targets such as the divergence of individual responses to identical stimuli among individual targets within a relevant population such as humans, animals, plants, landscapes, etc.;
- *Inference* effects relate to systematic and random errors in modelling inducing problems of drawing extrapolations or logic deductions from small statistical samples, from animal data or experimental data onto humans or from large doses to small doses, etc. All of these are usually expressed through statistical confidence intervals;

- *Indeterminacy* results from genuine stochastic relationship between cause and effects, apparently non-causal or non-cyclical random events, or badly understood non-linear, chaotic relationships;
- *System* boundaries allude to uncertainties stemming from restricted models and the need for focusing on a limited amount of variables and parameters;
- *Ignorance* means the lack of knowledge about the probability of occurrence of a damaging event and about its possible consequences.

The first two components of uncertainty qualify as epistemic uncertainty and, therefore, can be reduced by improving existing knowledge and advancing current modelling tools. The last three components are genuine uncertainty components and can be characterized, to some extent, by using scientific approaches, but cannot be completely resolved. The validity of the end results is questionable and, for risk management purposes, additional information is needed, such as a subjective confidence level in risk estimates, potential alternative pathways of cause–effect relationships, ranges of reasonable estimates, loss scenarios and others. Examples of high uncertainty include many natural disasters, such as earthquakes, possible health effects of mass pollutants below the threshold of statistical significance, acts of violence – such as terrorism and sabotage – and long-term effects of introducing genetically modified species into the natural environment.

2.1.4 Interpretative and Normative Ambiguity

Interpretative and *normative ambiguity* relates to divergent or contested perspectives on the justification, severity or wider 'meanings' associated with a given threat (Stirling 2003; Renn 2008). *Interpretative ambiguity* denotes the variability of (legitimate) interpretations based on identical observations or data assessments results, e.g. an adverse or non-adverse effect. Variability of interpretation, however, is not restricted to expert dissent. Laypeople's perception of risk often differs from expert judgments because it is also a response to qualitative risk characteristics such as familiarity, personal or institutional control, assignment of blame, and others. Moreover, in contemporary pluralist societies diversity of risk perspectives within and between social groups is generally fostered by divergent value preferences, variations in interests and very few, if any universally applicable moral principles; all the more, if risk problems are complex and uncertain. That leads us to the aspect of *normative ambiguity*. It alludes to different concepts of what can be regarded as tolerable referring e.g. to ethics, quality of life parameters, distribution of risks and benefits, etc. A condition of ambiguity emerges where the problem lies in agreeing on the appropriate values, priorities, assumptions, or boundaries to be applied to the definition of possible outcomes. Examples for high interpretative ambiguity include low dose radiation (ionizing and non-ionizing), low concentrations of genotoxic substances, food supplements and hormone treatment of cattle. Normative ambiguities can be associated, for example, with passive smoking, nuclear power, pre-natal genetic screening and genetically modified food.

Most risks are characterized by a mixture of complexity, uncertainty and ambiguity. Passive smoking may be a good example of low complexity and uncertainty, but high ambiguity. Nuclear energy may be a good candidate for high complexity and high ambiguity, but relatively little uncertainty. Endocrine disrupters could be cited as examples for high complexity, uncertainty and ambiguity.

2.2 Conceptual Design of an Integrative Risk Governance Model

The risk governance concept[2] that we propose consists of four consecutive phases (IRGC 2005):

- Pre-assessment
- Appraisal
- Characterization/evaluation
- Management

Risk communication accompanies all four phases. Each phase specifies activities that constitute important elements for good governance. This simple concept is in line with almost all other competing concepts and ensures the compatibility with professional codices and risk governance legislation. Moreover, it has transformed the linear structure more commonly found in other contemporary conceptions of risk governance into an open, cyclical, iterative and interlinked process, as shown in Fig. 2.1.

The four phases correspond to the two major challenges of risk governance: generating and collecting knowledge about the risk, and making decisions about how to mitigate, control or otherwise manage it. These two challenges are illustrated by the two activities portrayed on the horizontal axis: appraisal and management. However, there are two additional phases in which knowledge and values are closely intertwined: pre-assessment and characterization/evaluation. These two phases are located on the vertical axis and constitute interfaces between knowledge and values. During the phase of pre-assessment, the problem is framed and defined, and the terms of reference are specified. This task needs to be governed by societal values (stating the goals, objectives and contextual conditions) and inspired by what we already know about the hazard (suspected impacts, exposure, persistence and others) (Zinn and Taylor-Gooby 2006). Similarly, when looking at all the evidence collected and condensed in the phase of characterization/tolerability judgement, a good understanding of this evidence, as well as a prudent judgement competence for making the necessary trade-offs between risk, benefits and other important impact categories, is

[2]The following concept of integrative risk governance collects, condenses and re-interprets different ideas, modules, project results, and publications on which the authors have worked on over the last decade. See e.g. WBGU (2000); Klinke and Renn (1999, 2002); Klinke et al. (2006); IRGC (2005); Renn and Walker (2007); Renn (2008).

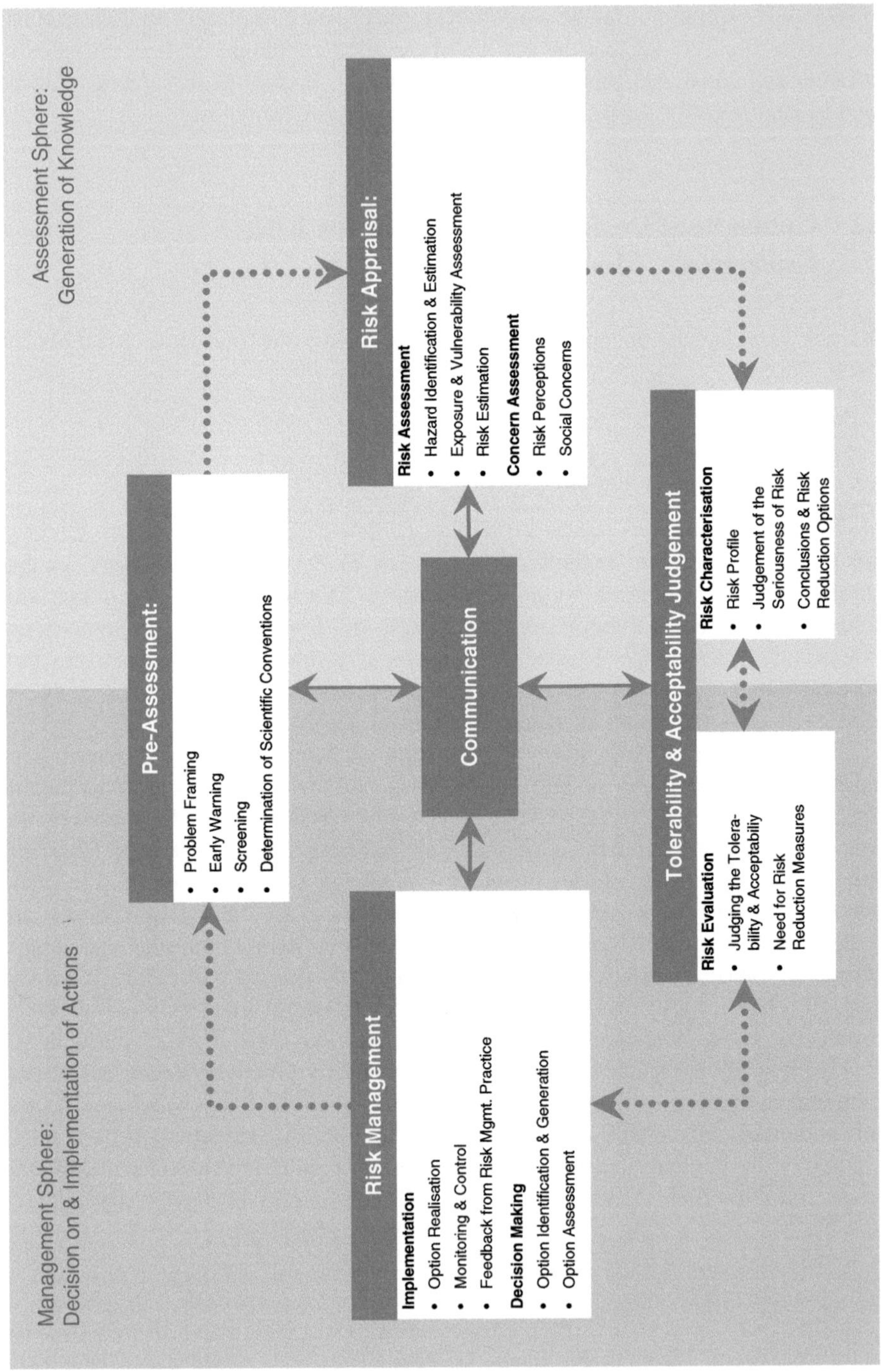

Fig. 2.1 IRGC Risk Governance Framework – general model (Source: IRGC 2005: 13)

essential for an effective governance process. This design of the four phases avoids the naive separation of facts here and values there, but also escapes the solipsism of post-modern relativity by honouring the analytical distinctions between the factual world and the world of values even if they clearly interact.

2.2.1 Pre-assessment

Risks are mental constructions, which are not real phenomena but the result of perceptions and/or interpretations by humans (Krohn and Krücken 1993; OECD 2003). The introduction of risk as a mental construct is contingent on the presumption that human action can prevent harm in advance. Risk as a mental construct has major implications on how risk is considered. Risks are created and selected by human actors. What counts as a risk to someone may be an act of God to someone else or even an opportunity for a third party. Although societies have over time gained experience and collective knowledge of the potential impacts of events and activities, one cannot anticipate all potential scenarios and be worried about all the many potential consequences of a proposed activity or an expected event. By the same token, it is impossible to include all possible options for intervention. Therefore societies have been selective in what they have chosen to be worth considering and what to ignore (Beck 1994). Specialized organisations have been established to monitor the environment for hints of future problems and to provide early warning of some potential future harm. This selection process is not arbitrary. It is guided by cultural values, by institutional and financial resources, and by systematic reasoning.

A systematic review of risk-related actions needs to start with an analysis of what major political and societal actors such as e.g. governments, companies, the scientific community and the general public select as risks and what types of problems they label as risk problems (rather than opportunities or innovation potentials, etc.). In technical terms this is called framing and encompasses the selection and interpretation of phenomena as relevant risk topics (Tversky and Kahneman 1981; van der Sluijs et al. 2003). The process of framing is mostly already part of the governance structure since governmental authorities (national, supranational and international agencies), risk and opportunity producers (e.g. industry), those affected by risks and opportunities (e.g. consumer organizations) and interested bystanders (e.g. the media or an intellectual elite) are involved and often in conflict with each other when framing the issue. What counts as risk may vary among these actor groups. Whether a consensus evolves about what requires consideration as a relevant risk depends on the legitimacy of the selection rule. The acceptance of selection rules rests on two conditions: (1) All actors need to agree with the underlying goal (often legally prescribed such as the threshold or maximum loading of specific chemicals in a water body); (2) They need to agree with the implications derived from the present state of knowledge (whether and to what degree the identified hazard impacts the desired goal).

2.2.2 Risk Appraisal

For politics and society to come to reasonable decisions about risks in public interest, it is not enough to consider only the results of (scientific) risk assessment. In order to understand the concerns of people affected and various stakeholders, information about both risk perceptions and the further implications of the direct consequences of a risk is needed and should be taken into account by risk management.[3]

Risk appraisal thus includes the scientific assessment of the risks to human health and the environment and an assessment of related concerns as well as social and economic implications (Renn and Walker 2007). The appraisal process should be clearly dominated by scientific analyses – but, in contrast to traditional risk regulation models, the scientific process includes both the natural/technical as well as the social sciences, including economics. The risk appraisal comprises two stages:

1. Risk assessment: experts of natural and technical sciences produce the best estimate of the physical harm that a risk source may induce.
2. Concern assessment: experts of social sciences including economics identify and analyze the issues that individuals or society as a whole link to a certain risk. For this purpose the repertoire of the social sciences such as survey methods, focus groups, econometric analysis, macro-economic modelling, or structured hearings with stakeholders may be used.

There are different approaches and proposals how to address the issue of risk appraisal. The German Advisory Council on Global Change (WBGU) has developed a set of eight criteria to characterize risks beyond the established assessment criteria (WBGU 2000; Klinke and Renn 2002). Some of the criteria have been used by different risk agencies or risk appraisal processes.

- *Extent of damage*: Adverse effects in natural units, e.g. death, injury, production loss, etc.
- *Probability of occurrence*: Estimate of relative frequency, which can be discrete or continuous.
- *Incertitude*: How do we take account of uncertainty in knowledge, in modelling of complex systems or in predictability in assessing a risk?
- *Ubiquity*: Geographical dispersion of damage.
- *Persistence*: How long will the damage last?
- *Reversibility*: Can the damage be reversed?
- *Delay effects*: Latency between initial event and actual damage.
- *Potential for mobilisation*: The broad social impact. Will the risk generate social conflict or outrage, etc.?
  - *Inequity and injustice* associated with the distribution of risks and benefits over time, space and social status.

[3]This includes the social mobilization potential, i.e. how likely is it that the risk consequences generate social conflicts and psychological reactions by individuals or groups?

- *Psychological stress and discomfort* associated with the risk or the risk source (as measured by psychometric scales).
- *Potential for social conflict and mobilisation* (degree of political or public pressure on risk regulatory agencies).
- *Spill-over effects* that are likely to be expected when highly symbolic losses have repercussions on other fields such as financial markets or loss of credibility in management institutions.

These four sub-criteria of the last category reflect many factors that have been proven to influence risk perception. The 'appraisal guidance' published by the UK Treasury Department in 2005 recommends a risk appraisal procedure that is similar to our proposal and includes as well both the results of risk assessment and the direct input from data on public perception and the assessment of social concerns (HM Treasury 2005).

2.2.3 Tolerability and Acceptability Judgment

Delineating and reasoning a judgment about the tolerability or acceptability of a given risk is one of the most controversial activities in the risk governance process. The term 'tolerable' refers to an activity that is seen as worth pursuing (for the benefit it carries) yet it requires additional efforts for risk reduction within reasonable limits. The term 'acceptable' refers to an activity where the remaining risks are so low that additional efforts for risk reduction are not seen as necessary. Judging risks according to their tolerability and acceptability leads to the well proven means of the traffic light model in form of a risk diagram with probabilities on the *y*-axis and extent of consequences on the *x*-axis (Fig. 2.2). In this variant of the model the red zone signifies intolerable risk, the yellow one indicates tolerable risk in need of further management actions and the green zone shows acceptable or even negligible risk.

To draw the line between 'intolerable' and 'tolerable' as well as 'tolerable' and 'acceptable' is one of the most difficult tasks of risk governance. The UK Health and Safety Executive developed a procedure for chemical risks based on risk–risk comparisons (Löfstedt 1997). Some Swiss cantons such as Basle County experimented with Round Tables as a means to reach consensus on drawing the two lines, whereby participants in the Round Table represented industry, administrators, county officials, environmentalists, and neighbourhood groups. Irrespective of the selected means to support this task, the judgement on acceptability or tolerability is contingent on making use of a variety of different knowledge sources. One needs to include the risk estimates derived from the risk assessment stage, and additional assessment data from the concern assessment within the appraisal stage.

Judgments on acceptability rely on two major inputs: values and evidence. What society is supposed to tolerate or accept can never be derived from looking at the evidence alone. Likewise, evidence is essential if we are to know whether a value has been violated or not (or to what degree). With respect to values and evidence we can distinguish three cases.

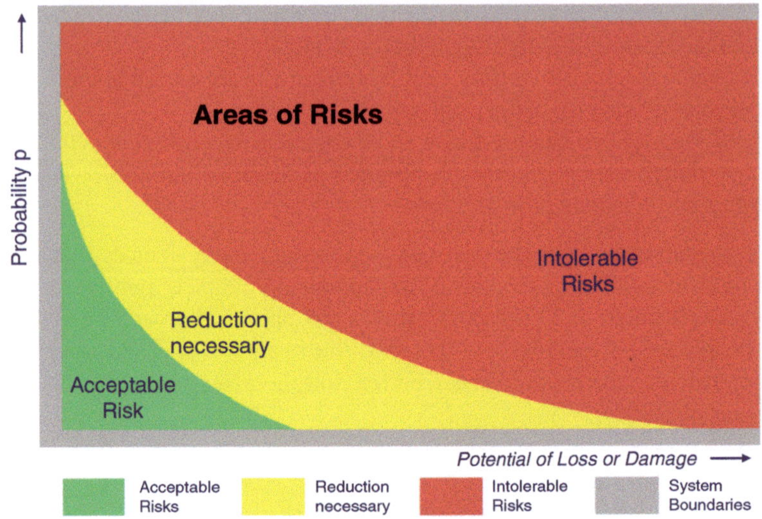

Fig. 2.2 Acceptable, tolerable and intolerable risks (Traffic Light Model) (Source: Renn 2008)

Interpretative ambiguity means that evidence is seen as ambiguous but not on values. In those cases where there is unanimous agreement about the underlying values and even the threshold of what is regarded as tolerable or acceptable, evidence in the form of risk estimates may be sufficient to locate the risk within the traffic light diagram. A judgement can then best be made by those who have most expertise in risk and concern assessments, in which case it makes sense to place this task within the domain of risk appraisal.

Normative ambiguity refers to the ambiguity on values but not on evidence. If the underlying values of what could be interpreted as tolerable or acceptable are disputed, while the evidence of what is at stake is clearly given and non-controversial, the judgement needs to be based on a discourse about values and their implications. Such a discourse should be part of risk management. In these cases, science is very familiar with the risks and there is little uncertainty and interpretative ambiguity about dose–effect relationships. Yet there is considerable debate whether the application is tolerable or not. One example may the use of phtallates in toys. All analysts are aware that the substance is potentially carcinogenic, but given the known exposure and the dose–response functions there is hardly any possibility for young children to be negatively affected. Yet the mere idea of having a carcinogenic substance in children's' toys has incited a fierce debate about the tolerability of such an ingredient in rubber toys.

Interpretative and normative ambiguity addresses a third case where both the evidence and the values are controversial. This would imply that assessment should engage in an activity to find some common ground for characterising and qualifying the evidence and risk management needs to establish agreement about the appropriate values and their application. A good example for this third case may be the interpretative and normative implications of global climate change. The Intergovernmental

Panel on Climate Change (IPCC) has gone through considerable efforts to articulate a common characterisation of climatic risks and their uncertainties. Given the remaining uncertainties and the complexities of the causal relationships between greenhouse gases and climate change, it is then a question of values whether governments place their priorities on prevention or on mitigation (Keeney and McDaniels 2001). To deal with similar issues regarding chemical pollution, efforts are in progress to initiate an International Panel on Chemical Pollution (IPCP).[4]

Since the last case includes both issues of the other two, the process of judging the tolerability and acceptability of a risk can be structured into two distinct components: risk characterisation and risk evaluation. The first step, risk characterisation, determines the evidence-based component for making the necessary judgement on the tolerability and/or acceptability of a risk; the step risk evaluation determines the value-based component for making this judgement.

The separation of evidence and values underlying the distinction between characterisation and evaluation is, of course, functional and not necessarily organisational. Since risk characterisation and evaluation are closely linked and each depends on the other, it may even be wise to perform these two steps simultaneously in a joint effort by both risk assessment and risk management. The US regulatory system tends to favour an organisational combination of characterisation and evaluation, while European risk management tends to maintain the organisational separation, e.g. in the food area (Löfstedt and Vogel 2001). The same is true for chemical regulation. ECHA, the European Chemical Agency, has the mandate to assess and characterize risks from chemical substances with the REACH regime. The management part of issuing regulation is left to the EU Commission.[5]

2.2.4 *Risk Management*

Risk management reviews the information and findings from risk appraisal (risk assessment and concern assessment) and the results and conclusions from tolerability and acceptability judgment (risk characterization and evaluation), in order to assess, evaluate and select appropriate risk management options. Starting point for risk management are three potential outcomes:

- In an *intolerable situation* the risk source such as a chemical or a technology needs to be refused or substituted.

[4]For the IPCP see the website http://www.ipcp.ch.

[5]An example for an institutionalized body to fulfil at least partially tasks and functions of tolerability and acceptability assessment and risk appraisal on the national level is the UK Chemical Stakeholder Forum. The forum consists of stakeholders from different associations such as chemical industry, business, environment, consumer protection as well as research institutes. They gather different perceptions and concerns, evaluate and prioritize different chemicals and propose risk management strategies in order to deliberate the government.

- In a *tolerable situation* the risks need to be reduced or handled in some other way within the limits of reasonable resource investments (ALARP, including best practice).[6]
- In an *acceptable situation* the risks are so minor – perhaps even regarded as negligible – that any risk reduction effort is unnecessary. However, risk sharing via insurances and/or further risk reduction on a voluntary basis presents options for action which can be worthwhile pursuing even in the case of an acceptable risk.

With regard to these outcomes risk management may either face a situation of unanimity, i.e. all relevant actors agree with how a given risk situation should be qualified, or a situation of conflict in which major actors challenge the classification undertaken by others. The degree of controversy is one of the drivers for selecting the appropriate instruments for risk prevention or risk reduction.

If risks are classified as tolerable, or if there is dispute as to whether they are tolerable or acceptable, risk management needs to design and implement actions that make these risks acceptable over time. Should this not be feasible then risk management, assisted by communication, needs at least to credibly convey the message that major effort is undertaken to bring these risks closer to being acceptable. This task can be described in terms of classic decision theory (Morgan 1990; Hammond et al. 1999).

The decision making process of risk management starts with identifying and generating risk management options. Generic risk management options include risk avoidance, risk reduction, risk transfer and possibly self retention. Whereas to avoid a risk means either selecting a path which does not touch on the risk (e.g. by abandoning the development of a specific chemical or technology) or taking action in order to fully eliminate a certain risk, risk transfer deals with ways of passing the risk on to a third party. Self retention as a management option essentially means taking an informed decision to do nothing about the risk and to take full responsibility both for the decision and any consequences occurring thereafter. Risk management by means of risk reduction can be accomplished by many different means such as:

- Technical standards and limits that prescribe the permissible threshold of concentrations, emissions, take-up or other measures of exposure
- Performance standards for chemical and technological processes such as minimum temperatures in waste incinerators
- Technical prescriptions referring to the blockage of exposure (e.g. via protective clothing) or the improvement of resilience (e.g. via immunisation or more indestructible constructions)
- Governmental economic incentives including taxation, duties, subsidies and certification schemes
- Third party incentives, i.e. private monetary or in kind incentives

---

[6]This can be addressed by private actors (such as corporate risk managers) or public actors (such as regulatory agencies) or both (public-private partnerships).

- Compensation schemes (monetary or in kind)
- Insurance and liability
- Co-operative and informative options ranging from voluntary agreements to labelling and education programs

All these options can be used individually or in combination to accomplish even more effective risk reduction. Options for risk reduction can be initiated by private and public actors or both together.

Afterwards the decision making unit assesses the risk management options with respect to predefined criteria. Each of the options will have desired and unintended consequences which relate to the risks that they are supposed to reduce. In most instances, an assessment should be done according to the following criteria:

- *Effectiveness*: Does the option achieve the desired effect?
- *Efficiency*: Does the option achieve the desired effect with the least resource consumption?
- *Minimisation of external side effects*: Does the option infringe on other valuable goods, benefits or services such as competitiveness, public health, environmental quality, social cohesion, etc.? Does it impair the efficiency and acceptance of the governance system itself?
- *Sustainability*: Does the option contribute to the overall goal of sustainability? Does it assist in sustaining vital ecological functions, economic prosperity and social cohesion?
- *Fairness*: Does the option burden the subjects of regulation in a fair and equitable manner?
- *Political and legal implementability*: Is the option compatible with legal requirements and the political culture?
- *Ethical acceptability*: Is the option morally acceptable?
- *Public acceptance*: Will the option be accepted by those individuals who are affected by it? Are there cultural preferences or symbolic connotations that have a strong influence on how the risks are perceived?

Measuring management options against these criteria may create conflicting messages and results. Many measures that prove to be effective may turn out to be inefficient or unfair to those who will be burdened. Other measures may be sustainable but not accepted by the public or important stakeholders. These problems are aggravated when dealing with global risks. What appears to be efficient in one country may not work at all in another country. Risk management is therefore well advised to make use of the many excellent guidance documents on how to handle risk tradeoffs and how to employ decision analytic tools for dealing with conflicting evidence and values (Wiener 1998; van der Sluijs et al. 2003).

Subsequently, risk management evaluates the risk management options, which is similar to risk evaluation since this step integrates the evidence on how the options perform with regard to the evaluation criteria with a value judgement about the relative weight each criterion should be assigned. Ideally, the evidence should come from experts and the relative weights from politically legitimate decision makers.

In practical risk management, the evaluation of options is done in close cooperation between experts and decision makers. As pointed out later, this is the step in which direct stakeholder involvement and public participation is particularly important and is therefore best assured by making use of a variety of methods (Rowe and Frewer 2000; OECD 2002; Renn 2008).

The final step in the decision making process should be the selection of risk management options. Once the different options are evaluated, a decision has to be made as to which options are selected and which rejected. This decision is obvious if one or more options turn out to be dominant (relatively better on all criteria). Otherwise, trade-offs have to be made that need legitimisation (Graham and Wiener 1995). A legitimate decision can be made on the basis of formal balancing tools (such as cost–benefit or multi-criteria-decision analysis), by the respective decision makers (given his decision is informed by a holistic view of the problem) or in conjunction with participatory procedures.

It is also the task of risk management to oversee and control the implementation process. In many instances implementation is delegated, as when governments take decisions but leave their implementation to other public or private bodies or to the general public. However, the risk management has at any rate the implicit mandate to supervise the implementation process or at least monitor its outcome.

2.2.5 Risk Communication

Given the arguments about risk perception and stakeholder involvement, it is essential to have an effective communication at the core of any successful activity to assess and manage risks. The field of risk communication initially developed as a means of investigating how best expert assessments could be communicated to the public so that the tension between public perceptions and expert judgement could be bridged. In the course of time this original objective of educating the public about risks has been modified and even reversed as the professional risk community realized that most members of the public refused to become ‘educated’ by the experts but rather insisted that alternative positions and risk management practices should be selected by the professional community in their attempt to reduce and manage the risks of modern technology (Plough and Krimsky 1987).

Risk communication is needed throughout the whole risk handling chain, from the framing of the issue to the monitoring of risk management impacts. The precise form of communication needs to reflect the nature of the risks under consideration, their context and whether they arouse, or could arouse, societal concern. Communication has to be a means to both ensure that (Renn 2008):

- Those who are central to risk framing, risk appraisal or risk management understand what is happening, how they are to be involved, and, where appropriate, what their responsibilities are.
- Others outside the immediate risk appraisal or risk management process are informed and engaged.

The first task of risk communication, i.e. facilitating an exchange of information among risk professionals, has often been underestimated in the literature. A close communication link between risk/concern assessors and risk managers, particularly in the phases of pre-assessment and tolerability/acceptability judgement, is crucial for improving overall governance. Similarly, co-operation among natural and social scientists, close teamwork between legal and technical staff and continuous communication between policy makers and scientists are all important prerequisites for enhancing risk management performance. This is particularly important for the initial screening phase where the allocation of risks is performed.

The second task that of communicating risk appropriately to the outside world, is also a very challenging endeavour. Many representatives of stakeholder groups and, particularly, members of the affected and non-affected public are often unfamiliar with the approaches used to assess and manage risks and/or pursue a specific agenda, trying to achieve extensive consideration of their own viewpoints. They face difficulties when asked to differentiate between the potentially dangerous properties of a substance (hazards) and the risk estimates that depend on both the properties of the substance, the exposure to humans, and the scenario of its uses (Morgan et al. 2001). Also complicating communication is the fact that some risks are acute, with severe effects that are easy to recognize, whereas others exert adverse effects only weakly but over a long period of time. Yet other risks' effects only start to show after an initial delay. Finally, it is no easy task to convey possible synergies of exposures to industrial substances with other factors that relate to lifestyle (e.g. nutrition, smoking, use of alcohol).

Effective communication, or the non-existence thereof, has a major bearing on how well people are prepared to face and cope with risk. Limited knowledge of and involvement in the risk management process can lead to inappropriate behaviour in emergency or risk-bearing situations (for example, when handling contaminated food or water or dealing with unknown chemicals). There is also the risk of failed communication: consumers or product users may misread or misunderstand risk warnings or labels so that they may, through ignorance, expose themselves to a larger risk than necessary. This is particularly prevalent in countries with high rates of illiteracy and unfamiliarity with risk-related terms. Providing understandable information to help people cope with risks and disasters is, however, only one function of risk communication. Most risk communication analysts list four major functions (Klinke and Renn 1999; OECD 2002):

- *Education and enlightenment*: inform the audience about risks and the handling of these risks, including risk and concern assessment and management.
- *Risk training and inducement of behavioural changes*: help people cope with risks and potential disasters.
- *Creation of confidence in institutions responsible for the assessment and management of risk*: give people the assurance that the existing risk governance structures are capable of handling risks in an effective, efficient, fair and acceptable manner (such credibility is crucial in situations in which there is a lack of personal experience and people depend on neutral and disinterested information). It should be kept in mind, however, that trust cannot be produced or generated,

but only be accumulated by performance, and that it can be undermined by the lack of respect for an individual within such an institution.

- *Involvement in risk-related decisions and conflict resolution*: give stakeholders and representatives of the public the opportunity to participate in the risk appraisal and management efforts and/or be included in the resolution of conflicts about risks and appropriate risk management options.

Although risk communication implies a stronger role for risk professionals to provide information to the public rather than vice versa, it should be regarded as a mutual learning process. Concerns, perceptions and experiential knowledge of the targeted audience(s) should thus guide risk professionals in their selection of topics and subjects: it is not the task of the communicators to decide what people need to know but to respond to the questions of what people want to know ('right to know' concept, see Baram 1984). Risk communication requires professional performance both by risk and communication experts. Scientists, communication specialists and regulators are encouraged to take a much more prominent role in risk communication, because effective risk communication can make a strong contribution to the success of a comprehensive and responsible risk management.

2.3 Conclusions

The starting point of this chapter was the insight that modern societies are in urgent need for a new inclusive and integrative framework promising to promote good risk governance, establish a more stringent approach to deal with serious, complex, uncertain and ambiguous risks, develop a more suited structure to cope with emerging systemic and global threats and provide a convincing and acceptable format for involving civil society in the decision-making process. Good governance seems to rest on the three components: knowledge, legally prescribed procedures and social values. It has to reflect specific functions, from early warning (radar function), via new assessment and management tools to improved methods of balanced risk evaluation, effective risk communication and deliberative participation. Criteria of good governance have been discussed in many different contexts. They need to be transferred to risk-related issues and put into operation so that best practices can be identified and recommended. Central items to be addressed are sound scientific expertise, adequate inclusion of public concerns, consistency and coherence in making trade-offs between risks and benefits, non-discrimination and proportionality in designing management options and assurance of thorough monitoring and independent oversight during implementation of management options. In addition, governance structures should reflect criteria such as transparency; effectiveness and efficiency; accountability; strategic focus; sustainability; equity and fairness; respect for the rule of law; and the need for the chosen solution to be politically and legally feasible, as well as ethically and publicly acceptable. Beyond the involvement of organized groups, the required framework needs to include procedures for general public participation and public dialogue and effective communication about risk issues.

In the modern pluralist world, most risks will need to be subject to such a robust governance approach if they are to be adequately managed.

Together with specialists and practitioners from different fields of risk analysis, we have tried to develop such a framework (Klinke and Renn 2002; IRGC 2005; Klinke et al. 2006; Renn and Walker 2007; Renn 2008). Drawing on an analysis of a selection of well-established approaches to what has traditionally been called 'risk analysis' or 'risk management', the new risk governance framework introduced in this chapter has been designed to offer both a comprehensive means of integrating risk identification, assessment, management and communication, and a tool that can compensate the absence of (or a weaknesses in) risk governance structures and processes (Bunting et al. 2007). Use of the framework promises to identify the key steps in the risk governance process, and the diagnosis of potential deficits, problems or shortcomings in governance institutions or procedures. In addition, it can assist in facilitating a thorough understanding of risk issues, identifying the stakeholders interested in (and concerned with) the risks and providing guidance for how, and when, to include stakeholders in the process.

What are the innovative features of the framework and how does it differ from those that were analysed in this volume?

The risk *governance* process is understood to include, but also to go beyond, the three conventionally recognized elements of risk analysis (risk assessment, risk management and risk communication). *Governance* thus includes matters of institutional design, technical methodology, administrative consultation, legislative procedure and political accountability on the part of public bodies, and social or corporate responsibility on the part of private enterprises. But it also includes more general provision on the part of government and commercial and civil society actors for building and using scientific knowledge, for fostering innovation and technical competences, for developing and refining competitive strategies, and for promoting social and organizational learning.

The framework builds upon the logical structure of four consecutive phases called pre-assessment, appraisal, characterization/evaluation and management. In addition, risk *communication* accompanies all four phases. Within each of the boxes, specific activities are listed that are deemed essential for meeting the requirements of good governance. The framework offers a truly interdisciplinary and multilevel governance approach. Most notably, it urges risk governance institutions to elicit not only knowledge about the physical impacts of technologies, natural events or human activities, but also knowledge about the concerns that people associate with this cause of risks. This concern assessment should not be confused with eliciting stakeholder feedback or providing platforms for participatory processes. It is, rather, a social science activity aimed at providing sound insights and a comprehensive diagnosis of concerns, expectations and worries that individuals, groups or different cultures may associate with the hazard or the cause of the hazard (Hyman and Stiftel 1988). This social science analysis should be submitted to the same kind of methodological scrutiny and peer review as any other natural science activity.

Parallel to this concern assessment, the framework provides input on all governance levels from stakeholders either by contributing additional knowledge or by inserting their values, interests and preferences into the evaluation of the risk itself and the selection of the most effective, efficient and fair set of management options. It promotes the idea of inclusive governance, which is seen as a necessary, although insufficient, prerequisite for tackling risks in both a sustainable and acceptable manner and, consequently, imposes an obligation to ensure the early and meaningful involvement of all stakeholders and, in particular, civil society (Jasanoff 1993).

How can the framework be used in the future? First, providing a unified, yet flexible, concept, it can assist risk researchers to conduct comparative analyses among and between different risk types, thus ensuring that resource distribution on risk management across risk sources and technologies follows a consistent and efficient pattern. Second, it may help risk governance institutions to structure their projects in line with the phases and components outlined in this volume. Third, the framework may be a worthwhile basis for diagnosing deficiencies in existing risk governance regimes around the world and may provide suggestions on how to improve them. These three functions are particularly pertinent to risk management for chemicals in order to develop the skills to harvest the benefits of chemical substances and processes as well as to master and limit the risks that are associated with their use.

References

Baram M. (1984) The right to know and the duty to disclose hazard information. *American Journal of Public Health* 74(4): 385–390.

Beck, U. (1994) The reinvention of politics: towards a theory of reflexive modernization. In: Beck, U., Giddens, A., and Lash, S. (eds.) *Reflexive Modernization: Politics, Tradition and Aesthetics in the Modern Social Order*. Stanford University Press, Stanford, pp 1–55.

Bunting, C., Renn, O., Florin, M.-V. and Cantor, R. (2007) Introduction to the IRGC risk governance framework. *The John Liner Review* 21(2): 7–26.

Functowicz, S.O. and Ravetz, J.R. (1992): Three types of risk assessment and the emergence of post-normal science. In: Krimsky, S., and Golding, D. (eds.) *Social Theories of Risk*. Praeger, Westport and London, pp 251–273.

WBGU – German Advisory Council on Global Change (2000) World in transition: strategies for managing global environmental risks. Annual Report. Springer, Heidelberg and New York.

Graham, J.D. and Wiener, J.B. (1995) *Risk vs. Risk*. Harvard University Press, Cambridge.

Hammond, J., Keeney, R., and Raiffa, H. (1999) *Smart Choices: A Practical Guide to Making Better Decisions*. Harvard Business School Press, Cambridge.

Hyman, E.L. and Stiftel, B. (1988) *Combining Facts and Values in Environmental Impact Assessment*. Westview Press, Boulder.

IRGC (International Risk Governance Council) (2005) *Risk Governance – Towards an Integrative Approach*, White Paper No 1, O. Renn with an Annex by P. Graham. IRGC, Geneva.

Jasanoff, S. (1993) Bridging the two cultures of risk analysis. *Risk Analysis* 13(2): 123–129.

Keeney, R. and McDaniels, T. (2001) A framework to guide thinking and analysis regarding climate change policies. *Risk Analysis* 6: 989–1000.

Klinke, A. and Renn, O. (1999) Prometheus unbound: challenges of risk evaluation, risk classification, and risk management. Working Paper No 153 of the Centre of Technology Assessment. TA-Academy, Stuttgart.

Klinke, A. and Renn, O. (2002) A new approach to risk evaluation and management: risk-based, precaution-based, and discourse-based strategies. *Risk Analysis* 22(6): 1071–1094.

Klinke, A., Dreyer, M., Renn, O., Stirling, A. and van Zwanenberg, P. (2006) Precautionary risk regulation in European governance. *Journal of Risk Research* 9(4): 373–392.

Krohn, W. and Krücken, G. (1993) Risiko als Konstruktion und Wirklichkeit. Eine Einführung in die sozialwissenschaftliche Risikoforschung. In: Krohn, W. and Krücken, G. (eds.) *Riskante Technologie: Reflexion und Regulation. Einführung in die Sozialwissenschaftliche Risikoforschung*. Suhrkamp, Frankfurtam Main, pp 9–44.

Laudan, L. (1996) The pseudo-science of science? The demise of the demarcation problem. In: Laudan, L. (ed.) *Beyond Positivism and Relativism. Theory, Method and Evidence*. Westview Press, Boulder, pp 166–192.

Löfstedt, R.E. (1997) Risk evaluation in the United Kingdom: legal requirements, conceptual foundations, and practical experiences with special emphasis on energy systems. Working Paper No. 92. Akademie für Technikfolgenabschätzung, Stuttgart.

Löfstedt, R. and Vogel, D. (2001) The changing character of regulation: a comparison of Europe and the United States. *Risk Analysis* 21(3): 393–402.

Morgan, M.G. (1990) Choosing and managing technology-induced risk. In: Glickman, T.S. and Gough, M. (eds.) *Readings in Risk. Resources for the Future*. Washington.

Morgan, M.G., Fishhoff, B., Bostrom, A. and Atmann, C.J. (2001) *Risk Communication: A Mental Model Approach*. Cambridge University Press, Cambridge, pp 17–28.

Mueller-Herold, U., Morosini, M. and Schucht, O. (2005) Choosing chemicals for precautionary regulation: a filter series approach. *Environmental Science and Technology* 39: 683–691.

OECD (2002) *Guidance document on risk communication for chemical risk management*. OECD, Paris.

OECD (2003) *Emerging systemic risks. Final report to the OECD futures project*. OECD, Paris.

Plough, A. and Krimsky, S. (1987) The emergence of risk communication studies: social and political context. *Science, Technology, and Human Values* 12: 78–85.

Renn, O. (2008) *Risk Governance. Coping with Uncertainty in a Complex World*. Earthscan, London.

Renn, O. and Walker, K. (2007) Lessons learned: a re-assessment of the IRGC framework on risk governance. In: Renn, O. and Walker, K. (eds.) *The IRGC Risk Governance Framework: Concepts and Practice*. Springer, Heidelberg and New York, pp 331–367.

Rowe, G. and Frewer, L. (2000) Public participation methods: an evaluative review of the literature. *Science, Technology and Human Values* 25: 3–29.

Stirling, A. (2003) Risk, uncertainty and precaution: some instrumental implications from the social sciences. In: Berkhout, F. Leach, M. and Scoones, I. (eds.) *Negotiating Change*. Edward Elgar, London, pp 33–76.

HM Treasury (2005) *Managing risks to the public: appraisal guidance*. HM, London. www.hm-treasury.gov.uk/media/0/B/Managing_risks_to_the_public.pdf. Cited September 2008.

Tversky, A. and Kahneman, D. (1981) The framing of decisions and the psychology of choice. *Science* 211: 453–458.

Van Asselt, M.B.A. (2000) *Perspectives on Uncertainty and Risk*. Kluwer, Dordrecht and Boston.

Van der Sluijs, J.P., Risbey, J.S., Kloprogge, P., Ravetz, J.R., Funtowicz, S.O., Corral Quintana, S., Guimaraes Pereira, A., De Marchi, B., Petersen, A.C., Janssen, P.H.M., Hoppe, R. and Huijs, S.W.F. (2003) RIVM/MNP guidance for uncertainty assessment and communication. Report No. NWS-E-2003-163. Copernicus Institute for Sustainable Development and Innovation and Netherlands Environmental Assessment Agency, Utrecht and Bilthoven.

Wiener, J., B. (1998) Managing the iatrogenic risks of risk management. *Risk: Health Safety & Environment* 9: 39–83.

Zinn, J.O. and Taylor-Gooby, P. (2006) The challenge of (managing) new risks. In: Taylor-Gooby, P. and Zinn, J. (eds.) *Risk in Social Science*. Oxford University Press, Oxford, pp 54–75.

Chapter 3
Communicating Chemical Risks: Beyond the Risk Society

Alison Anderson

Abstract The news media play a potentially crucial, yet often under-theorised, role in communicating chemical risks. Since research has tended to be restricted by traditional disciplinary boundaries, it has not always been able to benefit from insights gained from a broader perspective and has remained very fragmented. This chapter argues that recent approaches point to a more complex understanding of risk reporting and reflect wider social changes, both in the nature of 'risk' and the structure and workings of the media. It argues that any analysis of the media reporting of risk must be placed within the broader context of the growing concentration and globalisation of news media ownership and the increasingly 'promotional culture' which we inhabit, highlighted by the rapid rise of the public relations industry in recent years and claims-makers that employ increasingly sophisticated media strategies. This is illustrated by considering examples such as the furor over the proposed dumping of the Brent Spar oil installation at sea in 1995 and the Prestige oil disaster of 2002. Key issues are highlighted concerning news values, the credibility of news sources and access to the news media, and the current organisation of news work.

Keywords Brent Spar • Chemicals • News media • Prestige disaster • Risk society

3.1 Introduction

> Greenpeace's victory in the 'Spar' campaign forced business to re-evaluate its assumptions about decision-making, the relationship with customers and with other outsiders such as environmental groups. Because Shell is so large – a giant multinational (the fifth largest company in the world and Europe's biggest) – the impact of the 'Spar' decision spread through industry's boardrooms like a seismic shockwave. (Rose 1998: 92)

A. Anderson (✉)
University of Plymouth, School of Applied Psychosocial Sciences, Faculty of Health, Drake Circus, Plymouth, Devon, PL4 8AA, UK
e-mail: aanderson@plymouth.ac.uk

J. Eriksson et al. (eds.), *Regulating Chemical Risks: European and Global Challenges*, DOI 10.1007/978-90-481-9428-5_3,

Over the past 3 decades a series of dramatic disasters involving chemical risks (notably Bhopal and Chernobyl) have led industry to reflect upon the potential power of the news media to influence public opinion and policymaking. Since the 1980s numerous risk communication projects have developed in response to growing demands for greater openness and accountability (see Renn and Kastenholz 2000). In 1995 Royal Dutch/Shell's plans to sink the Brent Spar oil platform in the North Atlantic Ocean sparked off a major controversy and provided Greenpeace activists with, what they described as, a 'heaven-sent opportunity… And a visual symbol of sea dumping and toxic pollution' (Rose 1998: 10–11). The battle over the disposal of the redundant oil installation strongly resonated with public concern in Europe about polluted oceans and mistrust of big business. The Brent Spar is widely acknowledged to have marked a classic case of risk communication 'gone wrong', forcing the oil industry to re-examine its approach towards decision-making and corporate social responsibility. Moreover, it graphically illustrates how news media framing of chemical risks can have a significant impact upon policy and regulatory frameworks.

The news media, especially television, constitute a central source of information about science (Nelkin 1995). However, they offer an inevitably selective account of reality. Far from simply mirroring the world 'out there', or representing a random response to issues and events, the news media are structured by a range of routine organisational practices and resource constraints. News, then, is produced within a political, economic, cultural and social context. Routine news conventions, pressures and constraints lead journalists to frame certain truth-claims as legitimate and credible, whilst ignoring, trivialising or marginalising others. Objective scientific 'facts' are often highly contested and have moral and political dimensions. The concept of framing refers to the processes of selection and emphasis which leads some aspects of a news story to be accorded greater salience than others, as part of a consistent overall narrative weaved around defining the nature of a problem, its causes and recommended solutions (Entman 1993; Miller and Riechert 2000). Behind the scenes there is an intense struggle among contending stakeholders, each of whom is likely to be vying to make their voices heard or seeking to deflect attention away from the issues at stake. The emergence of particular dominant frames may prove crucial for the degree of support (or lack of) elicited from policymakers within the wider regulatory context (Petersen et al. 2009; Hornig Priest 2001). While the media do not determine public opinion, once they have framed the parameters of an issue during the early stages of debate it can prove very hard for other news actors to dislodge it (Anderson et al. 2005b; Bakir 2006; Nisbet and Huge 2007; Nisbet and Lewenstein 2002; Nisbet et al. 2003).

This chapter seeks to illuminate critical issues concerning the shifting role of the news media in communicating chemical risks. It begins by examining early approaches to researching risk and the media, before going on to discuss more recent developments. Key issues are raised concerning credibility, legitimacy and trust. Finally two illustrative case studies are presented (the Brent Spar in 1995 and the Prestige oil disaster of 2002) in order to elucidate what lessons can be learnt from prior crises.

3.2 Early Approaches to Researching Science and the Media

Within the science communication literature linear models often assumed a one-way flow of information from scientists (who publish their findings mainly via peer-reviewed scientific journals) to journalists, who then translate or popularise them for lay readers/audiences (Gregory and Miller 1998; Logan 1991). The 'information deficit' model viewed risk reporting un-problematically in terms of how far it was seen to accurately reflect expert opinion. The problem was judged to arise from an 'information gap' between scientific experts and lay publics who were often perceived to be ignorant about basic scientific 'facts'. The latter were seen as empty vessels that simply needed to be filled with scientific knowledge. However, this approach has come under growing criticism and new concerns have focused upon citizen involvement and moving public engagement 'upstream' (Schanne and Meier 1992; Wilsdon and Willis 2004). Linear models are widely acknowledged to have under-estimated the complexity of science communication processes, which involve multiple feedback loops, and to have downplayed scientific uncertainty and ambiguity (Lewenstein 1995a).

Scientists frequently complain that news media accounts of their work are 'biased' or 'inaccurate'. Distortions or misrepresentations of science are often explained in terms of the 'two cultures' of science and journalism. The notion of 'two cultures' implies that journalists and scientists occupy two distinctive worlds governed by different outlooks and aims. Journalists' attempts to bridge these cultures, to popularise science knowledge for lay readers/audiences, it is argued, leads to an over-simplification or distortion of the 'truth'. Studies based upon this 'diffusion' model of science popularisation often conclude that biased or inaccurate news reporting is a consequence of insufficient or inadequate information. According to this line of thought it follows, therefore, that media coverage can be 'improved' by better 'dissemination' of scientific or technological information (Lewenstein 1995a: 347).

However, the notion of the 'two cultures' of science and journalism has been criticised on a number of grounds (e.g., Dunwoody 1999; Hilgartner 1990; Lewenstein 1995a, b). It assumes that media institutions are linked with other institutions in a functionally interdependent way and that the communication process is relatively straightforward once the 'correct' information is presented to audiences (Nelkin 1995). Also, it denies the gradual evolution of a shared culture between science and journalism (Dunwoody 1999: 73–77). In the US a steadily rising number of science reporters have received scientific training and many are likely to work alongside scientists (Lewenstein 1995a: 345; Petersen et al. 2009). A shared culture implies a shared language. Scientists have been found to use 'popular' metaphors and imagery in journals such as 'Science' and 'Nature', which specialist journalists frequently use as key sources (Petersen 1999). The notion of popularisation ignores the multiple circuits of communication (e.g. personal contacts, emails, television news) between scientists and journalists and the fact that stories are often 'broken', long before the journey from peer review and science journals has been completed (Lewenstein 1995b; Petersen et al. 2009). Moreover, it glosses over the ways in which scientists, and other news sources, may attempt to control the news at various points in the production cycle (through news releases, press conferences, staged 'leaks' and so on). Indeed, some estimates suggest that more than half of all news sources are source generated so scientists are often able to package their work in a strategic fashion (Anderson et al. 2005a; Nisbet and Lewenstein 2002).

News coverage is the outcome of a series of behind-the-scenes battles between news sources, each seeking to offer their own definition of the issues at stake. As Ericson et al. observe: 'News is a product of transactions between journalists and their sources. The primary source of reality for news is not what is displayed or happens in the real world. The reality of news is embedded in the nature and type of social and cultural relations that develop between journalists and their sources' (Ericson et al. 1989: 377).

Routine reporting of science/technology is often mediated through the voice of the 'expert' and numerous studies have found that official sources (such as politicians and heads of government agencies, or scientists attached to trusted institutions) tend to gain advantaged access to the news media (see Manning 2001). Reporters typically approach these sources first before seeking out alternative viewpoints. This may explain the general tendency for risk reporting in general to offer information with a reassuring rather than an alarming tone (Petts et al. 2001; Schanne and Meier 1992). However, in the case of 'accidental' news, the spontaneity of the event may open up the potential for traditional frames to be challenged by 'alternative voices', especially where official sources choose to remain silent, or their response is delayed due to cumbersome bureaucratic procedures (Anderson 2002; Molotch and Lester 1975). Although environmental pressure groups may lack the finance, status and PR personnel advantages of official sources, they are frequently able to react to media demands much more rapidly (Anderson 1997; Kitzinger 1999). But gaining coverage is only half the battle since this offers no guarantee that a news sources' claims will be presented as credible or legitimate. As Ryan points out: 'The real battle is over whose interpretation, whose framing of reality, gets the floor' (Ryan 1991: 53). In many cases information that can be attributed to an official source is likely to be treated as more reliable and trustworthy compared to information from news sources which are perceived to have an 'agenda', such as environmental pressure groups (Anderson 1997).

Where chemical regulation is concerned, the news media potentially exert a significant influence upon both publics and policymakers. However, in contrast to early science communication models, it is important to recognise that this influence is far from linear – it is complex, dynamic and governed by a number of feedback loops. Assessing the extent to which media reporting of risks is 'accurate' and 'balanced' is extremely complex and far from straightforward. This debate reflects differing underlying philosophical perspectives on the ideal role of the media in reporting risk and whether objectivity is seen as possible or even desirable (Anderson 1997; Anderson 2006; Lichtenberg and MacClean 1991). Media coverage of chemical risks is the product of social and cultural processes and constructs particular versions of reality that do not simply mirror a single truth. Yet those who accuse the media of routine sensationalism tend to assume that sources of official information on risk operate purely on a scientific basis.

3.3 Which Risks Attract Attention, Why and Under What Conditions?

Rather than focus upon the extent to which the media exaggerate or under-estimate the 'objective' scientific evidence concerning risks, as Kitzinger (1999: 62) observes: '...the key question is which risks attract attention and how, when, why and under what conditions?' It is widely recognised in the social sciences that coverage of chemical risks is highly selective and it is not necessarily the most serious risks that attract the greatest attention. Risks that can cause the loss of life or serious injury of large numbers of people in one go – such as nuclear accidents – are far more likely to gain media coverage than those which have a gradual effect over a lengthy period of time – such as asbestos exposure (Greenberg et al. 1989; Singer and Endreny 1987). As the case studies discussed later on demonstrate, the newsworthiness of particular chemical risks is affected by a number of factors including their proximity to home, socio-economic factors, and the symbolic and visual dramatisation of an incident. Journalists and broadcasters typically possess a set of underlying ideas, a sort of 'sixth sense' about what constitutes 'news'. These, for the large part unquestioned news values, have a profound influence on a multitude of daily decisions concerning the selection, packaging and ordering of potential news items. They are influenced by a range of assumptions about the target audience and the editorial identity of the media outlet, as well as conventional story-telling mechanisms.

Research demonstrates that the reporting of chemical risks tends to closely revolve around 'events' (Molotch and Lester 1975). In particular, it thrives upon unexpected, dramatic disasters with a strong 'human interest' emphasis. In part this reflects the fact that much news coverage is based around a 24 hour cycle (see Anderson 1997). So unless a gradually developing problem is perceived to have come to a climax, it will often tend to be neglected in favour of the more immediate story. Since news quickly becomes stale the more unusual or sudden the event, the more likely it is to gain novelty value and grab headline attention. While this may increase public awareness of particular risks, the downside of event-centred coverage is that it tends to give readers/audiences the impression that blame can be put down to isolated instances where individuals or corporations have failed in their responsibilities, rather than associated with wider structural issues (Hannigan 2006). For example, in the case of the Exxon Valdez oil spill (1989) coverage tended to be framed around the allegation that the disaster was caused by the drunken state of the Captain, Joseph Hazelwood. This downplayed other possible explanations concerning the oil industry's poor capacity to clean up large spills in areas such as the Prince William Sound, or cuts in funding affecting maritime safety standards (Hannigan 2006). Indeed, frequently news media frames continue to hold sway long after the event, even when new evidence has emerged concerning wider causes (Darley 2000).

Routine chemical risks coverage, as with risk reporting in general, tends to be focused around policy events (such as parliamentary hearings, regulatory decisions, the release of white papers and so on). As Kitzinger (1999) observes, if there is little activity on the policy front then an issue is often likely to gain limited media exposure. The analysis of previous science controversies highlights how scientists tend to be the key news sources during the early stages when an issue is located within an administrative policy arena (Bauer and Gaskell 2002; Nisbet et al. 2003). Less advantaged groups may seek to re-orientate the framing of the debate away from narrow technical terms by creating 'pseudo events' that dramatise the issues and connect them with emotionally laden symbols (Anderson 1997; Gorss and Lewenstein 2005). Media attention does not typically peak until competing interests force the expansion of the issue and it shifts to more overtly political arenas (Nisbet et al. 2003). In the case of emerging debates over the potential risks and benefits of nanotechnologies, the issues are still largely bounded within the administrative policy arena. Although a spike of news media attention occurred in 2003 and 2004 in the UK national press, largely associated with HRH Prince Charles' intervention in the debate – and leading to the issues gaining coverage by political and general reporters – this was relatively short-lived (see Anderson et al. 2005b; Anderson et al. 2009).

Simply because a dramatic event involving chemical risks occurs, it does not mean that it will automatically command the same level of media interest across the globe since the newsworthiness of chemical risks is strongly influenced by their proximity to home. While the Exxon Valdez oil spill attracted sustained, intense media interest in the US – particularly given that it occurred in the setting of Prince William Sound which is widely regarded as symbolizing Alaska's 'unspoilt' natural heritage – it gained relatively little attention in the international media (Birkland and Lawrence 2002; Mazur 1998;Wheelwright 1994; Wilson 1992).

Finally, conflict and controversy also constitute important elements of a news story. Confrontational items involving 'goodies' and 'baddies' are seen as having particular audience appeal (see Anderson 1997). Greenpeace have become particularly adept at tuning into these news values. As Chris Rose observes:

> What Greenpeace are very good at is they've invented, if you like, a sort of morality play [...] You've got to have the pictures, it doesn't matter what they're talking about, you've got to have the pictures. So that takes Greenpeace straight out of the editorial system of gatekeepers... and it puts them in the same sort of news as the royal family/entertainment news... if you can't deal with it in those terms, and their formula, they can't really campaign on it, which is one reason they've stuck with boats on the high seas and are therefore not affected by things like trespass law. Issues are simplified, they're global problem issues and they're David and Goliath, a sort of pantomime I suppose. (Interview, Chris Rose, Media Natura, 24th January 1990 cited in Anderson 1997)

3.4 Risk Society

The discussion in the preceding sections has illustrated some of the principal ways in which chemical risks are socially constructed and selectively mediated, serving some interests over others. In this context, the work of German social theorist,

Ulrich Beck, has been particularly influential. He argues that, in contrast with the natural hazards that dominated pre-industrial society, today's 'manufactured risks' have unpredictable effects that permeate far beyond a specific geographical location, or a particular point in time. According to Beck, risks can 'be changed, magnified, dramatised or minimised within knowledge, and to that extent they are particularly open to social definition and construction' (Beck 1992: 23). Beck contends that the institutions which are involved in managing and assessing levels of risk, the 'relations of definition', play a critical role in informing public awareness and knowledge of dangers. Beck's broad-brush, macro-level approach draws attention to the key role played by the media in articulating competing rationality claims. However, while Beck's work has undoubtedly made a very valuable contribution towards understanding the nature of contemporary society, it has been critiqued as over simplified, unevenly developed and contradictory in places (Cottle 1998: 25; Mythen 2004: 76).

As Mythen (2007) comments, the risk society thesis needs to be assessed in the context of its over-arching purpose, which was to illuminate key aspects of the modern condition and stimulate debate. However, it is argued here that the gaps need to be fleshed out by empirical research to reveal the complexities and subtleties that may not be apparent from broad brush theorising. It is useful, therefore, to consider some of the blind spots with this theory. In particular, Beck makes bold claims about media influence without providing empirical evidence, and presents the media as monolithic (Anderson 1993: 51; 1997: 188). Research has demonstrated that the news media are differentiated, occupying their own particular market niches, and governed by a range of economic and political constraints (Anderson 1997; Hargreaves et al. 2003). Indeed, representations of chemical risks differ according to different media formats that are affected by their own particular restrictions and practices. For example, risk reporting in popular red top UK newspapers tends to focus more on the 'human interest' angle and the experiences of ordinary people compared with the elite press (Murdock et al. 2003) and television formats tend to favour risk items with strong visual appeal (Anderson 1997, 2006). Some chemical risks are likely to resonate more closely with publics than others which may be perceived to be less immediate or meaningful (Kitzinger 1999). As we shall see in Chapter 4 of this volume, there are also important divergences in how chemical risks are reported by different countries, yet Beck pays little attention to national and political cultures and contexts (Cottle 1998: 17–18). The media-centric focus of Beck's analysis directs attention away from considering the processes involved in news production and is: 'conspicuously silent … on the institutional field in which 'relations of definitions' compete for public recognition and legitimation' (Cottle 1998: 18). As will be demonstrated later, official sources of information do not automatically gain advantaged access to the media and it is necessary to consider a whole array of different claims-makers who deploy particular strategies towards voicing their views.

What is needed is careful empirical analysis of source–media relations which builds upon and tests this macro-level theory (Anderson 1997; Cottle 1998). The aim of this final section of the chapter, therefore, is to examine two case studies of the communication of chemical risks, the Brent Spar in 1995 and the Prestige oil disaster of 2002. These case studies illustrate the importance of considering the news media strategies of sources that are vying for media attention. The discussion considers how the analysis of the media reporting of chemical risks must be placed within the broader context of the growing concentration and globalisation of news media ownership and the increasingly 'promotional culture' in which we inhabit, highlighted by the rapid rise of the public relations industry in recent years. Globalisation and the growing reliance upon new media technologies, it is contended, have transformed the news media and the ways in which social actors target their activities.

3.5 The Brent Spar

Greenpeace had been campaigning about pollution of the oceans since the early 1980s but the row over the decommissioning of the Brent Spar oil storage and loading buoy in 1995 provided a highly visual 'crystallizing moment' to draw public attention to the issue (Jordan 2001; Rose 1998). After 3 years of research Shell/Royal Dutch came to the decision that disposal at sea was the best option for the Brent Spar, although it failed to consult more widely – including taking into account the views of environmental pressure groups. This decision was challenged by Greenpeace on environmental grounds who argued that it could set a dangerous precedent for industry to view the ocean as a dumping ground. Greenpeace activists began their occupation of the Spar on 30 April 1995 and in the coming months attracted considerable news media attention to the issue of deep-sea disposal in Northern Europe (Hansen 2000; Bakir 2006). After 3 years eventually a combination of pressures led to European regulatory change in 1998, when OSPAR voted for a full ban on dumping steel offshore oil rigs and platforms in the North Sea. The topsides of the Brent Spar were taken away for recycling and the hull sections of the installation reused in a harbour development in Norway.

The battle between Greenpeace and Royal Dutch/Shell is widely regarded as a landmark case in risk communication. From the start of its campaign Greenpeace labelled the Spar as toxic and radioactive and their use of the term 'dumping' signalled that Shell was behaving in an environmentally irresponsible manner. Consumer boycotts of Shell petrol stations were triggered across Northern Europe, though Greenpeace itself did not call for such action. There were threats of violence and an actual incident of fire-bombing at a petrol station in Germany and petrol sales were reduced by around 20% in Shell's 1,728 German stations (Löfstedt and Renn 1997). Such was level of public outcry that Shell Germany was sent over 11,000 letters in protest at their plans (Löfstedt and Renn 1997). The issue of deep-water disposal particularly resonated with the German public since they already had long-standing concerns over pollution of the North Sea and, given that there are no oil reserves in Germany, the country had no strong political ties with the oil industry (Bakir 2006; Jordan 2001). In an about-turn on 20th June 1995 Shell bowed to consumer pressure and announced that they had decided to abandon deep-sea disposal, though they still maintained that deep-water disposal was the best option on technical and environmental grounds. It took another 4 months before an independent scientific assessment was conducted. Shell's U-turn was seen as signalling that their case had been flawed. Moreover, the use of pro-Shell scientists in the media in the immediate weeks after the announcement (rather than independent scientists) only served to increase the level of distrust in their claims (Bakir 2006).

As with the Prestige oil spill, the row over the Brent Spar was of wider European interest. The campaign would not have been successful had there not been sufficient public outcry and political opposition generated in Germany and the Netherlands, since the controversy generated barely any protest in the UK (Bennie 1998; de Jong 2005; Jordan 2001). This was, in large part, down to Greenpeace's successful strategy of targeting the news media in Northern Europe (Anderson 2003). Fortuitously it also took place at a time where a number of important political events were happening, including a conference in June 1995 discussing the future of the North Sea, and the G7 meeting. Undoubtedly, the visually symbolic nature of the conflict was a key factor. Video news releases of two Greenpeace vessels being attacked with high pressure water cannons made dramatic visual images. At the beginning of the campaign 20 journalists were invited to board the Greenpeace vessel and live satellite pictures were beamed back direct to news desks. The pressure group spent around £350,000 on media out of a total campaign budget of £1.3 million, which included satellite equipment, filming costs and the hiring a helicopter that would pick up film from the deck of the platform and itself take more pictures (Pearce 1996). According to Chris Rose (the then UK director of campaigns for Greenpeace): 'Without a doubt, those images, though very expensive to get, changed the story for us. News-wise, not a lot was happening. But they were great pictures and they ran round the world' (Pearce 1996: 86).

During the campaign Shell's slow response and reactive approach was seen as signalling their guilt (Bakir 2006). Acting on the defensive, the company took a top-down approach to communication and often acted evasively frequently responding with 'no comment' (Palmer 2001 cited in de Jong 2005). Their communications strategy was hindered by their internal structure and weak co-ordination of the companies in different countries. This often meant that they lacked one clear 'voice' with the German office saying one thing and the UK office saying another (Löfstedt and Renn 1997). Key documents were treated as confidential and there was limited dialogue with external stakeholders. Shell had no website and tended to communicate via telephone or in person (de Jong 2005). Press releases tended to narrowly reflect Shell's interests rather than engage with the wider issues under debate. Furthermore, they were often issued after events had moved on and so, by the time they reached journalists, they were out-of-date (de Jong 2005). By contrast, Greenpeace had a very well developed website enabling them to communicate directly with their supporters during the campaign and offering them regular updates and a diary of events (de Jong 2005). They were one of the first environmental groups to develop an internal email system back in the late 1980s helping them to rapidly spread information within the organisation (Pearce 1996).

Moreover, Greenpeace were widely perceived by the public to have the moral high ground (Bennie 1998). While Shell/Royal Dutch communicated the issues on purely rational/technical grounds Greenpeace also sought to appeal directly to people's hearts through the moral dimension. When an independent audit of the Spar was carried out by Det Norske Veritas (DNV) revealing that Greenpeace's estimate of the oil content was too high, they pre-empted the publication of the findings by publicly apologising for the mistake 6 weeks before they were released. This transparent and open approach contrasted with Shell's tendency towards secrecy and closure. Following Greenpeace's admission public opinion polls showed that public support for their case had not been badly dented, though editors were quick to castigate the hand that had fed them (Anderson 1997; Rose 1998). Some complained that reporters had too readily accepted Greenpeace video news releases and had failed to cover the conflict objectively. At the Edinburgh International Television Festival, David Lloyd, senior commissioning editor for Channel 4 News, stated:

> On Brent Spar we were bounced. This matters – we all took great pains to represent Shell's side of the argument. By the time the broadcasters had tried to intervene on the scientific analysis, the story had been spun, far, far into Greenpeace's direction [...] When we attempted to pull the story back, the pictures provided to us showed plucky helicopters riding a fusillade of water cannons. Try and write the analytical science into that to the advantage of the words. (Cited in Rose 1998: 158)

Shell was also highly critical of the extent to which the news media had relied upon Greenpeace footage. The then Shell chief executive, Dr Chris Fay, complained: 'Like all good spin doctors, they knew how to manage the debate with a flow of simplistic allegations – the daily 'curve ball'. They knew that activists in rubber boats among the massive ironmongery of the North Sea made good television; David and Goliath' (see Goddard 1996).

However, Shell under-estimated the importance of trust in risk communication. Following the Brent Spar affair it hired a PR firm and spent millions of pounds in seeking to change its public image. The company engaged in wider consultation and adopted a two way dialogue model of communication (Bakir 2006). As discussed below, trust in sources again proved to be a major issue during the communication of the Prestige oil disaster. Moreover compared with previous oil spills which received a high profile within the media, such as the Exxon Valdez, official versions of events were much more easily challenged through the role of the international press and the Internet.

3.6 The Prestige Oil Disaster

> Never in the history of Spain has an environmental disaster aroused such public outcry, exerted such a political impact, or elicited such media coverage as the Prestige oil spill (WWF 2002).

The Prestige oil spill in 2002 was one of the worst oil spills to have affected the European coastline, sparking considerable levels of public concern and news media attention. Concerns over the handling of the crisis subsequently led to the publication of a Good Practice Guide for communicating risk with regards to accidental marine pollution incidents (AMPERA 2007). A content analysis of newspaper reporting of the Prestige accident in the UK, France, and Spain between November 14 and 26, 2002 was performed, taking into account the totality of coverage over the period and the number of front page items that led on the story (see Anderson and Marhadour 2007). The sample mostly included UK and Spanish newspapers, but two French newspapers were also included to enable further cross-cultural comparisons to be made. The articles were retrieved via Lexis-Nexis search facilities, with the exception of the two regional Spanish newspapers (where the newspaper's own archives were used). A variety of different framings of the Prestige oil spill were offered in national and regional newspapers in Spain, France, and the UK. The local Spanish press was found to provide the most sustained coverage and geographic propinquity to the accident was a good predictor of the frequency and intensity of reporting, though there are clearly a variety of different factors at work. Regional Spanish newspapers focused upon implications for the local economy rather than the effects on wildlife; whereas national newspapers in Spain, France, and the UK framed the oil spill largely in terms of its environmental impacts and the political controversy over who was to blame. Other research suggests that government sources gained the most advantaged access to the Spanish news press (Agraso et al. 2003). These findings echo those of an earlier study by Molotch and Lester (1975) into press coverage of the 1969 Santa Barbara oil spill, which found that official sources dominated coverage and that the greater the geographical proximity to the accident the more high profile and intense the news coverage.

From the start, the authority's response to the Prestige crisis was slow, uncoordinated and disorganised and Spanish and Galician government research centres made little contact with university marine scientists. As with the Brent Spar, a number of communication errors were made (AMPERA 2007; Freire et al. 2006). Communication tended to be top-down and different government spokespersons provided contradictory or ambiguous messages. Also, there was no independent expert commentator to justify the government's actions (AMPERA 2007). According to marine scientist, Ana Vilas Paz:

> The central government spokesman's press conferences were always denied in less than 12 hours, by the facts announced in some private TV channels or by Portugal and France. It appears that the government, because of the impossibility of controlling the situation, chose to hide the information by deceitfully reassuring messages, developing obstacles for the media, lies, and even censorship (forbidding for example the flights over the sinking zone). This kind of system doesn't work when it is perfectly evident what's going on and people can contrast it with false official information. The government even prohibited the public TV workers from using the term 'oil spill.' (Vilas Paz 2004)

The response from university scientists was also slow and generally ineffective, hampered by institutional structures and the absence of open-access databases on ecosystems and marine resources (Freire et al. 2006). However, international environmental pressure groups including Greenpeace, Friends of the Earth (FOE), World Wide Fund for Nature (WWF) and the International Fund for

Animal Welfare (IFAW) were quick to convey information about the crisis on their websites and produce their own reports (e.g., see Caballero 2003). A major coordinating role was played by the World Wildlife Fund (Spain) (WWF), who rapidly responded to the disaster by making various scientific documents available, issuing media releases and developing a crisis group to oversee communication and strategies on conservation policy (Anderson and Marhadour 2007; WWF Spain 2003). Pressure groups drew attention to the weaknesses of the Spanish government in handling the crisis. Established and newly formed environmental organisations effectively mobilised people to take part in mass protests. It was estimated that in December 2002 around 200,000 people gathered in protests on the streets in North West Galicia, claiming that information about the oil spill had been censored by the authorities (Anderson and Marhadour 2007). The Internet, alongside traditional media, proved to play an important role in conveying information. Marcus Fernández, assistant director of Código Cero (a Galicia based website) commented:

> Without the Web, the satellite photographs of the affected zone wouldn't have been shown publicly, people wouldn't have been coordinated quickly to act as volunteers in the most affected zones, information wouldn't have gotten to the volunteers about how to handle the sludge he said. Definitely, the (role) of the Internet has been decisive. (Scheeres 2002)

Particularly for activists the Internet has increasingly provided a key source of alternative perspectives challenging the official version of events. It is also likely to have affected news coverage as news agencies have a potentially crucial agenda-setting influence upon other media. Even where local publics tend to access local/regional news media as their main source of information, activists across the globe can coordinate protests and dispute official accounts with growing sophistication and speed. However, one of the difficulties associated with the pressures of round the clock news coverage is selective and intermittent reporting. While the international press may be more likely to frame such disasters in terms of their environmental impact, their attention tends to be short-lived and rarely includes an in-depth consideration of the role of institutional and structural factors influencing the shipping industry, such as the globalisation of powerful oil corporations (Anderson and Marhadour 2007).

3.7 Conclusion: Beyond the Risk Society

What can be learnt from this discussion of the news coverage of crises over chemical risks in contemporary Europe? Both of the examples considered here highlight the need for the ongoing cultivation of trust among the stakeholders involved in communicating chemical risks. As Bakir observes: '…social trust in risk communicators cannot be assumed but must be cultivated and maintained with key audiences prior to, and during, risk communication' (Bakir 2006: 82). Risk communication must be proactive, begin at an early stage and genuinely listen to the views of stakeholders and publics. Conflicting messages, poor co-ordination and a lack of transparency all tend to breed distrust. Since the mid 1990s levels of trust in policymakers and regulators among European publics has increasingly declined, whilst survey responses suggest that environmental NGOs have scored the highest in terms of trust (see Petry et al. 2006; Selin 2007). In the UK the handling of the crises over genetically modified (GM) food and crops and Bovine Spongiform Encephalopathy (BSE) significantly contributed to low levels of trust in risk regulators. Despite misgivings over the impact of REACH legislation on competitiveness some see this as presenting an opportunity to restore consumer trust in the chemicals industry. However, it remains to be seen whether citizen concerns over the use of harmful chemicals (in food, for example) will be amplified and stigmatised by the news media. Environmental NGOs have identified a number of loopholes and flaws with the legislation, and are pressing for safer alternatives to be found to potentially stigmatizing 'substances of very high concern', such as persistent and bio-accumulative substances, used in everyday consumer products (Chemical Reaction 2007a, b).

Once the news media have framed an issue in the early stages it can prove very hard for other news actors to shift the parameters of debate. The examples of the Brent Spar and the Prestige demonstrate how environmental NGOs are often able to speedily draw international attention to an issue which resonates with underlying public concerns. The intensification of the speed of modern communication networks places great pressure upon journalists to provide information in rapid sound bites. Where there is considerable uncertainty and ambiguity over the science, the news production process becomes especially fraught. Scientists are often reluctant to comment on chemical risks where the evidence is uncertain. At the same time the Internet provides a range of competing interpretations, including challenges to official accounts that can often be accessed at speed. Scientists need to develop a greater appreciation of the workings of the media (particularly the mismatch between scientific and journalistic timescales), while journalists need to cultivate a better understanding of the constraints affecting how scientists operate. Both news and science are always a social production involving contending claims about 'truth'. With the growth of the public relations industry and the politics of 'spin', journalists can benefit considerably from a deeper understanding of how relations between news sources and the media shape the reporting of chemical risks (Anderson et al. 2005a; Davis 2000, Miller and Dinan 2000).

Future case studies, it is hoped, will further illuminate the shifting dynamics of source-media relations in the communication of chemical risks and lead to improvements in reporting practices. Further work needs to systematically examine the kinds of discursive strategies that are used by stakeholders in their efforts to frame issues within the news media. In doing so, there is a need to go beyond a media-centred approach to consider the complexities and contingencies of social processes and the wider play of political power.

References

Agraso, M.F., Eirexas, F. and Jiménez-Aleixandre, M.P. (2003) The Prestige in the Galician press in 2002: a preliminary analysis. *Quark, Ciencia, Medicina, Communicacion y Cultura* 28: 44–56.

AMPERA (2007) Risk communication in accidental marine pollution: good practice guide for an effective communication strategy. http://otvm.uvigo.es/investigacion/informes/documentos/AMPERA_Publication_Nr2_RiskCommunication.pdf. Cited September 2008.

Anderson, A. (1993) The production of environmental news: A study of source-media relations. Unpublished PhD, University of Greenwich.

Anderson, A. (1997) *Media, Culture and the Environment*. UCL, London.

Anderson, A. (2002) The media politics of oil spills. *Spill Science and Technology Bulletin* 7(5): 7–16

Anderson, A. (2003) Environmental activism and news sources. In: Cottle, S. (ed.) *News, Public Relations and Power*. Sage, London.

Anderson, A. (2006) Media and risk. In: Walklate, S. and Mythen, G. (eds.) *Beyond the Risk Society: Critical Reflections on Risk and Human Security*. Open University/McGraw Hill, Maidenhead, pp 114–131.

Anderson, A. and Marhadour, A. (2007) Slick PR? The media politics of the Prestige oil spill. *Science Communication* 29(1): 96–115.

Anderson, A., Petersen, A. and David, M. (2005a) Communication or spin? Source -media relations in science journalism. In: Allan, S. (ed.) *Journalism: Critical Issues*. Open University Press, Buckingham.

Anderson, A., Petersen, A, Wilkinson, C. and Allan, S. (2005b) The framing of nanotechnologies in the British newspaper press. *Science Communication* 27(2): 200–220.

Anderson, A., Petersen, A., Wilkinson, C. and Allan, S. (2009) *Nanotechnology, Risk and Communication*. Palgrave Macmillan, Houndmills.

Bakir, V. (2006) Policy agenda-setting and risk communication: Greenpeace, Shell and issues of trust. *Press/Politics* 11(3): 67–88.

Bauer, M.W. and Gaskell, G. (2002) *Biotechnology: The Making of a Global Controversy*. Cambridge University Press, Cambridge.

Beck, U. (1992) *Risk Society: Towards a New Modernity*. Sage, London.

Bennie, L. (1998) Brent Spar, Atlantic oil and Greenpeace. *Hansard Society Series in Politics and Government* 7: 89–102.

Birkland, T.A. and Lawrence, R.G. (2002) The social and political meaning of the Exxon Valdez oil spill. *Spill Science and Technology Bulletin* 7(5): 17–22.

Caballero, M.J. (2003) The Prestige disaster: one year on. Greenpeace Spain Report, November. http://www.greenpeace.org/multimedia/download/1/346545/0/Prestige_report.pdf/

Chemical Reaction (2007a) Navigating REACH: an activist's guide to using and improving the new EU chemicals legislation, November. http://www.chemicalreaction.org/

Chemical Reaction (2007b) My voice – how you can demand better protection of human health and the environment from hazardous chemicals, May. http://www.chemicalreaction.org/

Cottle, S. (1998) Ulrich Beck, 'risk society' and the media: a catastrophic view? *European Journal of Communication* 13(1): 5–32.

Darley, J. (2000) Making the environment news on the Today programme. In: Smith, J. (ed.) *The Daily Globe: Environmental Change, the Public and the Media*. Earthscan, London.

Davis, A. (2000) Public relations, news productions and changing patterns of source access in the British national media. *Media, Culture and Society* 22(1): 39–59.

De Jong, W. (2005) The power and limits of media-based international oppositional politics – a case study: the Brent Spar conflict'. In: de Jong, W., Shaw, M. and Stammers, N. (eds.) *Global Activism, Global Media*. Pluto, London.

Dunwoody, S. (1999) Scientists, journalists, and the meaning of uncertainty. In: Friedman, S., Dunwoody, S. and Rogers, C.L. (eds.) *Communicating Uncertainty: Media Coverage of New and Controversial Science*. Lawrence Erlbaum, New York.

Entman, R.M. (1993) Framing: towards clarification of a fractured paradigm. *Journal of Communication* 43(4): 51–58.
Ericson, R.V., Baranek, P.M. and Chan, J.B.L. (1989) *Negotiating Control: A Study of News Sources*. University of Toronto Press, Toronto.
Freire, J., Fernández, L. and Muiño, R. (2006) Role of the Spanish scientific community in the initial assessment and management of the environmental damages caused by the Prestige oil spill. *Marine Policy* 30(4): 308–314. http://otvm.uvigo.es/investigacion/informes/documentos/archivos/marine_policy_freire.pdf
Goddard, L. (1996) Shell chief knocks media over simplistic coverage. *PR Week*, 24th May.
Gorss, J. and Lewenstein, B. (2005) The salience of small: nanotechnology coverage in the American press, 1986–2004. Paper presented at International Communication Association Conference, 27 May 2005.
Greenberg, M.R., Sachsman, D.B., Sandman, P.M. and Salome, K.L. (1989) Network evening news coverage of environmental risk. *Risk Analysis* 9(1): 119–126.
Gregory, J. and Miller, S. (1998) *Science in Public: Communication, Culture and Credibility*. Plenum, New York.
Hannigan, J.A. (2006) *Environmental Sociology: A Social Constructionist Perspective*. Routledge, London.
Hansen, A. (2000) Claims-making and framing in British newspaper coverage of the 'Brent Spar' controversy. In: Allan, S., Adam, B. and Carter, C. (eds.) *Environmental Risks and the Media*. Routledge, London.
Hargreaves, I., Lewis, J. and Speers, T. (2003) *Towards a better map: science, the public and media*. Report prepared for the ESRC.
Hilgartner, S. (1990) The dominant view of popularization: conceptual problems, political uses. *Social Studies of Science* 20: 519–539.
Hornig Priest, S. (2001) Cloning: a study in news production. *Public Understanding of Science* 10: 59–69.
Jordan, G. (2001) *Shell, Greenpeace and Brent Spar*. Palgrave, Basingstoke.
Kitzinger, J. (1999) Researching risk and the media. *Health, Risk and Society* 1(1): 55–69.
Lewenstein, B.V. (1995a) Science and media. In: Jasanoff, S., Markle, G.E., Petersen, J.C. and Pinch, T. (eds.) *Handbook of Science and Technology Studies*. Sage, Thousand Oaks.
Lewenstein, B.V. (1995b) From fax to facts: communication in the cold fusion saga. *Social Studies of Science* 25: 403–436.
Lichtenberg, J. and MacClean, D. (1991) The role of the media in risk communication. In: Kasperson, R. and Stallen, P. (eds.) *Communicating Risks to the Public: International Perspectives*. Kluwer Academic, London.
Löfstedt, R.E. and Renn, O. (1997) The Brent Spar controversy: an example of risk communication gone wrong. *Risk Analysis* 17(2): 131–135.
Logan, R.A. (1991) Popularization versus secularization: media coverage of health. In: Wilkins, L. and Patterson, P. (eds.) *Risky Business: Communicating Issues of Science, Risk and Public Policy*. Greenwood, New York.
Manning, P. (2001) *News and News Sources: A Critical Introduction*. Sage, London.
Mazur, A. (1998) Global environmental change in the news: 1987–90 vs. 1992–6. *International Sociology* 13(4): 457–472.
Miller, D. and Dinan, B. (2000) The rise of the PR Industry in Britain: 1979–98. *European Journal of Communication* 15(1): 5–35.
Miller, M. and Riechert, B.P. (2000) Interest group strategies and journalistic norms: news media framing of environmental issues. In: Allan, S., Adam, B. and Carter, C. (eds.) *Environmental Risks and the Media*. Routledge, London, pp 45–54.
Molotch, H. and Lester, M. (1975) Accidental news: The great oil spill as local occurrence and national event. *American Journal of Sociology* 81: 235–260.
Murdock, G., Petts J. and Horlick-Jones, T. (2003) After amplification: rethinking the role of the media in risk communication. In: Pidgeon, N., Kasperson, R.E. and Slovic, P. (eds.) *The Social Amplification of Risk*. Cambridge University Press, Cambridge, pp 156–178.

Mythen, G. (2004) *Ulrich Beck: A Critical Introduction to the Risk Society*. Pluto Press, London.
Mythen, G. (2007). Reappraising the risk society thesis: telescopic sight or myopic vision? *Current Sociology* 55(6): 793–813.
Nelkin, D. (1995) *Selling Science: How the Press Covers Science and Technology*. WH Freeman, New York.
Nisbet, M. and M. Huge (2007) Where do science debates come from? Understanding attention cycles and framing. In: Brossard, D., Shanahan, J. and Nesbitt, T.C. (eds.) *The Public, the Media and Agricultural Biotechnology*. CABI, Oxford, pp 193–230.
Nisbet, M.C. and Lewenstein, B.V. (2002) Biotechnology and the American media: the policy process and the elite press, 1970 to 1999. *Science Communication* 23(4): 359–391.
Nisbet, M., Brossard, D. and Kroepsch, A. (2003) Framing science: the stem cell controversy in the age of press/politics. *Harvard International Journal of Press/Politics* 8(2): 36–70.
Pearce, F. (1996) Greenpeace mind-bombing the media. *Wired* 2(5): 84–88.
Petersen, A. (1999) The portrayal of research into genetic-based differences of sex and sexual orientation: a study of 'popular' science journals, 1980 to 1997. *Journal of Communication Inquiry* 23(2): 163–182.
Petersen, A., Anderson, A., Allan, S. and Wilkinson, C. (2009) Opening the black box: scientists' views on the role of the mass media in the nanotechnology debate. *Public Understanding of Science* 18(5): 512–530.
Petry, T., Löfstedt, R.E. and Dietrich, D. (2006) Science and politics: from science to decision making. *Regulatory Toxicology and Pharmacology* 44(1): 1–3.
Petts, J., Horlick-Jones, T. and Murdock, G. (2001) The social amplification of risk: the media and the public. *Contract Research Report 329/2001*. HSE Books, Sudbury.
Renn, O. and Kastenholz, H. (2000) Risk communication: chemical products risks. OECD background paper. http://www.bfr.bund.de/cm/238/background_paper.pdf
Rose, C. (1998) *The Turning of the 'Spar'*. Greenpeace, London.
Ryan, C. (1991) *Prime Time Activism: Media Strategies for Grassroots Organising*. South End Press, Boston.
Schanne, M. and Meier, W. (1992) Media coverage of risk. In: Durant, J. (ed.) *Museums and the Public Understanding of Science*. Science Museum Publications, London.
Scheeres, J. (2002) Oil's not well in Spain. Wired, 17th December. http://www.wired.com/news/politics/0,1283,56877,00.html
Selin, H. (2007) Coalition politics and chemicals management in a regulatory ambitious Europe. *Global Environmental Politics* 7(3): 63–93. http://muse.jhu.edu/journals/global_environmental_politics/v007/7.3selin.pdf
Singer, E. and Endreny, P. (1987) Reporting hazards: their benefits and their costs. *Journal of Communication* 37(3): 10–26.
Vilas Paz, A. (2004) Prestige: Never again. http://www.archipelago.gr/defaulten.asp
Wheelwright, J. (1994) *Degrees of Disaster, Prince William Sound: How Nature Reels and Rebounds*. Simon and Schuster, New York.
Wilsdon, J. and Willis, R. (2004) *See-through Science: Why Public Engagement Needs to Move Upstream*. Demos, London.
Wilson, A. (1992) *The Culture of Nature: North American Landscape from Disney to the Exxon Valdez*. Blackwell, Oxford.
World Wildlife Fund (2002) The Prestige catastrophe: oil spill off Spain's NW coast. http://www.panda.org/news_facts/newsroom/crisis/spain_oil_spill/index.cfm
World Wildlife Fund Spain (2003) Prestige oil spill WWF Spain activities: progress report. http://www.wwf.dk/db/files/prestige_ en_oliekatastrofe_i_gali.pdf. Cited May 2003.

Chapter 4
Framing Chemical Risks in Sweden and Poland: Journalists' Narratives and Media Texts

Annika Egan Sjölander, Katarzyna Wolanik Boström, and Kenneth Ögren

Abstract How are chemicals framed in the press in Sweden and in Poland? We have conducted interviews with journalists representing local press, tabloids and national newspapers in order to grasp the professionals' own narratives about chemicals and also the range of diversity within journalism. What at first can appear as a marginalized topic, chemicals, partly because it is not an established journalistic genre, has turned out to have many faces. All news treating additives in food production and every report relating to medicines, such as the growing resistance towards antibiotics among the population, is part of the discourse, not to mention accidental releases of hazardous substances, etc. Secondly, it is a central part of the study to understand how these dominant themes are textually constructed in the press coverage. The news and media debate about chemicals are not only a central information source for the majority of citizens; the mass media also influence stakeholders, opinion-leaders and decision-makers in society. By and large the results indicate that the types of frames that are used by journalists in these two countries have a lot in common, even if the content of the media texts and the specific national contexts differ substantially between Sweden and Poland.

Keywords Chemicals • Framing • Journalists • Narratives • News media

4.1 Introduction

Framing has come to be a central and contested concept for media and communication scholars (Scheufele 2006:65, D'Angelo 2002:870, Weaver 2007:143, Scheufele 1999:103). In this article we want to investigate how chemicals are framed both in

---

A.E. Sjölander (✉), K.W. Boström, and K. Ögren
Umeå University, Institutionen för kultur- och medievetenskaper, SE 901 87, Umeå, Sweden
e-mail: annika.egansjolander@kultmed.umu.se; katarzyna@kultmed.umu.se; kenneth.ogren@kultmed.umu.se

J. Eriksson et al. (eds.), *Regulating Chemical Risks: European and Global Challenges*,
DOI 10.1007/978-90-481-9428-5_4,

journalists' narratives about their work and in the results of it, namely published media texts. We apply a comparative approach and have chosen to study the contemporary press in Sweden and Poland. As we see it, policy-making in relation to chemicals is influenced not only by expert-based risk assessments and related political processes but also by media framings.[1] Framing refers to interpretation and selection processes by journalists in the newsrooms in the production of news and other media texts. According to Robert M. Entman (1993:52):

> To frame is to *select some aspects of a perceived reality and make them more salient in a communicating text, in such a way as to promote a particular problem definition, casual interpretation, moral evaluation, and/or treatment recommendation* for the item described (italics in original).

The framing process is active, even if not necessarily deliberative or consciously manufactured by the individual news reporter. It is part of the common routines among the professionals and facilitates the process of sampling from a large amount of available information (Entman 1993, Johansson Lönn 2005:47f). Furthermore, the concept of framing is based on the assumption that the way in which an issue is characterized in news reports can have an influence on how it is understood by its audiences (Scheufele and Tewksbury 2007:11). We are convinced that the mass media discourse on chemicals and the risks related to the widespread use of them plays a significant role for a large portion of the population in Sweden and Poland. At least the media do, in so far as the public devotes interest to the subject, particularly when there is a crisis involving the spread of hazardous chemicals (Jarlbro 2001:8f, Nohrstedt and Nordlund 1993). The traditional news media still comprise a primary source of information for the majority of citizens, even after the broad introduction of Internet (cf. Carlsson 2008). In addition, when it comes to these kinds of complex, abstract subjects on which the individual has little personal experience to base his or her understanding, the mass media is more likely to be the only source of information for the citizen (Nelkin 1995, Nordlund 1995). Another reason for the pivotal role of the mass media is that politicians, experts and other decision-makers in society systematically use the media as an arena to express their views (Nord and Strömbäck 2004, Larsson 2005, Miczynska-Kowalska 2007). It also operates as a potential meeting point with the general public and provides decision-makers with a sense of public opinion on different subjects, which in turn affects their work and positions (Johansson 2000).

Another crucial point of departure in our work is the recognition that environmental problems, such as the spread of chemicals in society, are socially constructed (Anderson 1997, Hansen 1991, 1994). This means that they are strongly

[1]These three spheres – science, politics and media – have been studied simultaneously in the interdisciplinary research programme *Regulating Chemicals in the Baltic Sea Area: Science, Politics and the Media*, led by Dr. Michael Gilek, Södertörn University College, with support from The Foundation for Baltic and East European Studies (Östersjöstiftelsen). Apart from the research group, we especially thank Vanni Tjernström, Maritha Jacobsson and our colleagues at the Department of Culture and Media, Umeå University for constructive comments on earlier drafts of the manuscripts, and Martin Shaw for revising the English.

related to the ever-changing social, political and cultural understanding of them, including media framing (Eder 1996, Petersen 1997, Djerf-Pierre 1996, Hermansson 2002). Environmental issues have never received a great deal of attention in media and communication studies, and as Alison Anderson (Chapter 3) points out, there is still a great deal of research left to do. Studies on chemicals and the media tend to focus on catastrophes and crises (cf. Mazyr 1984, Jarlbro et al. 1997, Löfstedt and Renn 1997, Anderson and Marhadour 2007). Our project aims to study the 'ordinary' daily discourse about chemicals in the news media, an area about which less is known.

4.2 Purpose

The purpose of this chapter is to present and discuss results from a study on how chemicals are framed in the contemporary press in Sweden and Poland. These two EU countries in the Baltic Sea Region have different capacities for managing the risks involved in the past and present use of chemicals. We study the content and form of media texts as well as the journalists' own narratives about their work. We regard the combination of these empirical materials as an advantage in the study of media framing.

National and regional/local newspapers have been studied in order to gain a broader understanding of the different types of journalism and debates about chemicals that appear in the daily press. A comparative approach has been applied in the analysis, both between different news types/genres and between the two countries. Furthermore, we explore which similarities as well as differences that can be relevant in influencing the media framing process in the Swedish and Polish contexts.

4.3 The Swedish and Polish Cases

The member states of the European Union relatively recently (June 2007) implemented a highly controversial directive called Reach, which stands for Regulation, Evaluation, Authorization of Chemicals (cf. Hermansson and Reuter 2006; Eriksson et al. 2010). Briefly stated, the aim of the law is to increase knowledge and control of the use of chemicals in society in general, including both new and 'old' substances available on the market. The primary goal is to protect human health and the environment (Regulation (EC) No 1907/2006, European Union 2006). Public databases to be shared by all of the EU countries, with risk assessments and other relevant safety information about chemical products are currently being compiled under the direction of the new European Chemicals Agency (Echa).[2] Henceforth, users of chemical substances, such as industries and companies, are responsible for providing safety information and for gathering the required

[2]Reach is discussed in greater depth in Chapters 5, 6 and 13 in this volume.

knowledge about the risks that derive from their use. Primary regulatory supervision in Sweden is undertaken by the Swedish Chemical Agency (Kemikalieinspektionen). The Polish equivalent is the Office for Matters of Chemical Substances and Preparations (Biuro do Spraw Substacji i Preparatów Chemicznych).

Both Sweden and Poland have implemented Reach without any strong national opposition, but the processes seem to have been rather different, if you look at the scientific, political and media framing (cf. Hermansson and Reuter 2006, Eriksson et al. 2010). For example, preliminary findings indicate that the largest Swedish morning newspaper, *Dagens Nyheter*, framed the new EU directive largely in terms of environmental and health aspects, whereas the Polish equivalent, *Gazeta Wyborcza*, had a much stronger focus on employment and economic impact of the new stricter legislation (Hermansson and Reuter 2006).

The need for stronger regulation of chemicals has been given a high priority in relation to the EU by Swedish politicians and other stakeholders from the Scandinavian countries for a long time (Eriksson et al. 2010, Government Bill 2004/05:150). Part of the reason for this proactive approach is that Sweden, in comparison to other nations, has had stricter legislation in place for several years. The situation in Poland, along with other post-communist states, is quite different (Andonova 2004, Andonova et al. 2007). The implementation of Reach in Sweden is closely related to the work with sustainable development and, more specifically, the aspiration for a so-called non-toxic environment (Swedish Parliamentant Decision 1998/99:183). Swedish politicians have vowed to solve all of the major environmental problems that we are facing today within a generation. 'The environment must be free from man-made or extracted compounds and metals that represent a threat to human health or biological diversity' (Government Bill 1997/98:145:135). Chemicals are clearly a part of these risks.

That there are differences between Sweden and Poland is not surprising considering the historical, welfare and cultural contexts within which these risk regulation processes take place. However, it is of interest to know more about the specific differences and similarities with regard to media framing of this risk regulation.[3] Poland has undergone dramatic changes since the fall of communism in 1989. The Polish press, radio and television are in themselves good examples of this radical and complicated shift towards democracy (Lara 2008, Obermayer 1994). Giorgi's (1995: ix) argues that

> the restructuring of the media sector is particularly illustrative of the tension and dilemmas intrinsic to the process of transition, which is not surprising given the inextricable connection of the mass media to the political system.

During the communist era, the Polish mass media were state-controlled and rigorously monitored. Only politically 'suitable' messages were passed on to the masses. News about chemical hazards and environmental contamination did not 'fit in' with the communist frame and were systematically excluded (Gerholm 1990). Despite censorship, oppositional underground media did exist, and the Catholic Church was

[3]Sweden joined the EU in 1995 and Poland in 2004.

able to publish its own newspapers. In comparison to other Eastern European countries, Poland 'saw by far the largest and most enduring opposition to communist rules' (Sparks 2008:12). The Polish press is now considered to be independent and is privately owned, mainly by foreign companies (Lara 2008, Sparks 2008).

The Polish Journalist Association (SPD, Stowarzyszenia Dziennikarzy Polskich[4]) is an organization of professionals that has implemented guidelines and ethical standards similar to Western democracies after the shift from communism. In 1995 a so-called *Media Ethics Charter*, emphasising the importance of the principles of truth, objectivity, honesty, etc., was also adopted by a number of Polish journalist and media organizations.[5]

Even if Poland is considered to be one of the more successfully democratized post-communist countries (Giorgi 1995:4), the actual freedom of the media is continuously contested, not only within the country but also internationally. Colin Sparks (2008) argues that the autonomy of journalists is low in Poland and that state or government intervention is still very high, despite changes in the society in recent years.

We are interested in examining whether these structural constrains are present in the Polish journalist's narratives about their work. Like Swedes, Polish citizens used to be characterized as a reading population. However, now the proportion of the population that reads a daily newspaper is slowly decreasing. 'Sweden and Norway are Europe's strongest newspaper markets with 85% of adults reading a newspaper every day' (Niiranen 2007:23). In contrast, Poland is the weakest with only 35% of adults reading a daily newspaper (ibid, see also Lara 2008, Weibull and Jönsson 2008). Still, the recent introduction of the tabloid Fakt was a spectacular success. The sale of national dailies has experienced a significant boost, and, as in Sweden some years ago, there is an impressive growth in the share of free newspapers. The biggest losses are suffered by the Polish regional/local press (Niiranen 2007:168170). Furthermore, access to and use of the Internet is significantly higher in Sweden than in Poland (Findahl 2007). However, in both countries traditional media institutions rank high on the list of websites that people visit (Lara 2008, Weibull and Jönsson 2008). We are interested in the role of the web in contemporary news production. Today's Swedish media landscape is, as all media markets, also changing but is relatively stable. Media freedom from censorship has a long history. The essential *Freedom of the Press Act* is part of the constitution and dates back as far as 1766. It is known to be the oldest in the world. The *Fundamental Law on Freedom of Expression* from the same era is also crucial for a functioning democratic media system (Asp 1986, Larsson 1998).

Poland is an interesting case, as it is a post-communist country. Before the fall of communism, the Polish People's Republic had a notorious reputation for huge – though not officially recognized – environmental problems. Journalists were supposed to be a vehicle for the dissemination of the socialist ideology to the masses, and the ideal journalist was to be a political activist that used his or her talents for the realization

[4](http://www.sdp.pl/history.php).

[5](http://ethicnet.uta.fi/poland/media_ethics_charter).

of the Party's goals. The main objective was not to inform but to shape a *proper image* of reality (defined by the Party). Consequently, the mediated world, expressed in Gerholm's (1990:128) words, was a strongly modified world. We suggest calling it a 'communist frame'.

The effect of censorship and factitiousness on Polish public opinion led to a popular conviction that whatever was published in an official newspaper could not be reliable. However, some critical articles were tolerated during the communist regime, especially in intellectual journals. In 1989, censorship was abandoned and the press was commercialized and privatized (Majcherek 1999). In 1990, all newspapers and journals in the state concern called *Prasa-Ksiazka-Ruch* were sold to private owners or handed over to associations started by editors and professional journalists (like Gazeta Wyborzca) or other political actors, such as unions. There has been a strong concentration of the Polish press ever since, with the result that many local and regional newspapers and journals have had to shut down.

4.4 Method and Materials

We apply a comparative approach throughout the study, both between countries and between different types of media genres. In their review of research on mass media coverage of technological and environmental risks, Dunwoody and Peters (1992) point out that cross-cultural collaboration is rare. That observation is also emphasized by Anderson and Marhadour (2007) in their recent study of how Spanish, French and British newspapers framed the Prestige oil spill in 2002. They argue that there 'is a need to move beyond studies at the national level to examine cross-cultural differences in the reporting of global environmental risks' (2007: 97).[6]

To get at good overview of contemporary daily press coverage of chemicals in Sweden and Poland, we have chosen to study texts from national newspapers, including one tabloid and one regional/local newspaper in each nation during an entire year (2007). A time frame of 12 months enables us to cover variations in interest for the subject and represents the kinds of content about chemicals that readers encounter in general. Because of difficulties in accessing media texts from Poland, we have extended the search period there to one and a half years (from January 2007 until June 2008).

The Swedish newspaper articles have been collected via digital archives for media texts: *Mediearkivet* and *Presstext*. In addition to *chemicals*, we have used a large number of key words in order to capture the varied media content that involves chemicals.[7] The great variety of topics that relates to chemical use and its risks is

[6]However, comparative studies, including cross-cultural analysis of news coverage, are common within the field of political communication. For an overview of this research tradition see Tjernström (2001).

[7]As a starting point in formulating the search profile, we consulted our interdisciplinary research group of scientists and took guidance from a recently published book (Johansson 2006) that maps out and discusses the diverse chemical field.

Table 4.1 Sample of Swedish newspapers for content analysis

Swedish newspapers	Frequency	Political orientation	Circulation 2007 (weekdays)
National press			
Dagens Nyheter (est 1864)	Daily	Independent Liberal	344,200
Aftonbladet (est 1830)	Daily	Independent Social Democratic	388,500
Regional/local press			
Västerbottens-Kuriren (est 1900)	6 days/week	Liberal	38,000

also shown in the results (cf. Table 4.3). The largest and leading national morning newspaper, *Dagens Nyheter* (DN), has been studied as well as the most widely read tabloid, *Aftonbladet* (AB). *Västerbottens-Kuriren* (VK) is our regional/local choice, a liberal newspaper (like the majority of newspapers in Sweden) and the largest in the region, with its head office in Umeå (112,000 citizens) (Table 4.1).

The material from the Polish press had to be retrieved from the web archives on the homepages of the respective newspapers via the Internet. Because of that, there is a higher degree of uncertainty with regard to validity, e.g. when it comes to knowing that these particular texts were actually published in the paper editions. Each newspaper tends to develop its own archival routines, which clearly makes comparisons more difficult (cf. Egan Sjölander 2007). However, since we are seeking to grasp general features in the reporting, these methodological problems have been manageable.

The Polish press that we have selected represents three different types. The national newspaper *Gazeta Wyborcza* (GW), founded at the end of communism in 1989, was for many years the largest newspaper. In 2003 the centre-oriented tabloid *Fakt* (F) was started and took the lead. GW is owned by Agora S.A. and has a social liberal profile. The smaller, though perhaps more prestigious, national *Rzeczpospolita* (RZ) has existed since 1944 and is owned today by Presspublica. RZ has a conservative-liberal profile and claims to be the 'most cited newspaper in Poland'. Another competitor, *Dziennik*, founded in 2006, is owned by Axel Springer Verlag (like Fakt) and has a conservative profile. Our representative of a regional/local newspaper is *Glos Pomorza* (GP), based in the northwest (the old Szczecin-Koszalin-Slupsk counties). GP was founded in 1975 and is one of the largest newspapers. It is owned by Media Pomorskie/Mekom Europe (Table 4.2).

Content analysis has been used to study the framing of chemicals in our sample of texts.[8] It is an established method in media and communication studies with roots in quantitative research (Berelson 1952, Holsti 1969, Krippendorf 1980). As a method it works well when analyzing a large number of texts (Hansen et al. 1998:100,

[8]Linne and Hansen (1990), Johansson Lönn (2005) and Anderson and Marhadour (2007) have also carried out content analysis when studying media framing.

Table 4.2 Sample of Polish newspapers for the content analysis

Polish newspapers	Frequency	Political orientation
National press		
Gazeta Wyborcza (est 1989)	6 days/week	Social liberal
Rzeczpospolita (est 1980)	6 days/week	Liberal-conservative
Dziennik (est 2006)	6 days/week	Conservative
Fakt (est 2003)	6 days/week	Centre-oriented
Regional/local press		
Glos Pomorza (est 1975/2007)	6 days/week	None

Bergström and Boréus 2005:45). Briefly described, a coding scheme is first formulated with variables or questions that should reflect the purpose of the study. For example, in our case we look at *who* the actors involved in the chemical discourse are. So, in interviews with journalists we try to establish whether an actor represents politicians, researchers/experts, environmental activists, ordinary citizens, etc. A systematic process of coding of the sampled texts is then carried out, and the results are then brought together. Often, as is the case here, content analysis is combined with more qualitative text analysis (cf. Egan Sjölander 2007, Falkheimer 2004, Djerf-Pierre 1996). As mentioned before, we have borrowed inspiration from the broad field of narratives (Wolanik Boström 2005) as well as discourse theory (Sjölander 2004). These two traditions place emphasis on contextual understanding and focus on *how* meaning, for example of chemical risks, is constructed in different kinds of texts/contexts.

The content analysis of the Polish media has been somewhat restricted as compared to that of the Swedish, mainly for practical reasons, such as difficulties in accessing the empirical material and language barriers.[9] The analysis of the Polish press has focused primarily on general patterns in the news flow, such as dominant content/themes and prominent frames. In addition, the search profile was more limited here, even if the number of newspapers studied was greater and the time period longer.[10]

The journalists' own narratives have been an equally important source for better understanding the process of media framing (cf. Larsson 1998). So-called *production* studies still have a marginal position, as compared to the more dominant traditions of

[9]Only one of the persons in our research team speaks Polish, and that person has, therefore, conducted all the interviews with the Polish journalists.

[10]The list of key words was *chemical*, *pollution*, *Reach*, *sweetener*, *heavy metal*, *lead* and *food additive.*

content and reception research (Ekström and Nohrstedt 1996, Löfgren Nilsson 1999). We have conducted eight interviews with different journalists, four from each country and one from each newspaper.[11] Together they represent a variety of journalism.[12] One thing all of them have in common, though, is that they write about the environment and/or chemicals.

Through interviewing we wanted to capture the journalists' narratives about the circumstances of their work, e.g. the national context, the specifics of the newspapers and journals, the interest in environmental issues in newsrooms, etc. as well as their own professional and personal experiences of reporting in this particular area. One significant advantage with the use of qualitative interviews is that it offers the journalists the opportunity to illuminate their points inductively and indirectly, without being too restricted by a predefined (interview) form (Kvale 1997).

4.5 Results

4.5.1 Chemicals in the Swedish Press

The search for chemical news and debates in the Swedish press has revealed 550 hits in the three newspapers studied (AB, DN and VK) during 2007, which corresponds to several hundred published articles and roughly one published text in each newspaper every other day (cf. Table 4.3). The leading national newspaper, *Dagens Nyheter*, has most coverage involving chemicals (229 hits) but hardly any debate about it (Table 4.3). The local newspaper (VK) reports much more seldom on these kind of topics (128 hits) (Table 4.3) but, on the other hand, has more debate (23%) (Table 4.4). *Discharge, pollution* and *emissions* is the most frequently occurring theme and dominates the media discourse on chemicals. Nearly half of all texts studied (267 hits) deal with subjects such as *air pollution, overuse of fertilizers, asbestos, hazardous waste, environmental toxins, dioxin, toxin, heavy metal, lead, mercury, nickel* and *cadmium* (Table 4.3). Overuse of fertilizers, e.g. in farming and forestry, is the most common subject followed by the category environmental tox

---

[11]On average, each interview has lasted about 1 h. Each of them has been recorded as well as transcribed (the Polish also translated into English). For practical reasons, the interviews with Polish journalists have been conducted over the phone. The results (as compared to the four earlier interviews, mostly conducted at the workplaces of the Swedish journalists) were not much different. We believe that this was due to the type of interview (semi-structured) and also to the subject (chemicals/environmental reporting), which all of the reporters had a lot of knowledge about and could easily relate to.

[12]The group of Swedish journalists work at Dagens Nyheter, Svenska Dagbladet (the second largest morning newspaper in Sweden), Aftonbladet and Västerbottens-Kuriren. From Poland we have interviewed one reporter from each of Gazeta Wyborca, Rzeczpospolita, Dzienik and Glos Pomorza.

Table 4.3 Chemicals in three Swedish newspapers 2007. Total number of texts from Presstext (DN) and Mediearkivet (AB and VK)

Themes/keywords	Aftonbladet (AB) n = 193[13]	Dagens Nyheter (DN) n = 229	Västerbottens-Kuriren (VK) n = 128	Hits total
Chemical*	55 (44%)	36 (32%)	30 (23%)	121
DISCHARGE/POLLUTION/EMISSION				
Air pollution	8	31	6	45
Overuse of fertilisers	11	31	13	55
Asbestos	2	11	3	16
Hazardous waste	2	8	4	14
Environmental toxins	10	30	7	47
Dioxin	3	6	5	14
Toxin	2	3	1	6
Heavy metal	9	17	8	34
Lead	18	23	2	43
Mercury	10	11	10	31
Nickel	2	12	10	24
Cadmium	3	8	2	13
PESTISIDES/INSECTICIDES				
Pesticides	19	16	12	47
DDT	2	2	1	5

FOOD ADDITIVES				
Artificial sweetener	11	2	2	15
Food preservative	11	7	-	18
Food poisoning	5	4	2	11
RISK/RISK MANAGEMENT				
Limit value	3	10	6	19
Risk analysis	–	1	2	3
REACH	5	2	1	8
Environmental crime	1	3	2	6
CHEMICAL INDUSTRY PRODUCTS				
Solvent	1	6	2	9
Freon	3	3	1	7
Phthalate	2	2	–	4
Impregnating agents	2	2	1	5
Flame retardant	1	3	1	5
Total	201	290	134	625

[a]The total number of texts for each newspaper is less than this total number of hits, since several themes/keywords may be found in a single text.

Table 4.4 Different aspects of the Swedish media discourse on chemicals

	News articles	Debate articles	Ref. to accidents or crime	Ref. to EU	News agency source	Incl. one or more interviews	Comments by experts	Comments by politicians
Aftonbladet	49	6	7	9	7	20	7	2
N = 55	89%	11%	13%	16%	13%	36%	13%	4%
Dagens	36	–	–	9	7	25	9	6
Nyheter	100%			25%	19%	69%	25%	15%
N = 36								
Västerbottens-	23	7	8	1	5	13	4	2
Kuriren	77%	23%	27%	3%	17%	43%	13%	7%
N = 30								
Total	108	13	19	15	19	58	20	10

ins and air pollution. Discharge and pollution risks deriving from heavy metals (34 hits), such as lead (43 hits), mercury (31 hits) and cadmium (13 hits), are also quite common subjects. *Pesticides/insecticides* and *food additives,* such as food preservative and artificial sweeteners, is the theme in less then ten percent of the sample. *Risk* and *risk management* issues, such as Reach, is even less common (36 hits), as is the theme *Chemical industry products* (30 hits) where we gathered material about solvents, phthalates and brominated flame retardants.

In order to gain a more detailed picture of this general overview of themes and topics, we have looked more closely at the sample of 121 texts retrieved when searching for *chemical** (Table 4.3).[13] We have studied different elements to get a better understanding of the characteristic of this media discourse (cf. Table 4.4). More than eight of ten texts about chemicals are news articles, either longer ones or shorter press items. Thus, news material clearly dominates this media discourse. Not a single debate article or letter to the editor was published in DN, and only a handful were printed in the other two newspapers, and even these are not very closely related to each other. Journalists' sources play a significant role in the news-making process, not least in science and environmental reporting (cf. Nelkin 1995, Anderson 1997). Material from news agencies, as in this case Swedish TT or international Reuter, is regularly used in news reporting. However, less than one-fifth of the sample made use of this kind of source.

In general, researchers and experts are more common journalistic sources here than politicians. However, none of these groups of stakeholders seem to play a prominent role as opinion leaders in the field (Table 4.4). The content analysis also reveals that officials – quite frequently in the role of specialists – from the three authorities Swedish Environmental Protection Agency (35 hits), Swedish Chemical Authority (31 hits) and Swedish National Food Administration (26 hits) are represented in this material.

[13]Texts with a weak connection to the subject, i.e. in which chemicals are mentioned only *en passant* or in which the dominant topic of the text is something very different, have been excluded from this part of the analysis.

There are some differences between the different types of newspapers that are worth noting. Researchers/experts provide comments twice as often in DN (25%) as compared to the tabloid and the regional/local press (both 13%). Furthermore, in the national press it is far more frequent for articles to contain more than one interview (as many as 69%), which, of course, is more time-consuming to produce. Both researchers/experts and politicians are interviewed much more often by DN reporters as compared to others, and a *toxic risk* frame is often used in these types of news reports, in particular when experts are involved. For example, in one article (DN 071111) an eco-toxicologist from the Swedish Chemicals Agency is interviewed about a toxic substance (nonylfenolethoxilate) that has been found in clothing and effects the water environment and that seems to pose a risk to the health of human beings, according to comments made by the expert. In another interview a university professor in forestry protection answers a reporter's questions about the efficient use of an insecticide (cypermetrin) (DN 070124). The forbidden and toxic substance, nonylphenol, is the subject of another news item (DN 070928) based on a report released by the Swedish Environmental Agency. In this, the director of the Swedish Chemicals Agency comments herself on results indicating that the nation is not living up to the standards for water quality established by the EU. In order to improve the situation, she promises to put pressure on industry and to review regulations.

Two of three texts published in DN are based on at least two interviews, and in 25 percent of these there is a reference to the EU. These results are in line with our expectations, namely that the national press has greater resources and that the capital is closer to the centre of the EU, Brussels, as compared to most other regions of Sweden (cf. Ekström and Nohrstedt 1996). The results also show that references to the EU are extremely rare in the local media. Only one of all of the 23 news items published in VK mentions the European Union. In fact, the same pattern can be seen with regard to Reach. This new EU legislation appears to be a non-topic in the entire nation, but certainly in the local press (cf. Table 4.3). Part of the explanation for this can be that news material about national and international affairs are not produced locally at all anymore but instead purchased (as entire 'ready-made' pages) from news agencies, like TT, as confirmed to us during the interview with the reporter from VK. Since there are no reporters at the local newspapers that regularly cover this area, competence in this area naturally decreases. Also, in addition to the absence of conflicts among the nation's stakeholders, which would certainly attract media attention, a reason for this silence is that all of the central political decisions regarding Reach were taken before the period under study began (Ekström and Nohrstedt 1996, Egan Sjölander 2007).

Accidents and crimes are, on the other hand, common in the framing of local chemical news. About a quarter of it referred to different accidents or crimes involving chemicals, such as a collision between a tanker loaded with chemicals and a Norwegian fishing boat (VK 071010) and unsafe and negligent handling of chemical transports at Umeå university (VK 070524) and a discharge of formaldehyde at the regional university hospital (VK 070321). This event-driven, reactive *accident/crime frame* appears to be non-existent in DN but is found in 13% of the tabloid news production. In addition to discharge, pollution and emission, Aftonbladet's take on chemicals seems to be largely focused on food additives and pesticides/insecticides. Topics, such as the benefits and risks of artificial sweeteners in grocery products (AB 070801) and the use of food preservatives, as in the Swedes' beloved traditional Christmas ham (AB 071220), became news in AB. A *consumer-oriented frame* is often used in these types of health stories, which are characterized by the reporters' will to serve their readers with practical advice in their everyday lives. Eide and Knight (1999:527) call this service journalism and connect its development with 'the popularization of journalistic idioms, focus and modes of address'.

The heavy metal lead is a chemical that makes news in all of the Swedish newspapers during the period under study, though in different ways. An *economic/local frame* is applied, for example, when the VK reporter presents the promising new findings that lead and silver have been found in the region (VK 070907). The high and steadily rising price on the world market for lead and other minerals is emphasized in the text, and future plans for exploitation are described. Even though the test results indicate high concentrations of lead in the bedrock, no comments are made about any risks from possible exposure to lead, e.g. risks to employees at the site. The example of the 'lead story' in Aftonbladet is rather different and makes clear use of a risk/toxic frame combined with a consumer frame. The heading 'Dangerous heavy metal in three out of four jewelleries' (AB 071003) refers to a new study that reveals that several of the major Swedish retailers sell low price products containing lead. The story was previously published in another magazine that had initiated an investigation (ICA-kuriren). The reporter starts off by concluding that 'It is well known everywhere that lead is dangerous'. The inspection authority Swedish Chemical Agency was, according to the original article, considering filing a police report, and a business representative from one of the companies involved (Indiska) issued an apology and confirmed that 'We have not been testing the levels of lead up until now. We haven't had full control. I will take action immediately'. A list of short facts, such as lead being a poisonous, carcinogenic element that cannot be destroyed, was also published in conjunction with the main AB text and was clearly part of the tabloids framing. In the main morning newspaper, lead is, for example, mentioned in a short news item about organic cosmetics. The framing is consumer-oriented, and the journalist promotes the introduction of more healthy lipsticks without ingredients such as lead. The source for this article was a press release from a company that sells these products.

4.5.2 Chemicals in the Polish Press

The search for media texts about chemicals in the Polish press revealed a corpus of 179 news and debate articles (including a few editorial comments). A clear majority of them was published in the national press (151 articles), not including the popular tabloid Fakt (only 7 hits). When analysing the major themes and recurrent topics involving chemicals in the Polish media discourse, a general picture not that different from the Swedish press emerges. For example, a lot of the coverage treats discharges of toxic substances and pollution risks from mercury, lead, sulphates, acid, calcium fluorite, etc. and accidents, such as students getting injured when experimenting with explosives, receiving short 1-day reports. The Polish journalists, as their Swedish colleagues, fairly often write more consumer-oriented stories about health risks with children's toys, smoking, unhealthy eating and so on.

The most striking difference with regard to content is that the Polish stories on discharge and pollution deal mostly with the problematic legacy from the (communist) past, when chemical waste was not managed properly (cf. Andonova 2004, Andonova et al. 2007, Gerholm 1990). An illustrative example of this, in which a *toxic risk* frame is used in the journalists' texts, is the long and thorough series of Gazeta Wyborcza articles after the discovery of polluted tap-water in the town of Gryfino, with 22,000 citizens. The newspaper reported on the latest developments almost daily for a 2-week period, which is extraordinary considering the short 'lifetime' of most news stories (Egan Sjölander 2007). Three of seven deep wells that provide water to households in the community had been contaminated with carcinogenic substances (trichloroethane and tetrachloroethane) usually found in industrial solvents. However, the source of the pollution was unknown. Soon after the news was released, a GW journalist described the strong public reaction:

> The town of 22,000 is in psychosis. The people call the municipality office and ask if a rash on a child's body might be due to the water? Is the water harmful to pregnant women? For what period of time can the chemicals still damage your kidneys? (GZ 080701)

The town's mayor plays a significant role in this news story. His actions and critique of the *Sanepid* (the Sanitary Inspection), which had done tests but then waited several days before informing the affected inhabitants, is a crucial part of the whole 'drama' until the problem is solved. Experts are also key figures in this news production. Professor E. Milchert from Szczecin University of Technology, for example, confirms in an interview that the toxic substances can damage the kidneys and the liver, and a colleague from the same university fears that the chemicals may have been buried some distance away and a long time ago (GW 080702, GW 080701). Gazeta Wyborcza also covered the event when geologists from the State Geological Institute arrived to the affected area to investigate underground water flows (GW 080702), including when the mayor told about showing them a place in the woods where the chemicals could have been dumped (GW 080703). In this extensive reporting, the reader becomes well acquainted with the town's predicament, and suspense is carefully maintained. The activities of the team of geologists are followed and firms and factories are checked (GW 080711). The patronising tone in the portrait of the citizens at the beginning is transformed over time. Some women are quoted later on in the newspaper as expressing both anguish and caution in relation to the reassuring messages of the officials (GW 080704). This story received more space than any of the other discoveries of pollution, unknown chemicals or toxic waste during the period under study. In the media framing of all of these events, it is implied that 'old' chemical storage, be it from communist times or from even before, is the main cause of the current problems and the alarming environmental and health risks that are attached to them.

An *economic and legal frame* is, on the other hand, applied in much of the news coverage related to the European Union, in particular to the Reach legislation. In contrast to the relative silence about the new chemical regulation in the Swedish press (cf. Table 4.3), the Polish journalists at the two largest national newspapers seem to pay quite a lot of attention to the implementation stage. The mass media take on the EU in this context is one of scepticism, and the views of the Polish companies affected dominate the news material and are at the centre of the journalists' writing. In one article the reporter addressed the readers directly:

> Importers of pens, cosmetics, duster clothes and soap – you have only one month left to prepare for the EU regulation on registration of chemical substances. The Reach regulation may be the fall of many firms, especially the small ones. (GW 080506)

Rezczpospolita published even more than Gazeta Wyborcza on the subject, emphasising the fear that the new EU rules might lead to many firms closing down, in addition to the risk of rising prices, e.g. on children's toys and paper products. The aim of many texts is to provide help and guidance regarding the 'extremely expensive' registration process. But the alternative, neglecting it, is described as even worse and something that might have 'disastrous consequences' (RZ 071220, RZ 080624). Long and detailed instructions on what should be registered and practical advice on how to do so are published together with links to websites where more information can be found, including places that offer courses on Reach.

This consumer-oriented style of reporting, in which journalists provide practical support in solving the everyday problems of their readers, in this case business representatives, is also common in other areas (cf. Eide and Knight 1999). For example, when a *health frame* is prominent, such as in articles about the risks derived from additives in food or toys for children, reporters regularly offer concrete advice about which consumer products to avoid in order to stay 'healthy'. The question of why ham does not taste the same as it used to, for example, was raised in a headline in Dziennik (070216). In the text that follows, the reporter directly addresses everyone that has bought ham that has started to smell the day after or bread that was not edible after a few hours. The problem, according to the article, is that the ham and the bread are fake products from profit-seeking mass producers, probably 'stuffed with as many chemicals as possible' (Dz 070216). The purpose of this is to increase the weight of the meat and, therefore, water with nitrate and substances that bind water are added. Two experts are interviewed, and the one from the Main Inspectorate of Trade Quality of Agricultural and Food Products states that after entry into the EU, all that they can do is to control that the products contain what they declare to contain. At the very end of the article the reader/consumer is given a list of the most commonly used additives that can be consulted before future purchases.

As in the Swedish press, we have identified a series of event-driven articles with an *accident/crime* frame. These stories are found in all of the newspapers, even if they do not represent a large portion of the total articles. One illustrative example, though, published in two parts in GW (080610, 080611) involves six officers from the Polish Royal Mail that were injured while handling a package that leaked chemicals. The staff experienced light burns and sore/irritated throats. The fire brigade is said to have secured the package and found perhydrol. The sender was an Internet dealer and, as it is prohibited to send chemicals by post, the prosecutor's office was informed.

4.5.3 Swedish Journalists' Narratives About Chemicals

What do the journalists themselves say about chemicals in the news and their experiences of environmental reporting in general? To start with, when interviewing the Swedish reporters, it was evident that the main topic on the environmental agenda is climate change. It overshadows all other risks, including chemicals. *'Not even the cases of chemical discharge are very big right now /.../ I have a feeling that some 10–15 years ago, there was much more written about dioxin pollution, and now it is climate'* (VK-reporter). The issue of global warming has the highest priority in the newsroom, and two of the informants, the tabloid journalist and the local reporter, have authored a popular series of articles about climate change in their respective newspaper. The environmental reporter from the national newspaper *Svenska Dagbladet* confirmed that:

> Right now the climate is our main focus, which gives other big questions, such as biological diversity, the Baltic Sea and chemicals less attention. Still, chemicals in food get covered by our consumption reporter, so we still have some control.

The journalist from *Dagens Nyheter* also stressed that environmental politics is an area of importance for the newspaper. When asked about what good news or a 'scoop' regarding chemicals would involve, several journalists answered that it would be the discovery of 'buried barrels' of toxins.

> If you can see that in some area there are many birth defects or that fish are deformed, that would be a scoop. Deception and buried poisonous waste, or a toxic discharge that some company that is in business today tries to dump. There is a very small chance of that, but it would be a scoop, and the larger the company, the bigger the scoop. And exporting toxic waste to some poor Third World country, e.g. from the electronics industry. All *deception* is a scoop, even if officials from the Swedish Chemicals Agency were involved in deception. (SvD-reporter)

There is a clear reference here to the big BT Kemi scandal in 1977, when hundreds of barrels of poisonous waste products were buried in the centre of the Swedish town Teckomatorp. The disclosure generated massive public attention and has since become a symbol for environmental crime (Mårald 2007). Another scandal involving chemicals that the DN and SvD reporters have covered extensively in the past is the misuse of Rhoca-Gil, containing acrylamide, when building a railway tunnel through the rock comprising the ridge Hallandsåsen. Workers at the site showed symptoms of poisoning, fish died and the groundwater was contaminated (as in Gryfino, Poland). When a (mass) product, previously perceived to be relatively safe, turns out to be dangerous, then it winds up high on the news agenda.

> There is always a scoop when a chemical considered harmless is found to be dangerous, or if you discover synergies that were not known before, and this might be the biggest problem we are facing right now, when we *really* are groping – what are the implications of all of those chemicals being out there, spreading in nature? It might be nothing, but it may have much *worse* effects than we can imagine! (SvD-reporter)

The EU legislation Reach was mentioned several times in the interviews with representatives from the national press (AB excluded) and described as a 'Swedish affair of the heart' (cf. Eriksson et al. 2010). Despite the long legislative process, the many political and administrative 'turns' and the complicated specialist vocabulary, Reach is considered to comprise very important information for the Swedish public. The readers have to be informed at an early stage as well so that they can have a chance to influence the democratic process, in the view of one of them. All of the professionals interviewed, regardless of the profile of their newspaper, also emphasized the importance of writing in a comprehensible, pedagogical manner. *'I imagine it has to be understood by someone without any prior knowledge'* (AB-reporter). No reader should be excluded. These professional ideals, one could even call them requirements, of informing and educating the general public are well established among journalists. Together with the ambition of critically reviewing the actions of decision-makers in society, they form the very core of the journalistic endeavour (Weibull 1991, Nygren 2008).

The kind of sources that journalists make use of varies substantially, depending partly on the editorial profile and partly on the individual reporter. Some highlight the use of other media, such as websites called the American Science Daily or the British Observer, and others rely more on professional and personal contacts with specific sources, relationships that have been established over a long period of time. Reports and results from the local university have a bigger chance of receiving attention in the local press compared to any other media. But even in this case, as many other studies have shown (Nelkin 1995, Johansson Lönn 2005), journalists do not find researchers all that easy to deal with:

> Researchers are just *hopeless*! Because they don't talk to our readers, they talk to other researchers /…/ And as we have an ability to make things simpler, they think it sounds too easy, it is not serious and trustworthy in relation to other researchers. [Officials from public authorities] have a very difficult language, but they don't put quite as much prestige in it. (…) They have more *general* knowledge, and the researchers have *their own*. (VK)

In general the interviews give an impression that it is not difficult to find experts to interview, and in relation to the chemical field, officials from the Swedish Chemical Agency seem to have high status as experts (cf. Section 4.6). Politicians are not very visible as vital sources for reporters in relation to chemicals, despite the Reach process in which national representatives took the lead. It is clear, though, that part of the input and inspiration for Swedish journalists comes from environmental organizations, such as the Swedish Society for Nature Conservation (cf. Anderson 1997).

4.5.4 Polish Journalists' Narratives About Chemicals

Turning our interest towards the narratives of Polish journalists about their profession, one general conclusion that can be drawn is that the ideals and practices of the reporters in this post-communist country have a lot in common with Swedish journalists (cf. Weibull 1991, Nygren 2008). The Polish reporters agreed that interest in chemicals among the readers was not so great and varied over time but that this changes if people are directly affected, for example if a child shampoo is found to contain dangerous phthalates or if it is no longer possible to swim in the sea because of the algae.

> When there is a *direct connection* to the people, then the news value is much stronger. I started to write about the Baltic Sea pollution only when the blue [-green] algae had become visible from the beach, but the pollution had started much earlier, hadn't it? So when it becomes *visible*, when the people are *touched* by it. Or when there are huge *expenses* for a water purification plant. (DZ-reporter)

A scoop in this area would be 'some *bad* news, I am afraid!', preferably some kind of disaster: 'I associate it with leakage from a tank, or a lot of dead fish in the Vistula (*a river*) – this is news that deserves to be placed on the first pages' (DZ-reporter). The local/regional reporter from *Glos Pomorza* defines a scoop as, for example, if it were revealed that there was some chemical waste from the West stored illegally somewhere close by:

> I know there were such stories in other newspapers, those that are closer to the border, where the transports are easier, it has not been revealed here yet (...) I think we will be dealing with this issue in a while, as it *will* happen here, because people make money on storing waste, and they do it illegally to reduce costs. (GP-reporter)

The *Gazeta Wyborcza* journalist, who specializes in economic and EU issues, is also interested in environmental protection: 'as it is obvious that the chemical industry has a huge and not always positive impact on nature, I got interested in chemicals'. A scoop on the topic – as far as the economy is concerned – could involve important changes in ownership in the chemical branch, e.g. a fusion or a privatization of important companies. However, he argues, chemicals are not perceived as a particularly prestigious topic by other journalists. Other topics can have a larger news potential than some 'regulations in the rubber branch'. Reach has been one of this journalist's main topics during the past few years. Possible implications for the Polish chemical industry have been the focus of attention (which in turn has similarities with news coverage that apply an economic-legal frame, see Section 4.7). Journalists are seen as having an immensely important role in spreading information, as very few experts and institutions in Poland – and still fewer chemical firms – realized the economic impact of the coming directives. The Reach process was received with much reservation and caution in the country, he remembers, mainly because it was expected to involve huge costs for the industry. This GW reporter's experiences also support the preliminary findings in Hermansson and Reuter's (2006) analysis of Gazeta Wyborcza's coverage of Reach, namely that it had a strong focus on employment and economic impact.

For the local/regional reporter from *Glos Pomorza*, stories in which chemicals are involved often begin with a reader contacting the newspaper, i.e. with a consumer having complains after purchasing poor quality food in the supermarket or observing 'strange' garbage being dumped in the countryside. The journalist also related that inhabitants are becoming more and more aware of the environment and of health issues, so they are more alert to possible dangers, including harmful chemicals. The important social role of the local/regional newspaper in the society is captured very well in the following description (cf. Sjölander 2004, Asp et al. 1997):

> But on the other hand, we have to respond to people's needs – if they want explanations, we have to provide them. This is the fate of a regional newspaper, it is close to the people. To be quite honest, it is a service company. People come, they have problems, they cannot cope... We have noticed that we have become some kind of old [*communist*] County Office – when people could not cope, they went to the Party secretary and said 'Please help' – and now they come to the newspaper and say 'Please help!' (GP-reporter)

Other media also comprise important sources for this journalist, alongside with researchers from the local college or MIR, the Fishing Institute in Gdynia. The editorial staff of the regional newspaper discusses the reporting of Rzeczpospolita, Gazeta Wyborcza and Dziennik on a daily basis. 'We treat these three main media as a general base for information' (GP-reporter). The main sources for the journalists at these national newspapers are somewhat different. Announcements by the Polish Government and the European Commision are important news material. News agencies, such as Reuters and Polish Press Agency, are regularily used by all of these professional, and the Rzeczpospolita journalist said that she writes articles based on Nature, Science and New Scientist on a weekly basis. This reporter also interviews scientist from Poland and abroad when they receive awards or publish something of importance. It is good to have 'a living voice', she concludes. This group of Polish journalists have no problem finding experts, and several of them attest to the advantages of the Internet in that part of their work. One challenge with researchers they argue however is to translate their advanced scientific jargon into everyday language so that all of the readers can follow the text. This is an experience that they share with their Swedish colleagues (cf. Section 4.8). Political sources are rare for these journalists. In comparison to the situation in Sweden, the main difference regarding sources is the (lack) of availability and readiness of Polish officials to participate in the news production. 'There is some kind of governmental commission for environmental protection, but it does not seem to be very active' according to one reporter. Another says, however, that there is no problem with comments and expert opinions on the level of the EU. Even if there is a crisis or a catastrophe, it might be difficult for Polish journalists to find a politician or state official that is willing to comment. 'Then the officials as a rule are defensive and even avoid any contact, since nobody wants to explain how it could happen in his region'. It is a similar situation with the police and the fire brigade in cases of leakage and pollution. 'They are not talkative either', in the experience of this journalist. This structure is perhaps part of what Sparks (2008) refers to when he claims that the Polish media is subordinated to elite groups in society. Environmental organizations, like Greenpeace, are, on the other hand, active in providing input to these journalists according to the interviews.

That is of no surprise since Greenpeace in particular has a long history of so-called news management (cf. Hansen 1994, Anderson 1997). Reach was a good example of this lobbying process, in which the environmental organizations were positive, and the representatives of industry were negative.

4.6 Conclusions and Reflections

This study on how chemicals are framed in the contemporary press in Sweden and Poland reveals great similarities as well as specific differences in the media texts and journalists' narratives. By and large the results indicate that the types of frames that are used by journalists in these two countries have a lot in common, even if the content of the texts and the specific contexts differ substantially.

The toxic risk frame, perhaps the most dominant in these media discourses, is frequently used by both Swedish and Polish journalists. Chemicals and their use are here mainly understood as a threat to humans and the environment that in turn might be in need of protection. The risks are often caused by irresponsible industries driven by greed, or ignorant governments or public bodies, like in the communist past. The accident/crime frame, as well as the consumer-health frame, are also common in the press in both countries. The human 'factor', in other words negligence and lack of information or imagination, is implied as an explanatory factor in the first type of frame. The individual's rights and vulnerability is often emphasized in the second one. Variations on an economic frame have also been identified in both countries. The Reach legislation is often associated with economical difficulties in Poland, leading to loss of employment within the nation's important chemical industry. In contrast, new discoveries of mineral findings in Sweden are described as future opportunities creating jobs in rural communities. These differences in framing can easily be linked to the specific national context (cf. Hermansson and Reuter 2006). The local/regional newspapers in both Sweden and Poland stand out from the other types with regard to the news production process since they, not surprisingly, have such a clear local/community focus. On a general level, the analysis of the interviews with journalists also highlights the fact that the professionals in the 'old' democracy with a high environmental profile, and the post-communist nation with newly formed media institutions and environmental problems deriving from the past, actually have a lot in common. Above all, they structure their work in relation to chemical risks in a similar manner. Also, they largely share the same type of professional ideals and journalistic norms (cf. Weibull 1991, Nygren 2008). Even so, having emphasized the commonalities, one should not disregard the diversity in media practices that we have studied.

The most striking contextual factor that creates differences and that influences the way chemicals are framed is Poland's problematic past when it comes to chemicals (cf. Andonova 2004, Andonova et al. 2007). The unwillingness of today's elected politicians and public officials to communicate with journalists and the general public regarding chemical risks echoes some of the censorship rules from the nation's communist era. News was not automatically published then, especially if they concerned chemical pollution. And this because the health effects of chemicals (especially PVCs) or accidents in chemical plants could threaten the overall picture of a country characterized by prosperity and excellent working conditions (Gerholm 1990:131). The focus on climate change on the other hand is not at all as strong in Poland as in Sweden. That also has an impact on the media framing. Environmental journalism seems to be equated with global warming, bio fuels, etc. in contemporary Swedish newsrooms. This hegemony of climate change clearly runs the risk to overshadow other environmental problems, such as chemicals and the work for a non-toxic environment (Swedish Parliamentant Decision 1998/99:183).

Taken together the content analysis has revealed few attempts at investigative journalism in relation to chemical risks. Dependence on researchers/experts perspectives is at the same time relatively strong in the studied media texts, especially in the Polish press where such specialist groups seem to render particularly high status among journalists. The relative absence of politicians' voices in the news and in media debates about chemicals and their use reinforces this expert-dependency, as does the lack of journalist-initiated coverage of chemical risks. The same conditions, finally, provide relatively good opportunities for lobby organizations from industry and environmental organizations to influence and frame the media discourse on chemicals in Sweden and Poland.

References

Aftonbladet (2007, newspaper articles on chemicals. Stockholm).

Anderson, A. (1997) *Media, culture and the environment*. London: UCL.

Anderson, A. & Marhadour, A. (2007) Slick PR? The media politics of the prestige oil spill. *Science Communication* 29(1): 96–115.

Andonova, L. (2004) *Transnational politics of the environment: The EU and environmental policy in Central and Eastern Europe*. Cambridge, MA: MIT.

Andonova, L. Mansfield, E.D. & Milner, H.V. (2007) International trade and environmental policy in the postcommunist world'. *Comparative Political Studies* 40: 782–807.

Asp, K. (1986) *Mäktiga massmedier. Studier i politisk opinionsbildning*. Stockholm: Akademilitteratur.

Asp, K., Johansson, B. & Larsson, L. (1997) *Nära nyheter. Studier i kommunjournalistik*. Göteborg: Institutionen för journalistik och masskommunikation, Göteborgs universitet.

Berelson, B. (1952) *Content analysis in communication research*. New York: Hafner.

Bergström, G. & Boréus, K. (2005) *Textens mening och makt. Metodbok i samhällsvetenskaplig textanalys*. Lund: Studentlitteratur.

Carlsson, U. (ed.) (2008) *Nordicom-Sveriges Mediebarometern 2007*. Göteborg: Nordicom Sverige.

D'Angelo, P. (2002) News framing as a multiparadigmatic research program: A response to Entman. *Journal of Communication* 52(4): 870–888.

Dagens Nyheter (2007, newspaper articles on chemicals. Stockholm).

Djerf-Pierre, M. (1996) *Gröna Nyheter. Miljöjournalistiken I televisionens nyhetssändningar 1961–1994*. Göteborg: Institutionen för journalistik och masskommunikation, Göteborgs universitet.

Dunwoody, S. & Peters, H.P. (1992) "Mass media coverage of technological and environmental risks: A survey of research in the United States and Germany". *Public Understanding of Science* 1: 199–230.

Dziennik (2007, newspaper articles on chemicals. Poland).

Eder, K. (1996) *The social construction of nature. A sociology of ecological enlightenment.* London: Sage.
Egan Sjölander, A. (2007) *Som natt och dag trots samma kärnas ursprung? Om olikheter och opinioner i nationella och lokala/regionala mediers hantering av kärnavfallsfrågan.* R-07 12, Stockholm: Svensk Kärnbränslehantering AB.
Eide, M. & Knight, G. (1999) Public/private service. Service journalism and the problems of everyday life. *European Journal of Communication* 14(4): 525–547.
Ekström, M. & Nohrstedt, S-A. (1996) *Journalistikens etiska problem.* Stockholm: Rabén Prisma.
Entman, R. (1993) Framing: Toward clarification of a fractured paradigm, *Journal of Communication* 43(4): 51–58.
Eriksson, J., Karlsson, M. & Reuter, M. (2010) "Technocracy, politicization and non-involvement: politics of expertise in the European regulation of chemicals". *Review of Policy Research,* 27(2): 167–185.
European Union (2006) *Regulation (EC) No 1907/2006 of the European Parliament and of the Council of 18 December 2006 concerning the Registration, Evaluation, Authorisation and Restriction of Chemicals (REACH).*
Fakt (2007–2008, newspaper articles on chemicals. Poland).
Falkheimer, J. (2004) *Att gestalta en region. Källornas strategier och mediernas föreställningar om Öresund.* Göteborg: Makadam Förlag.
Findahl, O. (2007) *Internet i ett internationellt perspektiv. Del 1: Sverige i Europa.* Hudiksvall: World Internet Institute.
Gazeta Wyborcza (2007–2008, newspaper articles on chemicals. Poland).
Gerholm, T (1990) "Att korrigera verkligheten. Om mediernas makt och vanmakt i 1970-talets Polen" In *Medier och kultur,* Hannerz, U (ed.). Stockholm: Carlsson, p 121–148.
Giorgi, L. (1995) *The Post-Socialist Media: What Power the West? The changing media landscape in Poland, Hungary, and the Czech Republic.* Avebury: Aldershot.
Glos Pomorza (2007–2008, newspaper articles on chemicals. Poland).
Government Bill (1997/98:15) *Svenska miljömål. Miljöpolitik för ett hållbart Sverige.* 7 maj 1998.
Government Bill (2004/05:150) *Environmental quality objectives: A shared responsibility. Summary of Government Bill 2004/05:150.*
Hansen, A. (1991) The media and the social construction of the environment. *Media, Culture and Society* 13: 443–458.
Hansen, A. (ed.) (1994) *The mass media and environmental issues.* Leicester: Leicester University Press.
Hansen, A., Cottle, S., Negrine, R. & Newbold, C. (1998) *Mass communication research methods.* London: Macmillan.
Hermansson, C. (2002) *Det återvunna folkhemmet. Om teve journalistik och miljöpolitik i Sverige 1987–1998.* No 251. Linköping: Linköping Studies in Arts and Science.
Hermansson, C. & Reuter, M. (2006) "Reach out, don't touch or battle? The EU chemicals directive REACH in daily press in Sweden and Poland." Södertörns Högskola. Unpublished report.
Holsti, O.R. (1969) *Content analysis for the social sciences and humanities.* Reading: Addison-Wesley.
Jarlbro, G., Sandberg, H. & Palm, L. (1997) *Ammoniakolyckan I Kävlinge: hur myndigheterna kommunicerade med medier och medborgare.* Stockholm: Styrelsen för psykologiskt försvar.
Jarlbro, G. (2001) *Forskning om miljö och massmedier: En forskningsöversikt.* Lund: Media and Communication Studies, Lund University.
Johansson, B. (2000) Tron kan försätta berg. Tredjepersons-effekter och politisk kommunikation. *Nordicom Information,* no 3–4, p. 77–86.
Johansson, B. (ed.) (2006) *Giftfri miljö – utopi eller verklig chans?* Stockholm: Forskningsrådet Formas.
Johansson Lönn, E. (2005) *Jobb(iga) nyheter. Om dagstidningars bevakning av arbetsmiljöfrågor.* Umeå Studies in Media and Communication, No. 8, Umeå: Institutionen för kultur och medier, Umeå universitet.

Krippendorf, K. (1980) *Content analysis: An introduction to its methodology.* London: Sage.
Kvale, S. (1997) *Den kvalitativa forskningsintervjun.* Lund: Studentlitteratur.
Lara, A. (2008) The Polish media landscape. In *European Media Governance. The National and Regional Dimensions*, Terzis, G. (ed.). Bristol: Intellect, pp. 399–410.
Larsson, L. (1998) *Nyheter i samspel. Studier i kommunjournalistik.* Göteborg: Institutionen för journalistik och masskommunikation, Göteborgs universitet.
Larsson, L. (2005) *Opinionsmakarna: En studie av PR-konsulter, journalistisk och demokrati.* Lund: Studentlitteratur.
Linne, O. & Hansen, A. (1990) *News coverage of the environment. A comparative study of journalistic practices and television presentation in Danmarks Radio and the BBC.* København: Danmarks Radio.
Löfgren Nilsson, M. (1999) *På Bladet, Kuriren och Allehanda. Om ideal och organiseringsprincioer i den redaktionella vardagen.* Göteborg: Institutionen för journalistik och masskommunikation, Göteborgs universitet.
Löfstedt, R. & Renn, O. (1997) The Brent Spar controversy: An example of risk communication gone wrong. *Risk Analysis* 17(2): 131–136.
Majcherek, J. A. (1999) *Pierwsza dekada III Rzeczypospolitej 1989–1999*, Warszawa: Presspublica.
Mårald, E. (2007) *Svenska miljöbrott och miljöskandaler, 1960–2000.* Hedemora: Gidlunds förlag.
Mazyr, A. (1984) The journalist and technology: Reporting about Love Canal and Three Mile Island. *Minerva* 22(1): 45–66.
Miczynska-Kowalska, M. (2007) Media w komunikacji spolecznej. Informacja, czy celowa manipulacja? In *Media w Polsce. Pierwsza wladza IV RP?* Sokolowski, M. (ed.).Warszawa: WAiP.
Nelkin, D. (1995) *Selling science. How the press covers science and technology.* New York: W.H. Freeman.
Niiranen, V. (ed.) (2007) *Europeans Read Newspapers 2007.* Brussels: ENPA (European Newspaper Publishers' Association).
Nohrstedt, S-A. & Nordlund, R. (1993) *Medier I kris: en forskningsöversikt över mediernas roll vid kriser.* Stockholm: Styrelsen för psykologiskt försvar.
Nord, L. & Strömbäck, J. (ed.) (2004) *Medierna och demokratin.* Lund: Studentlitteratur.
Nordlund, J-E. (1995) *Politisk verklighet och opinionsbildning. Om dagspressens betydelse i regionala frågor.* Härnösand: Mitthögskolan.
Nygren, G. (2008) *Yrke på glid: om journalistrollens de-professionalisering.* Institutet för mediestudier. Stockholm: Sim(o).
Obermayer, H.J. (1994) Eastern European publishers face ethical dilemmas. *Journal of Mass Media Ethics* 9(2): 94–99.
Petersen, L. K. (1997) *Miljøbevidsthed på MTV. Om naturen I massekulturen.* København: Forlaget Sociologi.
Rzeczpospolita (2007–2008, newspaper articles on chemicals. Poland).
Scheufele, D. A. (1999) Framing as a theory of media effects. *Journal of Communication* 49(1): 103–122.
Scheufele, B. (2006) Frames, schemata, and news reporting. *Communications* 31: 65–83.
Scheufele, D. A. & Tewksbury, D. (2007) Framing, agenda setting, and priming: The evolution of three media effects models. *Journal of Communication* 9: 20.
Sjölander, A. (2004) *Kärnproblem. Opinionsbildning i kärnavfallsdiskursen i Malå.* Umeå Studies in Media and Communication No. 7, Umeå: Institutionen för kultur och medier, Umeå universitet.
Sparks, C. (2008) Media systems in transition: Poland, Russia, China. *Chinese Journal of Communication* 1(1): 7–24.
Swedish Parliamentant Decision 1998/99:183.
Tjernström, V. (2001) *Europa norrifrån. En nordisk komparativ studie av europeisk politisk kommunikation.* Umeå Studies in Media and Communication, No. 4b, Umeå: Institutionen för kultur och medier, Umeå universitet.

Västerbottens-Kuriren (2007, newspaper articles on chemicals. Umeå, Sweden).
Weaver, D. H. (2007) Thoughts on agenda setting, framing, and priming. *Journal of Communication* 57: 142–147.
Weibull, L. (1991) *Svenska journalister. Ett grupporträtt.* Stockholm: Tiden.
Weibull, L. & Jönsson, A-M. (2008) The Swedish media landscape. In *European Media Governance. The National and Regional Dimensions*, Terzis, G. (ed.). Bristol: Intellect, pp. 169–180.
Wolanik Boström, K. (2005) *Berättade liv, berättat Polen. En etnologisk studie av hur högutbildade polacker gestaltar identitet och samhälle.* Etnologiska skrifter nr 34. Umeå: Institutionen för kultur och medier, Umeå universitet.

Chapter 5
REACH: What Has Been Achieved and What Needs To Be Done?

Sven Ove Hansson and Christina Rudén

Abstract REACH is intended to increase the speed and efficiency of the risk assessment process and to make producers and importers of chemicals responsible for this process. In this contribution, the REACH data requirements are evaluated against the background of accepted data requirements and criteria for hazard assessment, classification and warning labelling. It is concluded that REACH will lead to increased availability of toxicological data, but not to the extent that would be needed to achieve a sound scientific basis for hazard assessment of all individual substances covered by the legislation. Amendments are proposed that would improve priority-setting and testing strategies in the REACH system.

Keywords Classification • Data requirements • REACH • Risk assessment • Risk management

5.1 Introduction

After extensive debates and preparations, the new European chemicals legislation, REACH, was finally adopted in December 2006.[1] The main parts of REACH came into force on the 1st of June 2008. Other parts will be introduced gradually, and the entire legislation will be implemented by the 31st of May, 2018.

C. Rudén (✉)
Royal Institute of Technology (KTH), Teknikringen 78B, SE 100 44, Stockholm, Sweden
e-mail: cr@abe.kth.se

S.O. Hansson
KTH, Skolan för arkitektur och samhällsbyggnad, Avdelningen för filosofi,
Drottning Kristinas väg 30, SE 100 44, Stockholm, Sweden
e-mail: soh@abe.kth.se

[1]REGULATION (EC) No 1907/2006 OF THE EUROPEAN PARLIAMENT AND OF THE COUNCIL of 18 December 2006 concerning the Registration, Evaluation, Authorisation and Restriction of Chemicals (REACH), Official Journal L396/17:1–849.

J. Eriksson et al. (eds.), *Regulating Chemical Risks: European and Global Challenges*,
DOI 10.1007/978-90-481-9428-5_5,

The new legislation aims at improved risk management of chemicals in Europe. Risk management has to be based on risk assessment. The lack of adequate risk assessments for the vast majority of general industrial chemicals was a major justification for the initiative by the Commission that ultimately led up to the new legislation (European Commission 2001). Unfortunately, it is not only risk assessments that are missing but also the toxicological data that is needed to make these assessments. The lack of such data was one of the major motivations for the work that led up to the construction of the new legislation.

Several studies have confirmed that toxicity data is insufficient or non-existent for a large proportion of the substances in use. In 1984, the National Academy of Sciences' National Research Council reported that 78% of the approximately 2,500 chemicals that were commercially used in high volumes did not even have 'minimal toxicity data' (National Research Council 1984). An update of this study 13 years later, performed by the Environmental Defense Fund (EDF), indicated that there had been no significant improvement; 75% of the top-volume commercial chemicals still lacked basic data from toxicity testing (Roe et al. 1997). A report from the European Commission showed that 72% of the 2,465 EU high production volume chemicals had less than a 'minimal dataset' (Allanou et al. 1999: 14). For low volume substances, the availability of data seems to be even lower. Estimates of the actual number of chemicals in use range between 30,000 and 100,000. For most of these substances, little or no toxicological information is publicly available. Obviously, well-informed risk management is impossible when no risk assessment can be made due to lack of data. Equally obviously, the creation of toxicity data for such a large number of substances is both time- and resource-demanding, and requires careful planning and optimization (Hansson and Rudén 2007).

It is against the background of these serious problems for the regulation of chemicals that the achievements of REACH have to be evaluated. In Section 5.2 we will further discuss the demands on risk assessment of individual substances and the consequent demands on toxicity data. In Section 5.3 we will assess the achievements of REACH in these respects. Based on this, we propose in Section 5.4 some further improvements of REACH.

5.2 Risk Management Criteria

Risk assessment is performed for the purpose of risk management, and regulatory testing of chemicals is performed for the purpose of such risk assessment. Therefore, our discussion must start with risk management. We need to ask: What should be the criterion for risk management decisions on the use of chemicals? Only when at least a tentative answer to that question is available is it possible to determine what demands to put on risk assessment and on data acquisition for risk assessment.

A traditional approach can be summarized as follows:

First Attempted Criterion
Refrain from all uses of chemicals that are known to be harmful.

There is a long tradition in risk management of acting only against proven effects. This approach still has active proponents.[2] However, there is also a long history of risk assessment failures showing that this strategy has a high cost in terms of human suffering and damage to the environment. The assumption that no known danger implies no danger is simply not tenable for chemical risks at our current state of knowledge (Harremoës et al. 2001).

A possible response to this problem is to radically reverse the above criterion, as follows:

Second Attempted Criterion
Refrain from all uses of chemicals that are not known to be harmless.

Unfortunately, this criterion is unrealistic since it is in practice impossible to prove that a substance is harmless (Hansson 1997). There are three major reasons why this is impossible. The first reason is that we cannot investigate the effects of a substance on all species and under all conditions. Exposure of untested species unavoidably introduces uncertainty into the analysis. Even if the test organisms are well-chosen and as representative as possible, there is a non-negligible possibility that other significant effects will unexpectedly turn up in other, untested organisms (Breitholtz et al. 2006).

The second reason is that even in one species, such as humans or a rodent studied with the purpose of predicting human toxicity, we cannot practically investigate all combinations of exposure routes and relevant endpoints. Indeed, testing for 'all relevant endpoints' is impossible in practice. Any realistic testing programme has to be based on a selection of endpoints. One pragmatic approach to endpoint selection is to refer to the classification and labelling directive and what data is needed for classifications according to the legislated toxicity criteria.[3] The classification and labelling rules refer to standardized tests that cover a standard set of endpoints and there are potentially significant effects/endpoints that are currently not covered by the standardized tests. Examples of such endpoints are those related to certain types of endocrine disruption and neuro-developmental effects.

The third (and perhaps least known) reason is that even when studying a single endpoint in a single species, surprisingly large harmful effects can be indetectable. This is perhaps best illustrated with an epidemiological example such as the following.

---

[2]For critical appraisals, see: Rudén and Hansson (2008); Ong and Glantz (2001).

[3]Then the following tests would be relevant: acute toxicity test, subacute/28-d study or a subchronic/90-d test, skin and eye irritation, corrosivity, skin sensitization, carcinogenicity, mutagenicicity, reproductive toxicity (2-gen study), ecotoxicity tests in fish, algae and Daphnia, evaluation of the potential for (or actual) bioaccumulation through the determination of log Pow (or BCF), and degradability.

Suppose that a subpopulation is subject to a chemical exposure that increases lifetime mortality in coronary heart disease from 10.0% to 10.5%. Statistical calculations will show that this difference is in practice indistinguishable from random variations. If this group is compared to an unexposed group in an epidemiological study, then there is no possibility to discover the increased incidence of lethal heart disease. More generally speaking, epidemiological studies cannot (even under favourable conditions) reliably detect excess relative risks unless they are greater than 10%. For the more common types of lethal diseases, such as coronary disease and certain types of cancer, lifetime risks are in the order of magnitude of about 10%. Therefore, even in the most sensitive studies, an increase in lifetime risk of the size 10^{-2} (10% of 10%) or smaller may be indistinguishable from random variations (Hansson 1999).

Such numbers should be compared to the level of concern, or in other words to how large health effects have to be in order to be subject to active risk management measures. In the 1960s and 1970s attempts were made to determine a level of 'acceptable risk'.[4] However, it was soon realized that this was an unrealistic project, since acceptability is value-based and since it depends not only on the risk but also on the associated benefits (Bicevskis 1982). There is therefore no general truth about what lethal risks are acceptable. However, judging by published proposals and by current practices it would seem reasonable to say that risks in the interval 10^{-4}–10^{-6} or larger are generally considered to be issues of concern. We therefore have a wide gap with a breadth of at least two to four orders of magnitude between those (probabilistic) risk levels that are scientifically detectable and those that are commonly regarded to be of concern (see Fig. 5.1). This knowledge gap is a major reason why the inference from 'no known risk' to 'no risk' is a dangerous one.

Although we cannot obtain conclusive evidence of harmlessness, we can in many cases obtain evidence of harmfulness that is sufficiently reliable to warrant risk management measures. With the exception of the very few cases when conclusive human experience is available, this evidence will have to be the outcome of toxicological testing. Based on this insight we might wish to retreat to the following criterion:

Third Attempted Criterion
Only use substances that have been adequately tested.

This is in our view reasonable as part of a criterion for risk management. Without toxicity data, knowledge-based risk management is not possibly. However,

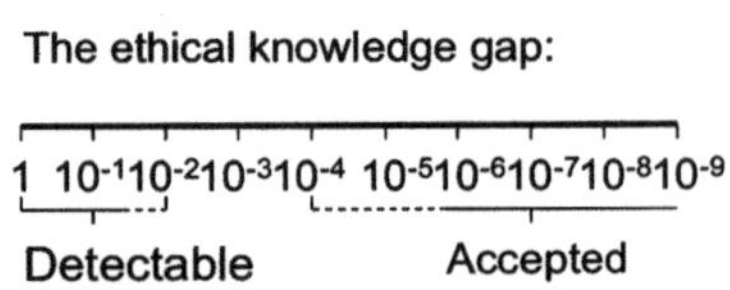

Fig. 5.1 The ethical knowledge gap

[4]For a summary of these efforts, see Philipson, Lloyd L (1983) Risk Acceptance Criteria and Their Development. Journal of Medical Systems 7(5):437–456.

as it stands the criterion is incomplete. We need to test, but we also need to take practical actions based on the data obtained. One possible way to amend the criterion is the following:

Fourth Attempted Criterion
Only use substances that have been adequately tested, with no sign of harmfulness detected.

However, the risk management strategy described here is too simplistic. There are two major ways to reduce the risks following chemical exposures: We should select as harmless chemicals as possible, and we should also arrange the handling and use of the chosen chemicals in a way that minimizes exposure. A risk management strategy that employs only one of these two strategies is bound to be inefficient. There are cases when a substance with harmful effects can be used safely in a specific process that has been tailored to preclude exposure. We can therefore amend the criterion as follows:

Fifth Attempted Criterion
Only use substances that have been adequately tested. Avoid harmful effects through a combination of substance selection and exposure reduction.

We are now approaching a reasonable criterion for chemicals risk management. However the criterion just stated does not take into account one of the factors that we emphasized above, namely that with any amount of testing we cannot be sure to have detected all harmful effects. There is no simple way to deal with this problem, but two strategies in combination will be efficient in many cases. One of these is to apply uncertainty factors (sometimes called 'assessment factors' or even 'safety factors') (Clausen et al. 2006). The need for of uncertainty factors can be seen from Fig. 5.1. If the gap between detectability and the accepted level is judged to be for instance 10^{-2}, then an uncertainty factor of 10^2 should in principle provide the desired level of protection. (Under the assumption of a linear dose-response relationship, this means that the NOAEL, No observed adverse effect level, is divided by 10^2 in order to arrive at an exposure limit.) The other method is to reduce exposure more generally, for instance by choosing technical measures that bring down exposures to all chemicals in a certain activity, rather than measures that will have effect only on selected substances with known harmful effects.

We can introduce these two strategies into our criterion as follows:

Sixth and Final Criterion
Only use substances that have been adequately tested. Avoid harmful effects through a combination of substance selection and exposure reduction, using uncertainty factors to compensate for uncertainty and indetectability. Whenever possible, exposures to substances not known to be harmful should also be reduced.

We will take this as our final criterion, for the present purposes. It should be clear enough what its implications are for risk assessment and data requirement: It requires that for all substances in use, there should be sufficient data for risk assessment of the substance.

What data are needed for a risk assessment depends of course on the level of ambition of that assessment. For an extensive risk assessment that covers many specific endpoints, a large number of studies are needed. Such extensive assessments are exemplified by the regulatory requirements for pharmaceuticals and pesticides. It would in practice not be realistic to demand such extensive data for the large number of untested industrial chemicals that are currently in use. We will therefore focus on minimal requirements. One useful approach is to identify the data that are needed before toxicologists and risk assessors consider it possible to perform an – albeit uncertain – assessment of the risks associated with exposure to a substance. Another useful, but more demanding, approach is to focus on the data needed to identify substances of particular concern. According to REACH 'substances of very high concern' (SVHC) are those classifiable for cancer, mutagenicity, or reproductive toxicity in categories 1 or 2 (according to the classification and labelling directive), substances classifiable as persistent, bioaccumulating and toxic (PBT), or very persistent and very bioaccumulating (vPvB), and substances of equivalent concern such as endocrine disruptors (criteria defined in the REACH legislation, Article 57). Basic questions are therefore: Will REACH provide us with sufficient information to perform the risk assessment necessary for determining whether a substance is classifiable with respect to the standard criteria for endpoints such as acute toxicity, irritation, sensitization, carcinogenicity, and ecotoxicity?[5] And, will REACH provide us with sufficient information to identify the substances pointed out as being of very high concern? These are the questions to which we will now proceed.

5.3 Data Requirements in REACH

An investigation of REACH will have to compare it to its predecessor, the previous European system for 'existing' and 'new' substances. This system was based on an inventory of all chemicals that were marketed or considered for marketing in the European Union as of the 18th of September 1981. The result of this inventory was registered in a database called EINECS (the European Inventory of Existing Commercial Chemical Substances), to which no additions were made after 1981. Substances listed in EINECS were called 'existing' chemicals, and those introduced thereafter were called 'new' chemicals. New substances had to be tested and notified before being introduced on the EU market (Council directive 92/32). Different test packages were applied depending on the amounts to be marketed

[5]The analyses are performed using the criteria in the European classification and labelling directive (67/548). These criteria will be replaced by a new directive that is based on the Globally Harmonized System (GHS) for classification and labelling developed by the United Nations (Reg. 1272/2008). The new rules will be implemented stepwise from June 2010 to June 2015.

annually. The analyses of these test packages, and the way in which they are used in test systems, is much facilitated by the fact that the test packages are (very close to being) linearly ordered in terms of inclusion. In other words, if we compare any two test packages (for different groups of chemicals), then one of them is a subset of the other. Therefore, we can represent the test packages as a series of increasing sets of test requirements (see Table 5.1).[6]

As can be seen in Table 5.1, according to the previous legislation, there were no general test requirements for ‘existing’ substances. The data requirements for ‘new’ substances marketed in quantities exceeding 1 t per year (per manufacturer) included mutagenicity in vitro, acute toxicity testing, skin and eye irritation, skin sensitization, and a 28-day toxicity study, and the required ecotoxicological data for these substances included biotic degradation and acute toxicity tests on fish, Daphnia and algae. Additional tests were required from 100 and 1,000 t respectively.[7] Throughout the years that this legislation was in force, the ‘existing’ substances were the vast majority. Over 100,000 substances were registered in EINECS while only about 5,000 were listed as ‘new’. The new substances thus represented about 5% of the total number of substances listed as either existing or new), and about 1% of the total production volume (European Commission 2001). The rest of the commercially available chemicals, representing about 99% of the total EU production volume, were to be risk assessed one by one according to defined principles (Commission directive 1488/94). A programme for risk assessments of the existing chemicals was initiated, but the pace of progress was far from impressive. In total 141 existing substances became prioritized for risk assessment. From 1993 to 2008, 137 ‘first draft’ risk assessments were published, conclusions were agreed for 118 substances, and results published for 56. (44 of these were deemed in need of further risk reduction).[8] For the prioritized existing substances a ‘base-set’ of data were required[9] but for substances not on the priority lists, no testing was mandatory.

In Table 5.1, the minimal data needed for important classifications are indicated with reference to the categories in the left hand column. As can be seen in the table, data sufficient for classifying according to acute toxicity and skin and eye irritation

---

[6]The major deviations from this orderly structure are the test requirements concerning mutagenicity and carcinogenicity, that form a special tiered approach, and the provisions for waiving certain test requirements.

[7]Council Directive 92/32/EEC of 30 April 1992 amending for the seventh time Directive. 67/548/EEC on the approximation of the laws, regulations and administrative provisions relating to the classification, packaging and labelling of dangerous substances Official Journal L 154, 05/06/1992 (Annex VIIA).

[8]See the newsletter of the existing substances programme on the homepage of the European Chemicals Bureau (ecb.jrc.it/existing-chemicals).

[9]The so called “base-set” included data on mutagenicity (*in vitro*), acute toxicity, skin and eye irritation, skin sensitization, a sub-acute/28-day toxicity study, acute toxicity tests on fish, *Daphnia* and algae and degradation data (Annex VII A of directive 92/32).

Table 5.1 Data requirements for different tonnage bands and substance categories in the previous legislation and within REACH

	Previous legislation				REACH						
	Existing	New >1 t[a]	New >100 t	New >1,000 t	<1 t	>1 t phase in	>1 t non-phase in	>1 t prio	>10 t	>100 t	>1,000 t
Chronic toxicity and carcinogenicity	No	No	(Yes)	Yes	No	No	No	No	No	No	(Yes)
Reproductive toxicity (one-generation)	No	No	Yes	Yes	No	No	No	No	No	(Yes)	(Yes)
Subchronic (90 d)	No	No	Yes	Yes	No	No	No	No	No	(Yes)	(Yes)
Subacute (28 d)	No	Yes	Yes	Yes	No	No	No	No	(Yes)	Yes	Yes
Acute toxicity second route	No	Yes	Yes	Yes	No	No	No	No	Yes	Yes	Yes
Acute toxicity one route	No	Yes	Yes	Yes	No	No	Yes	Yes	Yes	Yes	Yes
Skin sensitization	No	Yes	Yes	Yes	No	No	Yes	Yes	Yes	Yes	Yes
Skin + eye irritation	No	Yes	Yes	Yes	No	No	Yes	Yes	Yes	Yes	Yes
Mutagenicity (in vitro)	No	Yes	Yes	Yes	No	No	Yes	Yes	Yes	Yes	Yes
Additional ecotoxicity tests incl. long-term	No	No	Yes	Yes	No	No	No	No	No	Yes	Yes
Acute toxicity: fish	No	Yes	Yes	Yes	No	No	No	No	Yes	Yes	Yes
Acute toxicity: Algae	No	Yes	Yes	Yes	No	No	Yes	Yes	Yes	Yes	Yes
Acute toxicity: Daphnia	No	Yes	Yes	Yes	No	No	Yes	Yes	Yes	Yes	Yes
Biotic degradation	No	Yes	Yes	Yes	No	No	Yes	Yes	Yes	Yes	Yes

[a]Data on acute toxicity, skin and eye irritation, skin sensitization, mutagenicity and biotic degradation were required for new substances from 10 kg.

were required for all new substances over 1 t per year but for no existing substances (with the exception of the 141 prioritized substances, as described above). Data sufficient to classify according to aquatic toxicity were required for new substances above 1 t but not for the existing substances.

In REACH all general industrial chemicals are regulated in a single system. Therefore, the large difference in test requirements between 'new' and 'existing' substance has been eliminated. An important feature of the previous system that has been retained is the use of different test batteries for chemicals depending on their production volumes. The higher the production volume, the more extensive test batteries are applicable. The rationale for using production volume as a priority setting tool is the assumption that the higher the production volume, the higher the potential for exposure, and the higher the exposure the higher the risk of adverse effects.

In REACH, all chemicals produced in 1 t or more per year (and manufacturer) must be registered in a central database. The general obligation to register substances implies that substances that are not registered are not allowed to be manufactured or imported into the EU. The registration includes among other things, a technical dossier containing data from whatever tests are mandatory for the substance in question (as determined by its production volume).

The testing requirements in REACH are also summarized in Table 5.1. For the substances produced in volumes from 1 to 10 t in general only very limited information is required, namely data on the substances' physico-chemical properties. Chemicals in this tonnage band (1–10 t) are categorized as (1) 'phase-in substances' i.e. substances that were regulated also in the previous legislation, or (2) so called 'non-phase in' substances, i.e. substances that are introduced on the market subsequent to the REACH implementation, or (3) 'prioritised phase-in substances' that are substances with wide-spread and diffuse use (consumer products) that fulfil certain prioritization criteria.[10] For 'non-phase in' substances and for 'prioritized phase-in' substances the following test requirements apply: acute (oral) toxicity, in vivo skin sensitization, one in vitro test for gene mutations in bacteria, acute toxicity to algae and Daphnia, and biotic degradation (Ready biodegradability) (REACH, Annex VII).[11] For substances produced in 10 t or more per year and manufacturer, additional data are required such as in vivo skin and eye irritation, two in vitro cytogenicity/mutagenicity tests using mammalian cells, acute mammalian toxicity (second route), a 28-day mammalian toxicity study, screening for reproductive toxicity, acute toxicity to fish and microorganisms (activated sludge respiration inhibition), data on hydrolysis, and an adsorption/desorption screening

[10] I.e. prioritization criteria according to REACH Annex III: "Indications" that the substance is classifiable as a carcinogen , mutagen or reproductive toxicant in categories 1or 2, substances with at least one classifications according to 67/548, substances that are PBT or vPvB (according to the REACH criteria).

[11] According to REACH, results from in vitro testing of eye and skin irritation are also required for substances produced in 1 tonne per year or more. However no such standardized in vitro tests are currently available in the OECD test guidelines. Furthermore, a standardization or validation process has not been initiated for any such test in the guidelines.

study (REACH, Annex VIII). The 28-day toxicity study and the screening for reproductive toxicity are not mandatory, and testing can be waived based on for instance the extent and nature of human exposures. When the production volume exceeds 100 t, further data can be required including information about fate and behaviour (for instance BCF and identification of degradation products), long-term toxicity to fish and Daphnia, fish reproduction (OECD 210, 212, or 215), sub-chronic toxicity to mammals (90 days exposure), developmental toxicity (OECD 414), and a two-generation reproductive toxicity study (REACH, Annex IX). For all the additional tests introduced in this tonnage band there are possibilities to waive testing based on criteria specified in the legislation. At the highest level, above 1,000 t, additional (long-term) effect data on sediment living organisms, earthworms, soil invertebrates, and higher plants can be required, as well as additional data on fish reproduction (REACH, Annex X). Again, for all the additional tests introduced in this tonnage band there are possibilities to waive testing based on criteria specified in the legislation.

As we can see from Table 5.1, REACH results in a reduction of test requirements for new substances and an increase in the requirements for existing substances. Data sufficient for classifying according to acute toxicity, skin and eye irritation, and aquatic toxicity are now required for substances with production volumes above 10 t. Data sufficient to classify according to chronic toxicity and carcinogenicity are not routinely required for any of the tonnage bands, but can be required case-by-case based on initial genotoxicity tests or for substances with a yearly production above 1,000 t.

These data requirements will however only be in force after industry has been given the time for implementation considered necessary by the legislator. Registration of test data shall be completed before 30th of November 2010 for substances produced in >1,000 t per year, before 31st of May 2013 for substances produced in >100 t, and before 31st of May 2018 for substances produced in >1 t. A comparison between the old and the new legislation that takes into account the implementation periods can be found in Table 5.2.

Table 5.2 A summary of major European data requirements before REACH and in REACH

Type of data	Was required before REACH for	Is required in REACH for
Data needed to classify according to *acute toxicity, skin and eye irritation*	All new substances >10 kg (before marketing)	Non-phase in, and prioritized substances >1 t (in year 2018)
Data needed to classify according to *aquatic toxicity*	New substances >1 t (before marketing)	All substances >10 t (in year 2013)
Data needed to classify according to *chronic toxicity and carcinogenicity*	New substances >10 t (before marketing)	Can be required case-by-case >1,000 t (in year 2010) or in lower tonnages based on indications of mutagenicity

At the end of Section 5.2 we concluded that in order to make a well-informed risk management of chemicals possible, REACH would have to provide us with the minimal information necessary for determining whether a substance is classifiable with respect to the standard criteria for endpoints such as acute toxicity, irritation, sensitization, carcinogenicity, and ecotoxicity. We can now answer the question whether this is achieved: for substances produced in yearly volumes above 10 t by at least one producer it will be achieved by the year 2018 for most of these endpoints (but not for carcinogenicity).

It is interesting to note in this context that according to the REACH regulation, the manufacturer should produce a safety data sheet for substances produced in volumes between 1 and 10 t that are classifiable as toxic according to the classification and labelling criteria that we have referred to above.[12] (For substances produced in volumes above 10 t, safety data sheets are always required.) On the other hand, as we have seen manufacturers are not required to obtain such data if it is not available. This can be seen as incoherence in the regulation.

5.4 What Needs To Be Done

REACH has provided a structure in which a well-informed chemicals risk management can be developed. In particular, it creates a legislative and regulatory framework for all substances in which the procurement of data for making reasonably reliable risk assessment is possible. But on the other hand, as we have seen, it does not require the creation of such data for all substances for which it is needed. This should be no surprise. The deficiencies in the previous system of chemicals regulation were so large that it would be unrealistic to believe that they could be solved in one single reform. It is only to be expected that there should be scope for improvement. A discussion is needed that identifies the most important of the potential improvements of the system, and in this spirit we would like to propose three important issues for the further development of REACH.

1. The most pressing remaining issue is that of generating sufficient information for the risk assessment of chemicals produced in low volumes. It is clearly an untenable situation that a large number of substances are continuously put on the market and used although the minimal data required to risk assess, classify and label them is not available. In our view a decision should be made as soon as possible that after the last time-limit for data requirements that has already been decided (year 2018) similar data requirements as those currently required for substances produced in over 10 t per year will be introduced for those produced in 1–10 t yearly.

[12] Or if they are categorized as PBT or vPvB, or identified as a SVHC (REACH, vol. I article 29, Annex XI).

Fig. 5.2 The proposed labelling symbol for insufficiently investigated chemicals

2. The introduction of REACH does not change the fact that the classification and labelling system does not discriminate between a substance that has been tested with negative outcome (no harmful effect detected) and a substance that has not been tested at all. In both these cases the substance will remain unclassified. As we argued in Section 5.2, practical risk management should take into account not only known harmful effects but also uncertainties. In order to make this possible, the classification and labelling system should be modified to include reports of lack of data. We propose the introduction of a labelling symbol to be used when basic toxicity information about a substance is lacking (Hansson and Rudén 2003) (see Fig. 5.2). Significant improvements in the reporting of uncertainties are also needed in safety data sheets.
3. Finally, whereas REACH has its focus on chemical substances and products, significant exposures of both humans and the environment are mediated by articles that include or have been treated with chemicals. To take just one example, we are exposed to brominated flame retardants through electronic devices, furniture, building materials etc. rather than through chemical products. Tracing chemical substances in articles, such as these, is admittedly a much larger undertaking than that of keeping track of the contents of chemical products. Probably, a less comprehensive system may have to be chosen for articles than for chemical products. Nevertheless, a strategy for the risk management of chemicals is severely incomplete unless it tackles the distribution of harmful substances through the wide varieties of articles in which they are used.

References

Allanou, R., Hansen, B.G., and van der Bilt, Y. (1999) *Public Availability of Data on EU High Production Volume Chemicals*. European Commission EUR 18996 EN, Institute for Health and Consumer Protection, Ispra.

Bicevskis, A. (1982) Unacceptability of acceptable risk. *Search* 13(1–2): 31–34.

Breitholtz, M., Rudén, C., Hansson, S. and Bengtsson (2006) Ten challenges for improved ecotoxicological testing in environmental risk assessment. *Ecotoxicology and Environmental Safety* 63: 324–335.

Clausen, J., Hansson, S.O. and Nilsson, F. (2006) Generalizing the safety factor approach. *Reliability Engineering and System Safety* 91: 964–973.

European Commission (2001) *White Paper – Strategy for a Future Chemicals Policy*. Brussels 27.2.2001, COM(2001)88 final.

Hansson, S.O. (1997) Can we reverse the burden of proof? *Toxicology Letters* 90: 223–228.

Hansson, S.O. (1999) The moral significance of indetectable effects. *Risk* 10: 101–108.

Hansson, S.O. and Rudén C. (2003) Improving the incentives for toxicity testing. *Journal of Risk Research* 6: 3–21.

Hansson, S.O. and Rudén C. (2007) Towards a theory of tiered testing. *Regulatory Toxicology and Pharmacology* 48(1): 35–44.

Harremoës, P. et al. (2001) *Late lessons from early warnings: The precautionary principle 1896–2000*. European Environment Agency, Luxembourg.

National Research Council (1984) *Toxicity testing – strategies to determine needs and priorities*. National Academy Press, Washington, DC.

Ong, E.K. and Glantz, A.S. (2001) Constructing 'sound science' and 'good epidemiology': Tobacco, lawyers, and public relations firms. *American Journal of Public Health* (91): 1749–1757.

Philipson, Lloyd L. (1983) Risk acceptance criteria and their development. *Journal of Medical Systems* 7(5): 437–456.

Roe, D., Pease, W., Florini, K., and Silbergeld, E. (1997) *Toxic Ignorance*. Environmental Defense Fund (EDF), New York, USA.

Rudén C. and Hansson S.O. (2008) Evidence-based toxicology – 'sound science' in new disguise. *International Journal of Occupational and Environmental Health* 14: 299–306.

Chapter 6
Improving the Value of Standard Toxicity Test Data in REACH

Magnus Breitholtz, Elin Lundström, Ulrika Dahl, and Valery Forbes

Abstract Worldwide, environmental risk assessment strategies are based on the assumption that measuring direct effects of single substances, using a few single species tests, in combination with safety factors correcting for extrapolation inconsistencies, can be used to protect higher levels of biological organization, such as populations and even ecosystems. At the same time, we are currently facing a range of pollution problems (Millennium Ecosystem Assessment Series 2005), of which some could at least indirectly be linked to the fact that this assumption may not be fully valid. Consequently, there is an ongoing scientific debate on whether current chemical control protocols are sufficient for protection of ecosystems, and numerous suggestions for improvements have been presented by the scientific community, e.g. alternative tests and testing strategies. On the other hand, few of these suggestions actually reach the regulatory world (or become implemented), and risk assessment today basically follows the same paradigm as 30 years ago. While the new REACH regime is exceptionally ambitious, this chapter observes several problems and gaps in this regulatory framework. We suggest measures and approaches which imply increased ecological realism and understanding in future regulatory work.

Keywords Environment • REACH • Risk • Standardization • Toxicity test

M. Breitholtz (✉), E. Lundström, and U. Dahl
Department of Applied Environmental Science, Stockholm University,
SE 106 81, Stockholm, Sweden
e-mail: magnus.breitholtz@itm.su.se; elin.lundstrom@itm.su.se; ulrika.dahl@itm.su.se

V. Forbes
Deparment of Environmental, Social and Spatial Change, Roskilde University,
Universitetsvej 1, PO Box 260, DK 4000, Roskilde, Denmark
e-mail: vforbes@ruc.dk

J. Eriksson et al. (eds.), *Regulating Chemical Risks: European and Global Challenges*,
DOI 10.1007/978-90-481-9428-5_6, © Springer Science+Business Media B.V. 2010

6.1 Introduction

On the 1st of June 2007, a new European chemicals regulation came into force (REGULATION (EC) No 1907/2006). REACH (Regulation, Evaluation, Authorization and registration of CHemicals), as the legislation is called, will be introduced in a stepwise process and fully implemented in 2018. The main aims of the regulation are exceptionally ambitious in the sense that REACH 'should ensure a high level of protection of human health and the environment as well as the free movement of substances, on their own, in preparations and in articles, while enhancing competitiveness and innovation. This Regulation should also promote the development of alternative methods for the assessment of hazards of substances.' Knowledge about potential risks of both old (i.e. introduced on the market before 1981) and new chemical substances (introduced on the market after 1981) should be increased, and if the producer cannot produce data on the substance it will not go on the market, i.e. producers will be required to generate their own risk assessment reports. However, the legislation only covers chemical substances produced above 1 t per producer and year on the European market. This means that a large number of chemicals will not be covered by the regulation, and for low-volume chemicals (produced between 1–10 t per producer and year) data will not be sufficient even for an initial characterization of inherent properties (Rudén and Hansson 2006). Currently, we have knowledge about human health and environmental risk for about 1,500 of the 30,000 chemical substances that eventually will be covered by REACH (Swedish Chemicals Inspectorate 2007). This means that within the next 10 years a great number of (eco) toxicological tests must be performed in Europe. Still, REACH assumes that ecosystems can be protected by measuring direct effects in simple toxicity tests, whereas more complex – indirect effects – may be equally important to consider (e.g. Fleeger et al. 2003). For example, up to a production volume of 100 t per producer and year, only acute toxicity tests with a micro alga, *Daphnia*, a micro-organism and a fish species will be required. In fact, REACH will generate fewer systematic data as compared to the so-called new substances in the former European chemicals legislation. On the other hand, for old substances, REACH will increase the data requirements since these substances up until now have been excluded from control, unless they have been identified as priority substances. In all, REACH is an important, albeit small, step towards ensuring sustainable use of chemicals in Europe.

It is clear that the main focus in developing REACH has been related to at what level (i.e. production volume) requirements for certain relatively simple standard tests will come into force for hazard identification and dose–response assessment of single chemical substances. Little or no effort has been focused on the actual relevance of the ecotoxicological tests that should provide data for these crucial regulatory processes. As mentioned above, it has traditionally been a slow process for scientists to influence the regulatory framework, although thousands of scientific papers have presented additional test methods (e.g. reproduction tests, tests for endocrine disruption, molecular and biochemical analyses, population genetics, etc.) to improve our ability to protect ecosystems.

It is generally required that standard test methods be used for generation of toxicity data in European risk assessment (European Commission 2003; within Europe it has been decided that risk assessment procedures should make use primarily of the OECD test guideline programme [personal communication with Yvonne Andersson, Swedish Chemicals Inspectorate]), and it is likely that the costly and elaborate process related to development of new standards has constrained the incorporation of new test methods. Although a range of factors may hamper a straightforward standardization process, we believe that the following three aspects have hindered the development of new standard methods more than others, and therefore should be highlighted: (1) lack of funding for development and inter-calibration activities, (2) standardization work has low status within academia, and (3) (too) slow process (a new standard may take up to 10–15 years to develop).

The lack of funding (1) is mainly related to the fact that standardization bodies, such as ISO and OECD, do not provide central funding to support promising scientific proposals, meaning that Member States have to provide funding for these activities. Indirectly, this also means that the standardization process may be biased; larger more wealthy nations are more likely to be able to finance test development. The low status that standardization work has within academia (2) reflects the fact that it is difficult to publish such work in international peer-reviewed journals as original research. The slowness of the standardization process (3) likely depends on both of the first two reasons (1 and 2) but also on the inherently bureaucratic process, in which a large number of countries, at different hierarchical levels, need to reach consensus. Due to this, and to the large number of chemicals to be tested within REACH, it is not realistic to expect that future standard test batteries will include a wide range of additional test methods and animal groups, which could be one way to improve the chemical control process (see e.g. Breitholtz et al. 2006a, b).

Instead, to increase ecological realism and understanding in future regulatory work, we believe that alternative approaches can improve integration of ecology with baseline data from currently used standard test methods without much extra labour or cost. Since most ecotoxicological test methods are normally performed under optimal testing conditions (e.g. related to food quality/quantity, salinity, oxygen, pH, DOC, temperature, etc.), which may not be applicable for predicting toxic effects in ecosystems, in Section 6.2 we point to some suboptimal testing conditions that may be considered in regulating chemical substances. In Section 6.3 we present selected population modelling tools, which can use standard test endpoints, such as mortality and reproduction, to extrapolate to likely impacts at the population – or even the community level. In Section 6.4 we provide some suggestions on improvements of REACH and summarize our ideas.

6.2 Suboptimal Testing Conditions

6.2.1 Background

An ecotoxicological test is performed with the goal of investigating the impact of a chemical substance on ecological systems. The endpoints of such tests vary widely but are always based on stress-induced responses. Sibly and Calow

(1989) have defined stress as an 'environmental condition that reduces Darwinian fitness when first applied'. Such environmental conditions typically occur when organisms are exposed to pollutants (e.g. metals and organic chemicals) or UV radiation, food quality/quantity is low, and/or tolerance limits related to e.g. temperature, salinity, oxygen saturation, hardness and pH are exceeded. Although most organisms have an ability to regulate e.g. internal toxicant concentrations by detoxification and/or storage in the body, such activities cost energy, which means that less energy is available for other crucial processes in the body. This in turn means that organisms living under optimal conditions may better be able to handle stressors, such as chemicals, than organisms living under suboptimal conditions (e.g. Heugens et al. 2001, van der Geest et al. 2002).

In order to clearly quantify toxic effects of a chemical, current standard toxicity testing is normally performed in such a way that the organisms are exposed to a chemical under otherwise constant and optimal conditions. However, environmental conditions differ widely between different parts of the world, and even within ecosystems, which could mean that native organisms, as compared to the organisms used in the laboratory tests, are either more tolerant because they have adapted to handling external stressors, or less tolerant, as a result of combined chemical and environmental stress. Standard tests available within the large international standardization bodies, such as OECD and ISO, naturally cover only a minimal fraction of ecosystem diversity; freshwater species are, e.g. used to protect the marine environment, (high) temperatures provided by guidelines are often only representative for tropical regions, single species are used as representatives for tens of thousands or even millions of other species, etc. We can only speculate as to what extent this lack of ecological coverage limits the reliability of current chemicals regulation, but there is reason to believe that it is of major significance since so many unique biological and ecological systems and functions are missing.

In the following we will highlight a number of relevant environmental factors, which either alone or in combination may alter uptake and toxicity of both organic and inorganic chemicals, and which we think may be important for consideration in future chemicals regulation. We do not imply that the list of factors is complete, or that all of these factors should be considered as equally important for any given situation. Instead, our aim is to highlight some weak points in the current European chemicals legislation, which in some way need to be considered if the ecological relevance of REACH is to be improved. We are however fully aware that the selection of which environmental factors should be considered in REACH is a challenging task, which will need substantial research efforts in the future. We are also aware that changing environmental factors may have other implications (e.g. changes in community sensitivity due to lowered species diversity) than altered uptake and toxicity of chemicals, which may be equally or more important for ecosystem stability and functionality. However, since our focus is on the use of standard toxicity test data, such indirect ecosystem effects fall outside the scope of this chapter.

6.2.2 Environmental Factors of Importance for Uptake and Effects

In brackish environments with low salinities, such as the Baltic Sea, only a few species are able to persist, and those that do, live close to their tolerance limits (Bonsdorff and Pearson 1999). Given that the Baltic Sea receives discharges from about 85 million people living in 14 countries in the Baltic Sea drainage basin (Hannerz and Destouni 2006), it could, from a European perspective, be important to consider combinations of e.g. stressful salinity conditions and the toxic effects of environmental pollutants. This is further highlighted by the fact that the effect of salinity may not only affect organisms directly, but may also interact with, and thus modify, the toxic compound (Heugens et al. 2001). For example, due to metal complexation, bioavailability of metals often decreases with increasing salinity, which leads to decreased toxicity (e.g. Chapman et al. 1998, Witters 1998, McGeer et al. 2002). Also in marine coastal areas, there is a risk of organisms being exposed to anthropogenic exposure and salinity fluctuations simultaneously (Forbes 1991, Heugens et al. 2001). Forbes (1991) for instance found that the gastropod Hydrobia ventrosa grew more rapidly at 23‰ compared to at 33‰, whereas at 13‰ the gastropods did not grow at all. Interestingly, the negative effects of cadmium on growth were greatest at the highest salinity, and at the lowest salinity any effects of cadmium were masked by the salinity effect. Menezes et al. (2006), found lower levels of lactate dehydrogenase in brown shrimp (Crangon crangon L.) when they were exposed to simulated diurnal salinity conditions, and they concluded that this was due to energy loss associated with an increased osmotic burden.

Also temperature may influence the action of a toxicant, either by altering metabolism/detoxification (e.g. inducing heat shock or cold hardiness proteins), thus changing sensitivity towards a toxicant, or by changing feeding activity and thereby toxicant uptake (Heugens et al. 2001). In temperate coastal systems, temperature may fluctuate by 20°C annually (Camus et al. 2004), which could make standardized tests performed at one specific temperature of somewhat limited usefulness. Heugens et al. (2003) have for instance shown that, in D. magna, acute toxicity of cadmium differed substantially between 10°C and 35°C, with higher toxicity at higher temperatures. Water temperature is additionally an important factor for oxygen saturation, which may also affect the physiological status of aquatic organisms. For instance, in a laboratory test, Gardeström et al. (2007) exposed dog whelks (Nucella lapillus) collected from the intertidal zone and exposed them to 16°C (ambient), 26.5°C and 30°C under normal and hyperoxic conditions, respectively. They did not observe any thermally induced mortality at 26.5°C, but the mortality rate was 40–50% at 30°C, which however was reduced to 10% if extra oxygen was provided. It seems as if the oxygen supply was setting the limit for the whole organisms' thermal tolerance.

In the study, tissue samples were also analysed for protein-related parameters clearly showing that the stress response of dog whelks exposed to increased water temperatures differed from those exposed to lower temperatures, but that increased oxygen availability alleviated these differences thus increasing the similarity between heat-shocked and control animal protein patterns. This implies a more stable protein metabolism and might explain the increased survival of heat-shocked individuals when extra oxygen is supplied. This in turn demonstrates the importance of adequate oxygen levels for handling stress, which is something that could be considered when performing standardized toxicity tests. Oxygenation is also important for redox conditions and hence for bioavailability of metals in sediments. In anoxic conditions an important partitioning phase for cationic metals is the formation of metal sulphides, which have low solubility and hence bioavailability (Di Toro et al. 1991). In oxic conditions many metals (e.g. Cd, Zn and Cu) are instead associated to organic matter and inorganic structures, such as oxides, hydroxides of e.g. iron and manganese and clay minerals (Turner et al. 2004). Airas et al. (2008) have also shown that a combination of low dissolved oxygen concentrations and polluted sediment (but not low oxygen levels alone) reduced biomass in an oligochaete (Lumbriculus variegatus) and increased mortality in an insect (Chironomus riparius). From this study, the authors concluded that standard sediment toxicity tests may not provide sufficient data for risk assessment of contaminated sediments at sites where the actual conditions differ largely from laboratory conditions.

It is well-known that pH is of importance for bioavailability and uptake of many chemicals, such as metals (e.g. Chapman et al. 1998), but pH may also be important for the same processes concerning organic substances. Nakamura et al. (2008) recently showed that acute toxicity and bioconcentration of the pharmaceutical fluoxetine was affected by pH in the Japanese medaka (Oryzias latipes). Toxicity increased with increasing pH and bioconcentration was lower at pH 7 and higher at pH 9, likely because of increase in nonionized forms with significantly higher hydrophobicity than the ionized forms at pH values closer to pKa.

Organic material (e.g. food, particles) may influence bioavailability and bioaccumulation of both metals and hydrophobic organic substances, either by reducing or increasing uptake (e.g. Fliedner 1997, McGeer et al. 2002, De Schamphelaere et al. 2004, Wilding and Maltby 2006, Thorsson et al. 2008). Klüttgen and Ratte (1994) found that the development of juvenile D. magna was inhibited by cadmium at low food concentrations, while a body length reduction was clear at higher doses of food. The brood size was inhibited by 69% at high food levels, whereas no effect was found at low food levels at the same cadmium concentrations. Other studies have found effects on food availability and metal toxicity. For example, in acute toxicity tests, Chandini (1988) found that the cladocerans D. carinata and Echinisca triserialis were more sensitive to cadmium as food levels decreased, and Koivisto et al. (1992) found that in five cladoceran species (D. magna, D. pulex, D. galeata, Bosmina longirostris, and Chydorus sphaericus), copper exposure at low food levels decreased survival compared to high food levels. Further, although a sufficient amount of food may be available, it may still be of too low quality, which may have a negative impact on growth and reproduction (Li et al. 2008, Dahl et al. 2009). The choice of food may thus be of significant importance for risk characterizations based on standard toxicity tests, especially when using reproduction or population growth data since test organisms used to derive such chronic data need to be fed during testing.

Not only may food availability during testing influence how organisms respond to toxicants, but feeding conditions used to maintain animal cultures may also influence test results. For example, Pieters and Liess (2006) have shown that maternal nutritional state may have a significant influence on offspring sensitivity to pollutants. In their study D. magna offspring from females raised under either low or high food conditions responded differently when exposed to the pesticide fenvalerate. Low maternal food conditions increased the offspring size at time of birth, reduced age at first reproduction and increased reproductive output, which jointly enhanced offspring fitness as estimated by the population growth rate (r). Results also showed that fenvalerate exposure in combination with low maternal food levels caused a strong decrease in acute sensitivity of young daphnids (neonates), which was generally also observed for chronic endpoints.

Although it might seem logical to expect that animals with a high energy status are more successful in dealing with stress than animals with a low energy status, this may not be the case. Smolders et al. (2005) exposed D. magna to different food concentrations and measured energy status and scope for growth in animals exposed to a stressor (in this case increased salinities). Exposure to higher salinity significantly decreased survival and reproduction, but interestingly this decrease was more pronounced in the highest food concentrations, which shows that the high energy status of the daphnids from the high food concentrations at the start of the exposure did not provide an increased capacity to cope with additional stress. The authors speculated that this increased sensitivity was the result of a change in life history from emphasizing survival at low food supply to emphasizing reproduction at high food supply. The studies by Pieters and Liess (2006) and Smolders et al. (2005) clearly show that different testing conditions may have a profound impact on the outcome of standard toxicity testing and that this outcome may not be consistent with generally accepted hypotheses.

6.3 Population Modelling

6.3.1 Available Tools

Since the early 1990s, mathematical modelling has been accepted as a useful tool for developing exposure scenarios in environmental risk assessment, but has not received the same attention for effect characterization, although several techniques are available. Forbes et al. (2008) have identified three main classes of population models; i.e. demographic models, energy budget models and individual based models. Demographic models describe individuals with regard to their survival and contribution to future generations (i.e. offspring) and can either be structured or unstructured. Structured models define individuals in certain classes based on age or size, whereas unstructured models treat all individuals within the population as identical. These general models can further be supplemented with stochastic events (demographic or environmental), and by adding spatial structure, meta-population models may be obtained. It is also possible to incorporate density dependence, but this can in certain model formulations be difficult (Forbes et al. 2008). Energy budget models do not include survival as a response; instead, these models handle intake and output of energy for individuals, relating it to growth rate and reproduction. Individual based models consider each individual in a population and describe the individual responses. Population level patterns emerge from the combined responses of the individuals of which the population is composed. Individual based models are most powerful when they include great detail about individual exposures and responses and when they incorporate spatially and temporally realistic habitat features.

Population models can be used in environmental risk assessment for different purposes. They can 'detect' (or diagnose the cause of) adverse effects on populations exposed to chemicals. Data on the population is used to detect changes in population attributes and relate them to disturbances. They can also 'project' the likely consequences on populations under a set of environmental conditions, such as exposure to chemicals (or other stressors), and provide decision-makers with information about how populations are doing (see also discussion on suboptimal testing conditions in Section 6.2). Lastly, population models can 'forecast' the future behaviour of populations which is based on understanding the environmental variability as well as the dynamic interactions of density and biological processes (Munns et al. 2007). The difference between projection and forecasting is that the projection is what *would* happen to the population (given certain hypothesis, e.g. different management decisions), and a forecast is something that *will* happen and is based on a deeper understanding and more data than a projection (Caswell 2001).

6.3.2 Standard Test Data To Be Used for Regulatory Modelling

In REACH, for chemicals produced up to 100 t, it is not possible to use population modelling as an effect characterization tool since the standard tests required do not measure reproduction. However, for chemicals produced between 100 and 1,000 t, a chronic test with *Daphnia* is required, and for chemicals produced over 1,000 t reproduction tests with earthworms and chironomids are mandatory. For substances produced between 100 and 1,000 t, chronic tests with fish are also required, but there is currently no true reproduction test available with OECD or any other large international standardization body (however, a two-generation test is under development within OECD; see Table 6.1), which means that it will not be possible to generate adequate population data for fish under current testing requirements.

Further, since the reproduction test with *Daphnia* only comprises asexual reproduction, this also means that investigations focusing on sexual reproduction will be lacking for substances produced up to 1,000 t per producer and year. For substances produced above this production volume, the reproduction tests using earthworms and chironomids may however be used to generate adequate population data (see Table 6.1 for OECD test guidelines concerned with reproduction available for risk

Table 6.1 Endpoints obtained from adopted and proposed OECD Test Guidelines suggested for use in REACH

Daphnia reproduction (OECD guideline 211)	**Earthworm reproduction (OECD 222)**
Mortality	Mortality
Offspring	Changes in behaviour
Length/volume of individuals	Fecundity (number of juveniles produced)
Time to production of first brood	Body mass
Number and size of broods per animal	Pathological symptoms
Number of aborted broods	
Presence of males and ephippia	
Chironomid toxicity (OECD 219)	**Fish reproduction, 2-generation (OECD proposal)[b]**
Mortality	Survival
Offspring production	Behaviour
Sex	Fecundity
Weight of individuals	Fertilization success
Mean development rate of emerged midges	Hatchability, larvae appearance and survival
	Appearance of adults
Harpacticoid copepod development and reproduction (OECD proposal)[a]	Gonad size and morphology, and biochemical- endpoints (VTG, steroids)
Mean development rates from nauplius to the copepodite and adult stages, respectively	
Fertilization success	
Total viable offspring production per mating pair	
Time to production of first clutch	
Time interval between successive clutches	
Aborted egg sacs	
Necrotic and infertile eggs	
Sex ratio	
Stage specific mortalities	
Abnormal behaviour	

[a] OECD (Organization for Economic Cooperation and Development). OECD Draft Guidelines for Testing of Chemicals. Proposal for a New Guideline. Harpacticoid Copepod Development and Reproduction Test, Paris, France (Version: 19th of February 2008 – Current version includes only the species Amphiascus tenuiremis).

[b] OECD (Organization for Economic Cooperation and Development). Draft proposal for a new guideline. Fish Two-generation Test Guideline, Paris, France (Version: 8th of November 2002).

assessment in REACH. The proposals on copepods and fish are not yet adopted as OECD test guidelines but here serve as relevant examples). In all, this indicates that population modelling can only be part of the testing required for a rather small set of substances in the current system.

Table 6.2 is modified from Menzie et al. (2007) and contains attributes of organisms that can be used to obtain attributes of populations used in population modelling. The table highlights which of these attributes that can be derived from standard tests recommended in REACH and presented in Table 6.1. In our view, this clearly illustrates that population models may easily be incorporated into the regular risk assessment procedures within Europe.

Table 6.2 Attributes that can be used in population modelling. The table is modified from Menzie et al. (2007). An asterisk (*) denotes endpoints that can be derived from OECD guidelines and proposals suggested for use in REACH

Attributes of organisms	
Demographics of individuals	**Physiological characteristics**
Mortality*	Individual growth rate*
Reproductive state and output (e.g. fecundity, births per female)*	Respiration rate
	Ingestion rate
Development rate (e.g. time for larval development, time to maturity)*	Metabolism and excretion
Age*	**Genetic characteristics**
Size*	Individual genotypes
Sex*	Presence of particular alleles
	Heterozygosity
Ecology, behaviour and exposure	**Organism condition**
Life history for individual*	Condition factors (weight and length relationships)*
Habitat and food preference or location in space	Morbidity*
Locomotion, dispersal, migration and spatial extent of an individual	Deformities*
	Tumours and other histopathological anomalies*
Individual environmental exposure	
Attributes of populations	
Abundance	**Population growth rate**
Population size*	Intrinsic rate of natural increase*
Population density*	Finite rate of population increase
Equilibrium abundance (steady-state)	Birth, death, immigration and emigration rates
Carrying capacity	
Age class distribution*	**Spatial distribution and habitat**
Size class distribution*	Spatial distribution across available habitat
Sex ratios*	Critical patch size
	Habitat requirements
Genetic structure and variation	
Genotypic frequencies	
Heterozygosity	
Genetic diversity	

6.4 Suggestions for Improvements of REACH

In the current chapter we have illustrated how varying physical environmental factors may have a profound impact on physiological and toxicological responses in a number of aquatic organisms. The influence of varying environmental factors may become even more important as a result of expected climate change, resulting in rapid alterations of biotic and abiotic factors, and thus these issues should be of concern for chemicals regulation (Schiedek et al. 2007). In this context, it is certainly welcomed that the European Union has adopted the new chemicals regulation REACH with its very ambitious aim to protect the environment. We are however concerned that REACH may fall short of its aim as long as it relies on relatively weak tools for hazard identification and dose–response assessment of chemicals. However, we believe that the European Union has the ability to improve the current regulation and testing guidelines.

Naturally, increased numbers of test animals, and financial costs for introducing a wide range of physical environmental conditions as an obligatory part of current standard guidelines, will severely limit the development and incorporation of additional standard tests. Hence, before suboptimal testing conditions can become a mandatory aspect of risk characterization procedures, it is crucial to identify which environmental factors are of major concern. Once key physical environmental factors have been established, it is further important to develop alternative approaches to minimize any extra costs (in terms of both economical values and animal welfare) associated with increased testing frequencies. In this context our proposed use of population models would help yield more relevant data in the sense that the models may be used to predict the outcome of varying physical environmental conditions on populations, communities or even ecosystems, without any further testing. Both demographic models that incorporate environmental stochastic events and individual-based models that incorporate spatially and temporally realistic habitat features may be useful in this context. In some cases more complex models may be needed, which needs to comprise assumptions and simplifications of biological and ecological interrelationships.

A suggestion for incorporating state-of-the-art knowledge (concerning e.g. behaviour, tolerance, distribution pattern, etc.) in environmental risk assessment, which obviously is difficult to accomplish (see Section 6.1), would be to make use of such knowledge when constructing population models based on available baseline test data. We are aware that to e.g. add descriptive information or non-standard scientific data may go against what is the general paradigm in risk assessment, but we strongly believe that such actions would not be more problematic than the simplifications of ecosystem functionality and the sometimes poor extrapolations between species used in current risk assessment guidelines. In our view, the mathematical models may instead strengthen potentially weak assumptions and increase the ecological realism of the standard testing.

Moreover, taking varying environmental factors into account in chemicals regulation doesn't necessarily mean that additional temperatures or oxygen levels must be tested over the full concentration range. Selected test chemical concentrations can be tested at NOEC and EC/LC50-values for preliminary between-treatment comparisons. Further, when there are suspicions that there may be large regional differences in susceptibility to a certain chemical (possibly based on expert judgment) a new chemical ought to be tested at a range of physical environmental factors relevant for at least two of the most extreme regions within a certain area (e.g. Europe). For instance, Scandinavian freshwater systems are often oligotrophic, weakly buffered and threatened by low pH, which means that they in many aspects differ significantly from freshwater systems of middle and Southern Europe. This in turn may, as described above, affect bioavailability and toxicity of many environmental pollutants, especially metals.

In this context it is important to improve the analysis of the extent to which sensitive organisms and ecosystems in such areas may need specific test methods and specific concern in environmental risk assessment of chemicals (Breitholtz et al. 2006a). In the future, it is therefore important to increase research efforts to elucidate potential consequences of varying physical and chemical environmental factors for toxicity of a wide range of chemical substances, in order to develop tools for hazard identification and dose–response assessment that include scientifically well-based combinations of species, endpoints and environmental factors. The battery of endpoints to select from should, as far as possible, comprise population level data (Forbes and Calow 1999, Forbes et al. 2001, Breitholtz et al. 2006a), possibly obtained by using population models.

6.5 Concluding Remarks

In this chapter we have pointed to some fundamental physical environmental factors that in our view are important to take into account in order to improve REACH (and likely other chemical legislations). We have also pointed to the potential of an increased use of mathematical population models to obtain more relevant data for environmental risk assessment. With increased knowledge about how various physical environmental factors (e.g. temperature, pH, salinity, O_2) on one hand influence toxicity and on the other may be taken into account in the process of environmental risk assessment, chances will improve to achieve a process that is efficient, cost effective, scientifically robust, and meets the demands of science-based precaution. Environmental risk assessment within REACH would thus become a more diverse but at the same time more adequate process than the one presented in the current version.

References

Airas S., Leppanen M., Kukkonen J.V.K. (2008) Effects of temperature and oxygen concentration in sediment toxicity testing. *Ecotoxicology and Environmental Safety* 70(3): 475–482.

Bonsdorff E., Pearson T.H. (1999) Variation in the sublittoral macrozoobenthos of the Baltic Sea along environmental gradients: A functional-group approach. *Australian Journal of Ecology* 24: 312–326.

Breitholtz M., Rudén C., Hansson SO., Bengtsson B-E. (2006a) Ten challenges for improved environmental risk assessment. *Ecotoxicology and Environmental Safety* 63: 324–335.

Breitholtz M., Eriksson J., Green N., Ruden C. (2006b) Testing and risk assessment of persistent and bioaccumulating chemical substances – Improvements within REACH? *Human and Ecological Risk Assessment* 12: 782–805.

Camus I, Davies P.E., Spicer JI, Jones M.B. (2004) Temperature-dependent physiological response of Carcinus maenas exposed to copper. *Marine Environmental Research* 58: 781–785.

Caswell H. (2001) *Matrix Population Models–Construction, Analysis and Interpretation*. Sinauer Associates, Sunderlands, MA.

Chandini T. (1988) Changes in food (Clorella) levels and the acute toxicity of cadmium to Daphnia carinata (Daphnidae) and Echinisca triserialis (Microthricidae)[Crustacea, Cladocera]. *Bulletin of Environmental Contamination and Toxicology* 41(3): 398–403.

Chapman P.M., Wang F.Y., Janssen C., Persoone G., Allen H.E. (1998) Ecotoxicology of metals in aquatic sediments: binding and release, bioavailability, risk assessment, and remediation. *Canadian Journal of Fisheries and Aquatic Sciences* 55(10): 2221–2243.

Dahl U., Rubio-Lind C., Gorokhova E., Eklund B., Breitholtz M. (2009) Food quality effects on copepods in toxicity tests. *Ecotoxicology and Environmental Safety* 72(2): 351–357.

De Schamphelaere K., Vasconcelos F., Tack F., Allen H.E., Janssen C. (2004) Effect of dissolved organic matter source on acute copper toxicity to Daphnia magna. *Environmental Toxicology and Chemistry* 23(5): 1248–1255.

Di Toro D.M., Zabra C.S., Hansen D.J., Berry W.J., Swartz R.C., Cowan C.E., Pavlou S.P., Allen H.E., Thomas N.A., Paquin P.R. (1991) Technical basis for establishing sediment quality criteria for nonionic organic chemicals using equilibrium partitioning. *Environmental Toxicology and Chemistry* 10: 1541–1583.

European Commission (2003) Technical guidance document in support of commission directive 93/67/EEC on risk assessment for new notified substances and commission regulation (EC) No 1488/94 on risk assessment for exiting substances. Luxembourg, ISBN: 92-827-8011-2.

Fliedner A. (1997) Ecotoxicity of poorly water-soluble substances. *Chemosphere* 35(1/2): 295–305.

Fleeger J.W., Carman K.R., Roger Nisbet M. (2003) Indirect effects of contaminants in aquatic ecosystems. *Science of the Total Environment* 317: 207–233.

Forbes V.E. (1991) Response of Hydrobia ventrosa (Montagu) to environmental stress effects of salinity fluctuations and cadmium exposure on growth. *Functional Ecology* 5(5): 642–648.

Forbes V.E., Calow P. (1999) Is the per capita rate of increase a good measure of population-level effects in ecotoxicology? *Environmental Toxicology and Chemistry* 18(7): 1544–1556.

Forbes V.E., Calow P., Sibly R.M. (2001) Are current species extrapolation models a good basis for ecological risk assessment? *Environmental Toxicology and Chemistry* 20(2): 442–447.

Forbes V.E., Calow P., Sibly R. (2008) The extrapolation problem and how population modelling can help. *Environmental Toxicology and Chemistry* 27(10): 1987–1994.

Gardeström J., Elfwing T., Lof M., Tedengren M., Davenport J.L., Davenport J. (2007) The effect of thermal stress on protein composition in dogwhelks (Nucella lapillus) under non-normoxic and hyperoxic conditions. *Comparative Biochemistry and Physiology A – Molecular and Integrative Physiology* 148(4): 869–875.

Hannerz F., Destouni G. (2006) Spatial characterization of the Baltic Sea drainage basin and its unmonitored catchments. *Ambio* 35(5): 214–219.

Heugens E.H.W., Hendriks A.J., Dekker T., van Straalen N.M., Admiraal W. (2001) A review of the effects of multiple stressors on aquatic organisms and analysis of uncertainty factors for use in risk assessment. *Critical Reviews in Toxicology* 31(3): 247–284.

Heugens E.H.W., Jager T., Creyghton R., Kraak M.H.S., Hendriks A.J., Van Straalen N.M., Admiraal W. (2003) Temperature-dependent effects of cadmium on Daphnia magna: accumulation versus sensitivity. *Environmental Science and Technology* 37(10): 2145–2151.

Klüttgen B., Ratte H.T. (1994) Effects of different food doses on cadmium toxicity to Daphnia magna. *Environmental Toxicology and Chemistry* 13(10): 1619–1627.

Koivisto S., Ketola M., Walls M. (1992) Comparison of 5 cladoceran species in short-term and long-term copper exposure. *Hydrobiologia* 248(2): 125–136.

Li J., Sun S., Li C., Zhang Z., Pu X. (2008) Effects of different diets on the reproduction and naupliar development of the copepod Acartia bifilosa. *Journal of Experimental Marine Biology and Ecology* 355(2): 95–102.

McGeer J., Szebedinszky C., McDonald G., Wood C. (2002) The role of dissolved organic carbon in moderating the bioavailability and toxicity of Cu to rainbow trout during chronic waterborne exposure. *Comparative Biochemistry and Physiology Part C* 133: 147–160.

Menezes S., Soares A.M.V.M., Guilhermino L., Peck M.R. (2006) Biomarker responses of the estuarine brown shrimp Crangon crangon L. to non-toxic stressors: temperature, salinity and handling stress effects. *Journal of Experimental Marine Biology and Ecology* 335(1): 114–122.

Menzie C., Bettinger N., Fritz A., Kapustka L., Regan H., Moller V., Noel H. (2007) Population protection goals. In: Barhthouse L.W., Munns W.R. Jr, Sorensen M.T. (eds) *Population-level ecological risk assessment*. CRC, Boca Raton.

Millennium Ecosystem Assessment Series (2005) Ecosystems and human well-being: current state and trends. Findings of the condition and trends working group, chapter 15. ISBN: 1-55963-228–3.

Munns W., Gervais J., Hoffman A., Hommen U., Diane N., Nakamaru M., Sibly R., Topping C. (2007) Modelling approaches to population-level ecological risk assessment. In: Barhthouse L.W., Munns W.R. Jr, Sorensen M.T. (eds) *Population-level ecological risk assessment*. CRC, Boca Raton

Nakamura Y., Yamamoto H., Sekizawa J., Kondo T., Hirai N., Tatarazako N. (2008) The effects of pH on fluoxetine in Japanese medaka (Oryzias latipes): acute toxicity in fish larvae and bioaccumulation in juvenile fish. *Chemosphere* 70: 865–873.

Organization for Economic Cooperation and Development (1998) Daphnia magna reproduction test. OECD Guidelines for Testing of Chemicals. Technical Guidance Document 211.

Organization for Economic Cooperation and Development (2004a) OECD Guidelines for the Testing of Chemicals, Sediment-water chironomid toxicity test using spiked water, OECD 219, adopted 13 April 2004.

Organization for Economic Cooperation and Development (2004b) OECD Guideline for the Testing of Chemicals, Earthworm Reproduction Test (Eisenia fetida/Eisenia Andrei), OECD 222, adopted 13 April 2004.

Pieters B.J., Liess M. (2006) Maternal nutritional state determines the sensitivity of Daphnia magna offspring to short-term Fenvalerate exposure. *Aquatic Toxicology* 76: 268–277.

Rudén C., Hansson SO. (2006) Improving REACH. *Regulatory Toxicology and Pharmacology* 44: 33–42.

Schiedek D., Sundelin B., Readman J.W., Macdonald R.W. (2007) Interactions between climate change and contaminants. *Marine Pollution Bulletin* 54: 1845–1856.

Sibly R, Calow P (1989) A life-cycle theory of responses to stress. *Biological Journal of Linnean Society* 37(1–2): 101–116.

Smolders R, Baillieul M, Blust R (2005) Relationship between the energy status of Daphnia magna and its sensitivity to environmental stress. *Aquatic Toxicology* 73: 155–170.

Swedish Chemicals Inspectorate (2007) REACH Factsheet. http://www.kemi.se/upload/Trycksaker/Pdf/Faktablad/FbReachMars08.pdf

Thorsson M.H., Hedman J.E., Bradshaw C., Gunnarsson J.S., Gilek M. (2008) Effects of settling organic matter on the bioaccumulation of cadmium and BDE-99 by Baltic Sea benthic invertebrates. *Marine Environmental Research* 65(3): 264–281.

Turner, A., Millward G.E., Le Roux S.M. (2004) Significance of oxides and particulate organic material in controlling trace metal partitioning in a contaminated estuary. *Marine Chemistry* 88: 179–192.

van der Geest H.G., Soppe W.J., Greve G.D., Kroon A., Kraak M.H.S. (2002) Combined effects of lowered oxygen and toxicants (copper and diazinon) on the mayfly Ephoron virgo. *Environmental Toxicology and Chemistry* 21(2): 431–436.

Wilding J., Maltby L. (2006) Relative toxicological importance of aqueous and dietary metal exposure to a freshwater crustacean: implications for risk assessment. *Environmental Toxicology and Chemistry* 25(7): 795–1801.

Witters H.E. (1998) Chemical speciation dynamics and toxicity assessment in aquatic systems. *Ecotoxicology and Environmental Safety* 41: 90–95.

Chapter 7
Testing in Aquatic Ecotoxicology: What Are the Scientific Conditions for the '3R' Concept?

Peter Pärt, Argelia Castaño, and Bengt-Erik Bengtsson

Abstract In this chapter, we will evaluate how and if a 3R based approach can be applied in testing of ecotoxicity of chemicals. The 3R approach (reduce-refine-replace) is a strategy to reduce or totally abolish the use of experimental animals in favour of alternative methods. We review the current status of alternatives in aquatic ecotoxicology and how well they perform in comparison with current *in vivo* methods. We will conclude that theoretically can alternative methods and approaches replace animal based testing but the way to reach this goal is long. A strong development of more sophisticated alternative methods is needed focusing on specific and physiologically/toxicologically relevant endpoints. We underline the importance to gain more information on toxic mechanisms of chemicals. New exciting biochemical techniques are waiting around the corner, e.g. in the genomics area and they need to be integrated in future test paradigms.

Keywords 3R • Aquatic • Ecotoxicology • Fish tests • REACH

P. Pärt (✉)
Institute of Environment and Sustainability,
European Commission, Joint Research Centre, Via E. Fermi 2749, I-21027 Ispra (VA), Italy
e-mail: peter.part@jrc.ec.europa.eu

A.Castaño
Centro Nacional de Sanidad Ambiental, Instituto de Salud Carlos III, Ctra. Majadahonda – Pozuelo, 28220 Majadahonda (Madrid), Spain
e-mail: castano@isciii.es

B.-E. Bengtsson
Department of Applied Environmental Science, Stockholm University, SE 106 91, Stockholm, Sweden
e-mail: bengt-erik.bengtsson@itm.su.se

J. Eriksson et al. (eds.), *Regulating Chemical Risks: European and Global Challenges*,
DOI 10.1007/978-90-481-9428-5_7,

7.1 Introduction

Europe is introducing a new regulatory system (REACH) for management of chemicals. The overriding objective is to ensure protection of human health and the environment and at the same time have a control of the flow of use of chemicals in the society. An important component in REACH is extended hazard assessment for chemicals in use of more than 1 t/year. Approximately 30,000 chemicals fall in this group. Information on toxicity and the biological behaviour of chemicals has therefore a central position in REACH. This need leads to an increasing demand on test methods. However, already in the starting instructions of REACH (Commission White Paper on a new Chemical legislation, EC 2001) it was stated that REACH should not lead to an increased use of test animals. Instead, REACH should stimulate the development of alternative methods and approaches. These include *in vitro* methods, *in silico* methods (QSAR, SAR) and the optimal use of information from one method to another, so called 'read across'. An intensive development is currently going on, particularly in the industry, to meet the REACH testing requirements in a cost-effective manner.

Information on aquatic toxicity is mandatory when introducing new chemicals on the market. An aquatic toxicity 'base-set' of tests is required by the European REACH legislation for chemicals produced or marketed more than 10 t/year. Base set testing builds on acute toxicity tests with algae, a crustacean (*Daphnia magna*) and fish. In general, there is a high reliance in the current chemicals regulations on *in vivo* data from fish tests. The following numbers are reported by ECETOC 2005. If base-set information were required for 30,000 chemicals (1–100 t production), the number of fish required for testing would be 1.3 million for level 0, and for level 1 (100–1,000 t production) approximately two million. The ECETOC report concludes that this must be considered as a conservative estimate as often tests will use more than the minimum specified in the OECD protocol in order to satisfy global regulatory demands. Moreover, the testing of these chemicals *in vivo* would be extremely costly and time demanding. In addition, over the years there has been an increasing awareness among the public that vertebrate animal testing is causing distress and pain in the test animals (Nagel 2002, Braunbeck et al. 2005) and that such activities might be in conflict with current animal welfare legislations. Since fish are vertebrates, fish tests are under objection and are under public pressure to be replaced by alternatives. To overcome resource and ethical problems, REACH promotes a change in testing strategy, to use *in vitro* methods for the initial hazard evaluation of chemicals. This will require the establishment and validation of *in vitro* test protocols.

However, we should keep in mind that in spite of a long history of testing we have been unable to foresee the environmental hazards of many anthropogenic substances, such as DDT, PCB, chlorinated paraffins, brominated flame retardants, PFOS and lately the pharmaceutical drug diclophenac (Oaks et al. 2004). Mankind and the environment pay a high price for these mistakes and we have to do a better job in the future. So the new challenge is: When we replace one test method with a new one, it should not just be 'as good' but better than what we have used before.

7.2 Why Fish Tests At All?

Fish tests have a long tradition in ecotoxicology. Historically occurrence of fish kills raised the public interest to find an explanation and ways to avoid similar events in the future. Fish is also an important food resource and has high recreational values. The ambition to protect fish is therefore socially well established in most cultures. From a more scientific point of view, chemicals sooner or later end up in the aquatic environment and fish play an important role (or roles depending on species and size) in aquatic food chains. A key piece of information required for a risk assessment is the concentration at which a chemical causes an adverse effect, in particular the so called NOEC (No-Observed-Effect-Concentration). Such information is used to derive PNECs (Predicted-No-Effect-Concentration). In environmental risk assessments, PNECs are compared with PECs (Predicted-Environmental-Concentrations) to establish whether there is any environmental risk (i.e. PEC/PNEC ratio is >1). PNECs can be estimated from acute or chronic data. The test used to establish PNEC for a chemical depends on the quantity of chemical being placed on the European market. For chemicals below 100 t/year, acute toxicity is enough, for chemicals above 100 t additional tests for sub-lethal endpoints are required. Therefore, test guidelines are covering both mortality and morbidity. The following fish based tests are currently in use or are in discussion to be included in test guidelines:

Acute Toxicity *The* acute fish toxicity test is the fish test in the 'base-set'. The test (OECD 203, 1992a) involves exposure of fish to a test substance for a preferred period of 96 h. A minimum of seven fish are required for each of at least five concentrations plus control(s). Acute toxicity is expressed as the median lethal concentration that kills 50% of the population (LC-50) over the given time period. The species of fish used, according to OECD test guideline 203 (OECD 203, 1992a), should be selected on the basis of practical criteria, such as their ready availability throughout the year, ease of maintenance, convenience for testing, relative sensitivity to chemicals, and any economic, biological, ecological or geographical factors.

Fish Prolonged Toxicity Test – 14 Day Study (OECD 204, 1984) is a variant of the acute fish test used when a longer exposure time is needed, for example when testing highly lipophilic and poorly water soluble substances, and/or when reporting of additional information is considered necessary. The principle of the test is that threshold levels of lethal and other observed effects and NOEC are determined at intervals during the test period. The test requires at least ten fish per concentration plus control(s).

The Fish Juvenile Growth Test (OECD 215, 2000) measures the effect of a chemical on the growth of a population of juvenile fish, which are in the exponential growth phase. The fish are weighed prior to commencement of the study and exposed for 28 days at sub-lethal concentrations of the test chemical. Fish are fed a ration based on the initial weight of the fish, which may be recalculated after 14 days. The fish are reweighed at the end of the test and effects on growth rates determined through regression analysis or through one-way analyses of variance followed by multiple range tests comparisons with control data to determine a NOEC and lowest observed effect concentration (LOEC).

Early Life Stages (ELS) Toxicity Test Tests with early life stages of fish are assumed to target a particularly vulnerable period in the life cycle of fish. The tests give information on both the chronic- and sublethal toxicity of substances (OECD Guideline 210, OECD 210, 1992b). The more recent OECD guideline 212 (OECD 212, 1998) aids in forming a bridge between lethal and sublethal tests. We will in a later chapter discuss the applicability of ELS tests as replacements of the acute fish *in vivo* test according to OECD test guideline 203.

Test for Endocrine Disruption in Fish Currently, there are no internationally validated methods to determine the potential of chemicals to impact the endocrine system. However, a number of methods have been proposed as screening assays to identify the potential endocrine activity as well as confirmatory tests to assess adverse reproductive effects. Candidate protocols are described in a detailed review paper (OECD 2004). The tests discussed are all based on endpoints related to reproduction. These include gross morphology (appearance, including secondary sexual characteristics and gonado-somatic index (GSI)); plasma or liver vitellogenin levels; and histopathology of excised gonads. Other end-points are fecundity, hatchability, growth and survival of eggs and larvae and behavioural parameters connected to spawning and nursing behaviour.

Fish Full Life Cycle Test (FFLC) The FFLC test is based on the US EPA guideline for fish full life-cycle toxicity testing (US EPA 1986). In summary, the test begins with embryos (P) less than 24 h old, which are continuously exposed throughout the development of the fish until the fish are sexually mature. Once the fish achieve maturity, they are assessed for reproductive behaviour and fecundity. During the reproductive phase, embryos (F1) obtained from the P fish are developed for a minimum of 28 days (post-hatch) to determine in-life biological effects. If required, the development of the F1 fish can also be progressed to determine histological and biochemical endpoints. Typically, the minimum duration of a FFLC test using fathead minnow is 250 days.

Fish Two-Generation Test The fish two-generation test guideline is a proposed redesign of the FFLC. It is intended to establish the effects on reproduction of parent and offspring exposed to a toxicant and to capture any transgenerational effects. The test also enables histological and biochemical endpoints to be determined. The duration of the test is a minimum of 180 days, although it may be considerably longer depending upon the species used, for example, with fathead minnow the test duration would be at least 3 weeks longer than the traditional FFLC. The principle of the test is that adult spawning fish (P) are exposed for a minimum of 21 days and embryos collected from these fish are used for the development of the next generation (F1). The F1 fish are then progressed to maturity and assessed for reproductive behaviour and fecundity. During the reproductive phase, embryos (F2) obtained from the F1 fish are developed for a minimum of 28 days to determine in-life biological effects.

Fish Bioconcentration Test (BCF_{fish}) A fish bioconcentration test may be required for chemicals with a log P_{ow}>2.7 (P_{ow} = n-octanol:water partition coefficient) and a production- or marketing volume > 100 t/year. There are two different methods to evaluate BCF according to the OECD guideline 305 (OECD 305, 1996). In the first fish are exposed until 'steady-state' and BCF is calculated from the concentration of chemical in the fish divided with the concentration in water. In the other methods, kinetic rate constants for uptake (K_1) and depuration (K_2) are measured during an exposure phase and a depuration phase. BCF is calculated as $BCF = K_1/K_2$ under the assumption that the accumulation of chemicals in fish is described by a one-compartment model. OECD 305 requires three groups of fish, two exposures and one control, minimum four fish/sampling occasion and at least five sampling sessions for the accumulation phase and four for the depuration phase. Minimally 108 fish are used per chemical tested.

7.3 Alternative '3 R' Based Approaches in Ecotoxicology

The 3Rs approach as formulated by Russel and Burch (1959) is outlining three strategies for how the number and the suffering of experimental animals used in research and testing can be minimised. The 3Rs stand for Replacement, Reduction and Refinement. Replacement means the substitution of conscious living higher animals by an alternative non-animal system, a less sensitive living species (bacteria, plants or invertebrates) or an *in vitro* system. Reduction means reduction in the numbers of animals used to obtain information of given amount and precision. Refinement addresses any decrease in the incidence or severity of inhumane procedures applied to those animals that still have to be used. We will here outline how this approach has been applied, or can be applied in testing the aquatic ecotoxicity of chemicals.

7.3.1 Reduction: Acute Threshold Approach

Today, hazard identification for the aquatic environment is based on three tests: algae growth inhibition test, Daphnia magna immobility test and fish lethality. The lowest EC/LC50 is selected to perform risk characterisation (PEC/PNEC). Hutchinson et al. (2003) proposed a strategy, the acute threshold (step) down approach (OECD 2010) to reduce the number of fish used in hazard identification of pharmaceuticals. The objective is to reduce the number of fish and to estimate PNEC by applying comparative threshold data obtained from the most conservative data from algae and/or *Daphnia* acute tests. Hutchinson et al. (2003) substantiated this principle for 91 pharmaceuticals. Approximately 80% of the tested pharmaceuticals had a LC-50 value in fish which was equal to or higher than the most sensitive algae or *Daphnia* test. Hutchinson et al. (2003) proposed that for the remaining 20% of the pharmaceuticals, it would be possible to extrapolate a comparable LC-50 value for the fish by employing a step-down factor of 3.2 to the EC-50 value derived from the most sensitive species. Building on this, Jeram et al. (2005) identified full data sets for 1,400 chemicals in the New Chemicals Database of the European Chemicals Bureau. For 85% of the substances, either the algae or *Daphnia* was the most sensitive species and accordingly was providing the basis for calculation of PNECs. For only 15% of the tested substances was fish the most sensitive. Jeram et al. (2005) proposes a procedure when first tests are preformed with algae and *Daphnia*, where after a one-concentration test is made with fish with the lowest EC-50 concentration from the *Daphnia*/alga test. If mortality occurs, a full LC-50 test is performed with fish or, alternatively, in a step-down fashion until a concentration is reached which does not affect the animals. Jeram et al. (2005) calculated that this procedure would save between 54% and 71% of the fish from being sacrificed. The ECVAM Scientific Advisory Committee (ESAC) has made a formal statement on the validation and endorsed the approach. At present a draft guideline is under revision and acceptance in OECD.

7.3.2 Replacement: In Vitro Cell Based Methods

7.3.2.1 Cytotoxicity

In vitro methods based on toxicity in cell cultures offers alternatives to testing with whole organisms. Many methods are using cultured fish cell lines and are using cytotoxicity (cell death) as the endpoint. The use of fish cells in assessing environmental hazard of chemicals and effluents is increasing in popularity and is gaining broader regulatory acceptance. Cell lines will probably become even more important in the future, as the new concepts of genomics and proteomics become incorporated into screening tests. In addition, cellular systems could provide the basis for automated high throughput technologies which will facilitate the screening and analysis of large number of samples in a standardised and reproductive way. A comprehensive review on the current and past use of fish cells in ecotoxicology is found in Castaño et al. (2003) and for general conditions for *in vitro* methodology and endpoints in ecotoxicological assessment the reader is referred to the exhaustive review of Schirmer (2006).

Two main types of fish cell cultures have been described and used in toxicological work, primary cell cultures and immortalised cell lines. Primary cultures are prepared from fresh tissues from the organism. The advantage is that the cells maintain many of their original properties and they provide a tool for studies of mechanisms at cellular level as they occur in the organism. The disadvantage is that the survival time of primary cultures is limited. Many primary cells do not divide in culture and if they do, they only will go through a few passages. Therefore, primary cultures are a ‘fresh preparation’ and animals have to be sacrificed in order to ensure a steady supply of cells. Primary cultures also have a ‘memory’ of the history of the organism, reflecting exposure to toxicants, nutritional status and other adaptive responses. In basic studies this could be an advantage because it gives an opportunity to analyse the mechanisms of adaptation and tolerance. However, in routine screening studies this could be a problem because of a large and irregular variation.

Immortalised cell cultures, or cell lines, overcome this because they are clones which have been propagated over several hundreds of cell generations. The obvious advantage is a continuous and steady supply of cells in the laboratory. More than 150 continuous cell lines have been established from fish. Most of them are either fibroblast-like or epithelial-like, and originate mainly from the tissues of *Salmonid* or

Cyprinid fish species. Established fish lines are grown basically in basal culture media supplemented with mammalian sera. They are generally anchorage dependent, and grow attached in conventional tissue culture ware. Cell lines resist freezing and can be stored for long periods in frozen conditions. They also retain their viability during long-term low temperature storage and grow normally after return to the optimum temperature. For example, the RTG-2 cell line, derived from rainbow trout gonad, is able to remain viable for 2 years at 4°C without any medium change (Wolf and Quimby 1969). Fish cells are sensitive for high temperatures and most piscine lines are unable to grow above 30°C. Growth and cellular functions are strongly temperature dependent and consequently temperature affects the cytotoxic responses.

These practical advantages of immortalised cells are however balanced by disadvantages regarding properties and metabolic capacities and how well they represent the functions in the original tissue. Some evidence indicate that chromosomal aberrations can occur in immortalised cells lines and that the cells become de-differentiated and loose many of their original properties and capacities. This is particularly important in biotransformation studies since immortalised cells seem to have reduced metabolic activity. Cell lines are regularly used in toxicity studies when the end-point has been basal cytotoxicity. Some representative fish cell lines used in different types of toxicity studies are shown in Table 7.1.

Cytotoxicity is a measure of toxicity to living cells as a result of toxic exposure (Table 7.2). Basal cytotoxicity has been defined as the adverse effects resulting from interference with structures and/or processes essential for cell survival, proliferation, and/or function common to all cells in the organism. Basal cell functions generally support organ-specific cell functions. Basal cytotoxicity data are expressed as IC50 (concentration affecting 50% of cells compared to the untreated control cells), which can be mathematically calculated from the concentration–effect curves. As a rule, cells are exposed to different concentrations of the chemical for a given period, after which the degree of inhibition of basal cell functions is measured by using different

Table 7.1 Examples of common fish cell lines used in *in vitro* assays. For exhaustive compilations, see Castaño et al. (2003) and Schirmer (2006)

Type of cell line	Origin and culture condition	Advantage
PLHC-1	Top minnow hepatocellular carcinoma (monolayer culture)	Retains certain metabolic activity. Easy to culture
RTG-2	Rainbow trout gonad (monolayer culture)	Easy to culture. Some metabolic capacity. Can be stored for long periods in refrigerated conditions. Good for cytotoxicity and genotoxicity studies. Has been used as hosts to develop reporter gene systems
R-1	Rainbow trout liver (monolayer culture)	Good correlation with *in vivo* results concerning liver toxicity
BF-2	Caudal trunk of blue gill sunfish (monolayer culture)	Easy to culture. More sensitive than RTG-2 for some groups of chemicals
CHSE-sp	Chinook Salmon (suspension culture and monolayer cultures)	The same sensitivity in suspension as in monolayer culture. Suspension cultures easier and more rapid than monolayer cultures

Table 7.2 Cytotoxicity assays. Commonly used endpoints. For more detailed compilations, see Castaño et al. (2003) and Schirmer (2006)

Assay	Endpoint	Advantages/disadvantages
Neutral red release (NRR)	Membrane integrity	More rapid than cell viability and cell growth inhibition. Not appropriate for all types of chemicals
Neutral red uptake (NRU)	Cell viability since live cells accumulate the dye in lysosomes	Good correlation with LC-50 in fish. Highly reproducible
MTT (mitochondrial reduction of triazolium salts	Cell viability based on active mitochondria	Sensitive
ATP	Active metabolism	Sensitive and representative for both oxidative phosphorylation and for glycolysis
LDH (lactate dehydrogenase leakage	Membrane integrity	Easy to measure. Fluorescent methods sensitive
Thymidine incorporation	Cell growth	Sensitive and integrative of many essential cellular functions
Protein content	Cell growth, cell detachment	Easy to measure

end points. Estimates of cytotoxicity are generally based on the uptake or exclusion of dyes and are many times an indication of the integrity of the plasma membrane or some intracellular organelles. Other cytotoxicity indicators are depending on an active mitochondrial respiration or on intact lysosomes. The two assays most frequently employed for the assessment of basal cytotoxicity in this study are the MTT assay and the neutral red uptake assay (NRU). Another approach in assessing cytotoxicity is to measure cell proliferation, by counting cells, by measuring cell protein or by following incorporation of radioactive (^{3}H-labelled) thymidine in DNA or the incorporation of the thymidine analogue BrdU (5-bromo-2′-deoxyuridine) which can be measured with antibody based assays.

The question is how well cell based *in vitro* test performs in comparison to the *in vivo* fish LC-50 test. Two aspects are important, the relative potency and the sensitivity. In this respect should the *in vitro* test be able to generate comparable results on the relative potency of toxicants, and at effect concentrations similar to those in fish *in vivo* bioassays. In addition, *in vitro tests* should not indicate false positive or false negative results (Segner 2004). A good and generally strong correlation has been established between *in vitro* cytotoxicity and *in vivo* LC-50 in fish (Table 7.3) (Castaño et al. 1996, Castaño et al. 2003) and it can be concluded that as far as relative potency is concerned, the *in vitro* tests perform well as the *in vivo* LC-50 test.

The problem with fish cell tests, as well as with mammalian *in vitro* test, is their lower sensitivity compared with animal *in vivo* tests. On the average, the fish cell cytotoxicity assays are one or two orders of magnitude less sensitive than the *in vivo* acute fish tests (Castaño et al. 2003). The low absolute sensitivity and possibility of generating false negatives compared to fish *in vivo* LC-50 test is the strongest criticism against using fish cell lines as alternatives to acute fish test.

Table 7.3 Correlation coefficients between EC-50 values of various chemicals obtained with established fish cell lines and LC-50 data obtained with the *in vivo* test (Modified from Castaño et al. 2003, the reader is referred to this publication for details and references)

Fish cell line	Endpoint	Fish species used for LC-50	Slope	n	r	r2	Chemical class
FHM	Neutral red	Golden orfe		49	0.89	0.79	Heterogeneous
GFS	Neutral red	Carp		18	0.69	0.48	Organophosphate pesticides
GFS	Neutral red	Carp		34	0.85	0.72	Pesticides
GFS	Neutral red	Fathead minnow		31	0.96	0.92	Narcotics, anilines, Phenols
GFS	Neutral red	Guppy		29	0.96	0.92	aldehydes & pesticides
GFS	MTT	Goldfish		7	0.96	0.93	chlorophenols
GFS	MTT	Medaka		15	0.92	0.84	chlorophenols
GFS	MTT	Guppy		8	0.96	0.92	chlorophenols
RTG-2	ATP content	Rainbow trout		26	0.97	0.94	Heterogeneous
RTG-2	Neutral red	Rainbow trout		26	0.98	0.96	Heterogeneous
RTG-2	Cell detachment	Rainbow trout		26	0.98	0.95	Heterogeneous
BG/F	Neutral red	Rainbow trout		4	0.98	0.96	Organ-mercurial compounds
RTG-2	cell attachment	Rainbow trout		9	0.92	0.85	Phenols, benzenes, anilines
RTG-2	MTT	Zebra fish		5	0.95	0.90	Heterogeneous
RTG-2	Neutral red	Zebra fish		4	0.99	0.98	Heterogeneous
PLHC-1	MTT	Medaka		9	0.80	0.64	Organo tin compounds
PLHC-1	Neutral red	Medaka		8	0.86	0.74	Organo tin compounds
BG/F	Neutral red	Platessa		4	0.99	0.98	Organo lead compounds
FHM	Total protein content	Golden orfe		25	0.90	0.81	Hetrogeneous

Another approach is to use other cell-based systems that may have a higher sensitivity but still are representative for fish toxicity (Castaño et al. 2003; Segner 2004). Since basal cytotoxicity reflects adverse effects on cell structures and processes that are intrinsic to virtually all cells, most cell systems should show a similar response, and respond similarly when toxicity is measured by various viability criteria. Mammalian cells that are cultured at higher temperatures and that proliferate faster than fish cells, may therefore be more sensitive and could provide a better *in vitro* system to predict acute fish lethality particularly if cell growth is considered as the endpoint (Castaño et al. 2003; Segner 2004). Castaño and Gómez-Lechón (2005) compared the sensitivity of fish and mammalian cell lines to 51 chemicals during 24 or 48 h exposure (Fig. 7.1).

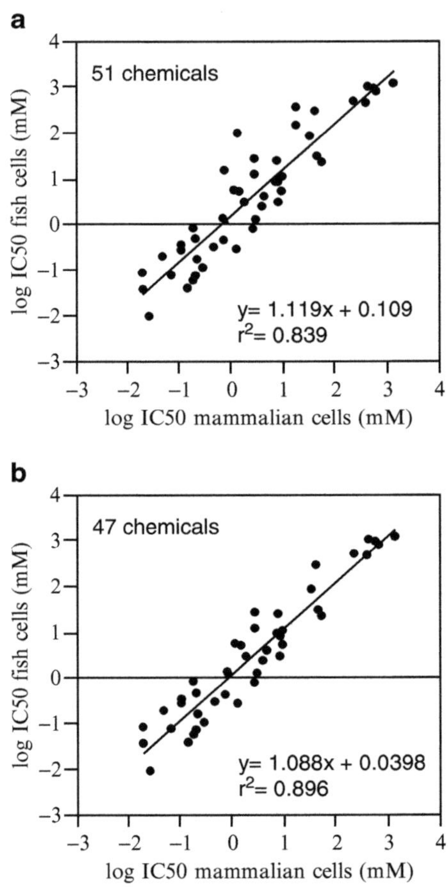

Fig. 7.1 IC-50 of chemicals from mammalian cells against IC-50 from fish cells lines after 24 h exposure. (**a**) 51 chemicals, (**b**) 47 chemicals (excluding paraquat, potassium chloride, dichloromethane, xylene) (From Castaño and Gómez-Lechón 2005)

The slope of the lines is very close to 1 which indicates that the fish cells and the mammalian cells have very similar responses to the tested chemicals. Since the intercept of the regression lines is close to 0 it indicates that fish and mammalian cells are equally sensitive to the tested compounds and consequently mammalian cells are not a better predictor of fish *in vivo* toxicity than fish cells.

This confirms that there are many fundamental similarities between fish- and mammalian cells with respect to cellular mechanisms and toxic responses. The conclusion is important from a hazard assessment point of view, because by accepting that results from mammalian cytotoxicity test as representative for fish, and in a longer range ecotoxicity, the knowledge base for assessments is increased.

However, fish cells also reflect a number of fish-specific traits that cannot be assessed with mammalian cells (Wolf and Quimby 1969; Castaño et al. 2003). An example is production of the yolk protein vitellogenin in the liver, which frequently is used as an *in vivo* biomarker for estrogen-like activity. Fish hepatocytes can be used as an *in vitro* screening method for estrogenicity with vitellogenin as a convenient endpoint. Mammalian hepatocytes lack this capacity. Fish cells have many practical advantages over mammalian cells: they can be incubated at room temperature (20°C) and in the ambient atmosphere, which means that specialised incubators are not needed; and they can be stored for long periods at 4°C, circumventing the need for freezing/ thawing the cultures. Fish cells can be exposed to various aquatic environmental samples at varying osmolarities, something that can be done only with mammalian renal cells. Due to the commonality of end points and simpler handling, fish cells could even replace the use of mammalian cells in some specific tests (for example, for the testing of non-sterile environmental matrices) (Castaño et al. 2003).

7.3.2.2 Sub-lethal and Mechanistic Endpoints

In vitro systems are excellent for analysing toxic mechanisms at cellular level. *In vitro* systems can also be used to screen for chemicals assumed to have a specific mechanisms of toxicity. Examples are genotoxic chemicals, (Becerril et al. 1999, Sánchez et al. 2000, Castaño and Becerril 2004) hormone active chemicals (endocrine disrupters) (Hornung et al. 2003, Ackermann et al. 2002), inducers of drug metabolising systems (P450 enzymes) (Pesonen and Andersson 1997; Huuskonen et al. 2000), and induction of vitellogenin or heat chock proteins. There are several advantages with cell lines targeting specific endpoints and therefore there is a rapid development in this area. They provide a methodologically easy and straightforward approach to analyse chemicals or environmental samples (extracts of sediments, tissues, air pollution particles etc) with respect to certain, toxicologically important, endpoints like oestrogen activity, induction of P-450 (CYP) enzymes or other properties. This has a large value when working with samples from unknown origin and complex environmental mixtures but also for single chemicals and chemical mixtures (Castaño et al. 2000). We are not going into depth on the different types and approaches that are currently used and explored. A comprehensive review can be found in ECETOC (2005).

7.3.2.3 Bioconcentration and Bioaccumulation

A considerable number of fish are sacrificed for data on the bioconcentration and bioaccumulation potential of chemicals. Currently intense research activities are ongoing to develop and define *in vitro* models for bioconcentration and to replace the *in vivo* based methods. Clearly, this is a delicate task since the bioconcentration and bioaccumulation process involves several functions in the body, which are on an organisational level above the individual cell. However, attempts have been made to define a BCF_{cell} (Segner and Cravedi 2001) by utilising the fact that BCF can be described as a ratio between absorption- and elimination rates. The elimination rate is related to the biotransformation potential of the chemical. For example, for a hydrophobic compound, the faster the measured *in vitro* biotransformation, the less likely will the substance bioaccumulate (Segner and Cravedi 2001). An advantage of *in vitro* methodologies for assessing biotransformation is that they are rapid and less expensive than *in vivo* tests. Measurements of absorption rates are technically more difficult but may be valuable as screening tools. In addition, absorption rates are also important in assessing the bioavailability of the chemicals either from body fluids to tissues or from the environment to the organisms. *In vitro* epithelial preparations of intestinal epithelial cells (CaCo-2 cell line) have been used to study intestinal absorption and bioavailability. Wood and Pärt (1997) developed a cultured gill epithelium based on primary cultures of rainbow trout gill epithelial cells and recently a similar preparation based on a continuous gill cell line RTGgill-W1 has been described (Lee et al. 2009). The advantage of these preparations is that they sustain exposure to water on the side of the epithelium, which normally faces water in the intact gill. Therefore, *in vitro* experiments can be made with environmentally realistic exposure conditions with the chemicals dissolved in water and in the way they occur in the nature. Cultured gill epithelia have until now been used in studies of basic gill physiology and for responses on toxic exposures, but not for direct absorption rate or bioavailability measurements (Pärt and Wood 2003).

7.3.3 *Replacement: Fish Egg/Embryo Systems*

Test with early life stages of fish have for long time been used to assess hazard of chemicals with the rationale to evaluate the effects on development and developmental stages. OECD has developed guidelines for early life stages test and there are protocols available for several species (OECD guidelines 210 and 212). These tests are of high ecotoxicological relevance since developmental stages are assumed to be particularly vulnerable for toxic insults. The effect of chemicals on fish embryos are evaluated from several endpoints. Besides overall lethality, endpoints such as coagulation of the egg, gastrulation, number of somites, movement, development of organs, pigmentation, heartbeat and circulation (Schulte and Nagel 1994) have been used. In Table 7.4 we report test protocols and end-points of three common species, Zebra fish (*Danio rerio*), Medaka (*Oryzias latipes*) and Fathead

Table 7.4 Comparison of test conditions for four species commonly used in embryo tests (Modified after Braunbeck et al. 2005 and Eriksson 2007)

	Zebra fish	Fathead minnow	Medaka	Stickleback
Origin of species	India, Burma, Malacka, Sumatra	Temperate zones of central North America	Japan, China, South Korea	Temperate and subarctic zones of the northern hemisphere coastal areas
Wet weight of adult fish	**Females**: 0.65 ± 0.13 g **Males**: 0.5 ± 0.1 g	**Females**: 1.5 ± 0.3 g **Males**: 2.5 ± 0.5 g	**Females**: 0.35 ± 0.07 g **Males**: 0.35 ± 0.07 g	**Females and males:** 1.5–2.5 g
Water quality	Salinity: 0 ppt, pH ~ 8	Salinity: 0 ppt, pH ~ 8	Salinity: 0 ppt, pH ~ 8	Salinity: 0–30 ppt, pH between 6 and 9
Male to female ratio for breeding	**4:2**	**2:4**	**15:15**	**1:1**
Embryo development at 25°C	**18 h**: Somite development	**22 h**: Somite development	**28 h**: Somite development	**24 h**: –
	21 h: Tail detachment **26 h**: Heart-beat visible **28 h**: Blood circulation **72 h**: Hatching	**25 h**: Tail detachment **27 h**: Heart-beat visible **30 h**: Blood circulation **160 h**: Hatching	Tail detachment **30 h**: Heart-beat visible **32 h**: Blood circulation **120 h**: Hatching	**48 h**: Somite development Tail detachment Heart-beat visible Blood circulation **120–144 h**: Hatching
Test type	26°C, 24-well plates (2 ml per cavity)	25°C, 24-well plates (2 ml per cavity)	26°C, 24-well plates (2 ml per cavity)	25°C, 96-well plates (280 μl per cavity)
Major toxicological endpoints at 25°C	**24 h**: Movement	**28 h**: Tail and somite development	**30 h**: Somite development	**48 h**: Movement
	Tail and somite development	**3 d**: Blood circulation	**78 h**: Blood circulation	**72 h**: Heart rate and pigmentation
	48 h: Heart rate, circulation, oedema and pigmentation	**4 d**: Blood circulation	**7 d**: Blood circulation	
			10 d: Blood circulation **14 d**: Blood circulation	**48 and 72 h**: Coagulated egg Oedema, eye and tail malformation, circulation, somites
	24 and 48 h: Coagulated egg, eye malformation			

Minnow (*Pimephales promelas*). The three-spined Stickleback (*Gasterosteus aculeatus)* is included because an embryo test has been recently developed (Eriksson 2007) and because this species is receiving increasing attention as a test species representative for European coastal and inland waters.

Use of embryos has several technical and practical advantages:

- Single fish embryos can be maintained in small volumes of test solution.
- Fish embryos can be cultured in microtitre wells and processed automatically on a standard microtitre plate reader. Hence, they can be used as a high throughput screening tool.
- Most fish embryos are inexpensive to maintain and easily bred in large numbers (e.g. single mating pairs of zebrafish produce between 100 and 200 eggs in one spawning).
- The majority of fish embryos are completely transparent and therefore development can be easily observed, enabling the possibility of determining specific organ toxicity, such as to the liver and the kidneys, as well as developmental teratogenicity through immunochemical techniques.
- Fish embryos can be used for determining the genotoxicity of a substance since approximately 90% of the genome is active during embryogenesis whilst only approximately 10% is functional during adult life.

The Fish Embryo Test (FET) is proposed as a promising alternative to the classical acute fish toxicity test. Today, FET is a routine test in whole effluent testing in Germany after official acceptance 2005 (Lammer et al. 2009), and the test for effluents has been standardised within the OECD. The crucial question in a 3Rs context is if an embryo is an animal. There are different definitions of what actually defines an 'animal' (Eriksson 2007), but currently the EU Directive 86/609/EEC is the norm in Europe (will be replaced by a new directive COM(2008)543/5 in 2009). An 'animal' is defined as any live non-human vertebrate including free-living and/or larval forms. Fetal or embryonic forms during their first two thirds of the development and which are not self sustaining are not defined as animals. During the last third of the developmental phase, the embryo is considered to realise pain and suffering and is included under the definition of an animal. Therefore, there is a potential to use embryos up to 2/3 of development and elutheroembryos (yolk sac dependent juveniles) as replacements for fish in the acute fish toxicity test without violating the 3R's principles.

Scholz et al. (2008) conclude that fish embryos represent an attractive model for environmental risk assessment (ERA) of chemicals since they offer the possibility to perform small-scale, high-throughput analyses. The authors suggest that toxic mechanisms may be studied and indications of adverse and long-term effects by adding new test applications to the FET. They also point at the need to estimate limitations of the test model and to what extent it can be optimised for regulatory purposes. In this context it deserves to point out again the importance of knowing the mode-of-action (MOA) of the actual chemical. Since the chorion acts as a protective shell around the embryo, some substances may be stuck on the surface of the egg or be unable to penetrate the eggshell. Testing of 'difficult' substances (e.g. large molecules/lipophilic) substances with eleutheroembryos might be an alternative, but might also be on the borderline of what is considered ethically acceptable in the near future?

In a recent paper Lammer et al. (2009) have made an in-depth analysis on how well FET perform in relation to the conventional acute fish toxicity test (OECD 203). They made a series of comparisons of chemical toxicity measured as fish embryo toxicity data (FET), including both embryo- (E) and elutheroembryo (EL) toxicity, and acute toxicity data for several fish species. A careful selection of high quality data was undertaken to make sure that proper comparisons would result and in total data for 73 chemicals was used. As can be seen in Fig. 7.2 the correlation is very strong indicating that FET can provide similar sensitivity as the acute test. Lammer et al. (2009) show also a strong correlation between E and EL data, but a small number of chemicals (e.g. selected polymers and higher molecular weight non-ionic surfactants) seemed to be more toxic to EL than to E, probably due to the protective effect of chorion in the latter category. The final conclusion that fish embryo tests are neither better nor worse than acute fish toxicity tests and offer a

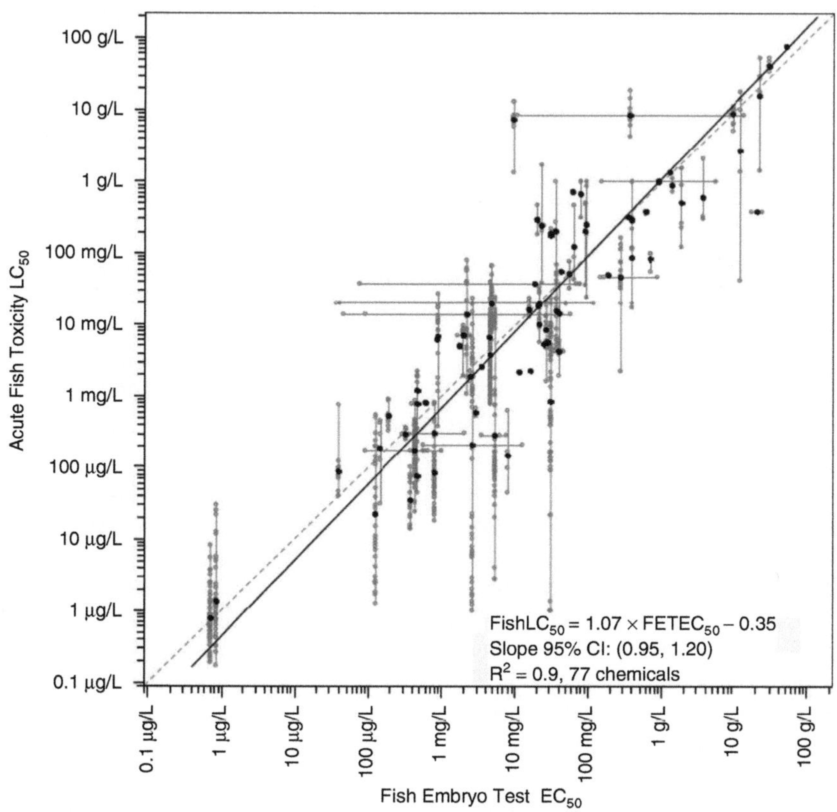

Fig. 7.2 Comparison of all available fish embryo toxicity data and acute fish toxicity data for 77 chemicals (Reproduced with permission from Lammer et al. 2009)

reasonable and realistic alternative to the acute fish *in vivo* fish test. They also remark that more research is needed to include new species which are more representative for the European environment than the currently used ones.

7.4 Current Trends in Ecotoxicological Testing

Today acute toxicant emissions are very rare and (usually?) unintentional. The exposure situations we want to protect the environment from are therefore best described as low level exposures over long periods of time. Following the precautionary principle, we want to guarantee (or make sure) that even a lifelong exposure, including all the sensitive development stages in the life of an organism, will not cause harm to the individual and the population. For practical and economic reasons, this is not possible to obtain for more than a few substances and a few organisms. Even less realistic is to get toxicity data for chemical substances in mixtures and to assess the ecologically realistic responses at a biological community or ecosystem level. In this perspective is REACH advocating a pragmatic and conservative approach based on short-term and rather simple methods. Facing that perhaps 30,000 chemicals have to be tested, this is probably the most feasible approach, but the results have to be interpreted and extrapolated with great caution. In addition it puts strong responsibility on the scientific community to develop representative testing methods and to refine the predictive models.

Already before the animal ethical aspects (e.g. the 3R-concept) had become an important issue for vertebrates used in ecotoxicology, the strive to extrapolate to chronic toxicity from shorter (acute) test periods was evident. The reason was very practical: to save time and money, but still with the ambition to make a judgment on possible effects of long-term or even life-long exposure to the chemical of concern. A series of full-life cycle experiments with the fathead minnow were conducted in the U.S. already in the 1960s with metals and pesticides (Mount 1968; Mount and Stephan 1967, 1969) from which safety or uncertainty factors were calculated based on the acute (96-h LC-50 values). However, as more data emerged, it became clear that different categories of substances with different physiochemical properties might exercise one mode of action at the acute exposure but another mode of action under chronic conditions in the same organism (e.g. Bengtsson 1974). This resulted in a high level of uncertainty when calculating reliable safety factors from acute toxicity data. Raimondo et al. (2007) determined the variability in acute to chronic toxicity ratios for data pairs of freshwater (FW) and saltwater (SW) fish. They found a median and maximum range of 9.0 and 645, respectively for FW fish ($n = 176$) and 5.2 and 772, respectively for SW fish ($n = 35$). These results indicate that when median or mean toxicity ratios are applied without considering mechanisms there is a high risk of underestimated chronic toxicity for some chemicals. Consequently, as much information as possible about the mode of action for each substance to be tested should be made available from the manufacturer to assist in the final hazard assessment. Barron et al. (2008) suggest from calculations using 140 matched fish acute toxicity ratio pairs that more frequent observations of mortalities in the acute test may increase the precision in transferring acute data to estimations of adverse chronic effects as long as the same mode of action may be assumed.

In tests over two generations, exposure of the parental generation often results in observable effects in the next (unexposed) generation. Thus Ji et al. (2008) found higher mortality and histopathological changes in the F1 generation of medaka exposed to PFOS. This phenomenon is a serious toxicological indicator per se, but it is far from certain that such effects are occurring at lower concentrations than were found hazardous already in the F0 generation. Accordingly, in Sweden a 2-generation test with the zebrafish has been applied during 2001–2007 to test the toxicity of 14 pulp mill effluents. In no case did the 2-generation tests result in higher sensitivity than was observed after one generation (T. Viktor pers. comm.). This may indicate that test resources may be better used than in resource-demanding 2-generation tests when estimating NOEC.

None of the three freshwater fish species presently used internationally in embryo tests are relevant to the European environment: The Japanese medaka lives in Asian paddy fields, the fathead minnow in central North America and the zebrafish has its origin on the Indian subcontinent. In a study at Stockholm University (Eriksson 2007) the three-spined stickleback was tested according to the zebrafish embryo test protocol as a more relevant alternative. Besides that the stickleback test requires 1 day more (i.e. 96 h) to complete due to slower development, the tests indicated that the species was equally useful. Beyond the fact that the three-spined stickleback occurs all over Europe and around the northern hemisphere in both fresh, brackish and marine waters, the three-spined stickleback has some other useful features: (a) genetically determined sex; (b) the occurrence of a unique marker for androgens: 'spiggin'; and (c) a full description of the genome. If we want to have a stronger connection between the field and the laboratory ('ecological realism'), this species seems to be a very promising candidate to be added to the OECD test arsenal (i.e. OECD 203, 210 and 215).

7.5 Conclusion: Has the 3R's Concept a Future in Ecotoxicology?

The present authors have worked professionally within ecotoxicology for a time corresponding to almost the length of one generation and during which this discipline has seen a tremendous development. The attitudes towards the environment and living organisms has gradually changed and become more humane among the public in Europe but also among the scientists themselves. In the early years and in the search for evidence we were focusing on producing convincing data to prove that mankind was polluting nature and ourselves. Statistically significant data required many replicates and many individuals. Animals were sacrificed in testing procedures for a 'good purpose'. Very few reacted – an animal life was clearly less worth than that of a human which justified the use of animals in protecting human life. However, attitudes have changed and there is a public pressure to reduce the use of animals in scientific experiments and in testing procedures. The 3Rs concept was presented by Russel and Burch (1959) but it took almost 50 years for these recommendations to have an impact on legislation. The REACH legislation has included animal alternative considerations to reduce tests with live animals and the transition from *in vivo* tests to alternative methods is encouraged. However, it was also stated as early as in 1986 (Directive 86/609/EEC) that an alternative method once it is 'practically and reasonably available' should replace methods requiring live animals. To promote the research and validation of alternative methods, the European Commission in 1991 created ECVAM – The European Centre for Validation of Alternative methods within the Joint Research Centre of the European Commission with the mission to facilitate this work.

The question to be answered is if alternative methods have the potential to provide realistic information for environmental risk assessment. The answer is a strong 'yes' but we are not there yet. The problem with the current *in vitro* cytotoxicity tests is their low sensitivity – they are one to two magnitudes less sensitive to a toxic challenge than to the intact animal. The conclusion is that the test and the end-point are not reflecting the critical toxic mechanisms in the intact organism. Therefore, to improve the situation we need an increased understanding of toxic mechanisms at an integrated level of a whole organism. We need also information from different groups of organisms in the environment and also to identify specifically vulnerable species or life stages. It is a shame, but we have still – although more than 50 years of toxicological research – very vague ideas and information about the actual and detailed cause of death in acute toxicity tests. There is a strong need to improve this situation, to increase our understanding of toxic mechanisms at the whole animal level. This information will be essential and on basis of this information we can start to develop and design 'more realistic' *in vitro* systems.

There is currently an intense and rapid development in the biotechnology area from which also the development of test methods will benefit. A number of engineered cell lines have been developed which express one specific physiological mechanism in combination with an easily measurable signal systems. These systems will gain importance to identify specific properties of chemicals, like hormone activity, induction of specific metabolic pathways or genotoxicity. Tests will be more specific and targeted. The rapid developments in the omics area will probably provide new tools to understand and interpret toxicological challenges on the individual and the cellular level, although here still a lot of basic work needs to be done before robust applications are available.

The issue is how to interpret results from highly refined *in vitro* systems to 'ecological relevance'. The leap is probably not as big as it seems in a first glance. We should not be blinded by some kind of ecological 'fundamentalism'. The ecosystem is built up by individuals. If the survival and fitness of the individuals is compromised the whole system will suffer. The current *in vitro* methods and particularly and hopefully the future methods have the full potential to assess impacts on, and through relevant legislation, protect the individuals. In this way we also achieve a safe protection of the environment and the ecosystem.

References

Ackermann, G.E., Brombacher, E. & Fent, K. (2002) Development of a fish reporter gene system for the assessment of estrogenic compounds and sewage treatment plant effluents. *Environmental Toxicology and Chemistry* 21: 1864–1875.

Barron M.G., Raimondo S., Russom C., Vivian D.N. & Yee S.H. (2008) Accuracy of chronic aquatic toxicity estimates determined from acute toxicity data and two time–response models. *Environmental Toxicology and Chemistry* 27: 2196–2205.

Becerril, C., Ferrero, M., Sanz, F. & Castaño, A. (1999) Detection of mitomycin c- induced genetic damage in fish cells by use of RAPDs. *Mutagenesis* 14: 449–456.

Bengtsson B.-E. (1974) The effect of zinc on the mortality and reproduction of the minnow, *Phoxinus phoxinus* L. *Archives of Environmental Contamination and Toxicology* 2: 342–355.

Braunbeck T., Böttcher M., Hollert H., Kosmehl T., Lammer E., Leist E., Rudolf M. & Seitz N. (2005) Towards and alternative for the acute fish LC-50 test in chemical assessment: the fish embryo toxicity test goes multispecies – un update. *ALTEX* 22: 87–102.

Castaño, A. & Becerril, C. (2004) *In vitro* assessment of DNA damage after short and long-term exposure to benzo(a)pyrene using RAPD and the RTG-2 fish cell line. *Mutation Research* 552: 141–151.

Castaño A. & Gómez-Lechón M.J. (2005) Comparison of basal cytotoxicity data between mammalian and fish cell lines: A literature survey. *Toxicology In Vitro* 19: 695–705.

Castaño, A., Cantarino, M.J., Catillo, P. & Tarazona, J.V. (1996) Correlations between RTG-2 cytotoxicity test EC-50 and *in vivo* LC-50 rainbow trout bioassay. *Chemosphere* 32: 2141–2157.

Castaño, A., Carballo, M., Llorente, M., De la Torre, A. & Muñoz, M.J. (2000) Evaluation of complex mixtures of fish populations using alternatives systems. *The Science of the Total Environment* 247: 337–348.

Castaño A., Bols N., Braunbeck T., Dierickx P., Halder M., Isomaa B., Kawahara K., Lee L.E.J., Mothersill C., Pärt P., Repetto G., Riego Sintes J., Ruffli H., Smith R., Wood C.M. & Segner H. (2003) The use of fish cells in ecotoxicology. *ATLA* 31: 317–351.

ECETOC (2005) Alternative testing approaches in environmental safety assessment. Technical Report No. 97. European Centre for Ecotoxicology and Toxicology of Chemicals, Brussels, pp. 1–182.

Eriksson K. (2007) Method development: Embryo testing with the three-spined stickleback (*Gasterosteus aculeatus*). Diploma work at the Department of Applied Environmental Science, Stockholm University, Stockholm, Sweden.

Hornung, M.W., Ankley, G.T. & Schmeider, P.K. (2003) Induction of an estrogen-responsive reporter gene in rainbow trout hepatoma cells (RTH 149) at 11° or 18°C. *Environmental Toxicolology and Chemistry* 22: 866–871.

Hutchinson T.H., Barrett S., Buzby M., Constable D., Hartmann A., Hayes E., Huggett D., Laenge R., Lillicrap A.D., Straub J.O. & Thompson R.S. (2003) A strategy to reduce the numbers of fish used in acute ecotoxicity testing of pharmaceuticals. *Environmental Toxicology and Chemistry* 22: 3031–3036.

Huuskonen, S.E., Tuvikene, A., Trapido, M., Fent, K. & Hahn, M.E. (2000) Cytochrome P4501A induction and porphyrin accumulation in PLHC-1 fish cells exposed to sediment and oil shale extracts. *Archieves of Environmental Contamination and Toxicology* 38: 59–69.

Jeram S., Riego Sintes J.M., Halder M., Fentanes J.B., Sokull-Kluttgen B. & Hutchinson T.H. (2005) A strategy to reduce the use of fish in acute ecotoxicity testing of new chemical substances notified in the European Union. *Regulatory Toxicology and Pharmacology* 42: 218–224.

Ji K., Kim Y., Oh S., Ahn B., Jo H. & Choi K. (2008) Toxicity of perfluorooctane sulfonic acid and perfluorooctanoic acid on freshwater macroinvertebrates (*Daphnia magna* and *Moina macrocopa*) and fish (*Oryzia latipes*). *Environmental Toxicology and Chemistry* 27: 2159–2168.

Lammer E., Carr G.J., Wendeler K., Rawlings J.M., Belanger S.E. & Braunbeck T. (2009) Is the Fish Embryo Toxicity test (FET) with the zebrafish (*Danio rerio*) potential alternative for the fish acute toxicity test? *Comparative Biochemistry and Physiolology C* 149: 196–209.

Lee, L., Dayeh, V., Schirmer, K. & Bols, N. (2009) Application and potential uses of fish gill cell lines: examples with Rtgill-W1. *In vitro Cellular and Developmental Biology (Animal)* 45: 127–134.

Mount D.I. (1968) Chronic toxicity of copper to fathead minnows (*Pimephales promelas*, Rafinesque). *Water Research* 2: 215–223.

Mount D.I. & Stephan C.E. (1967) A method for establishing acceptable toxicant limits for fsih-malathion and butoxyethanol ester of 2,4-D. *Transactions of the American Fisheries Society* 96: 185–193.

Mount D.I. & Stephan C.E. (1969) Chronic toxicity of copper to fathead minnow (*Pimephales promelas*) in soft water. *Journal of the Fisheries Research Board of Canada* 26: 2449–2457.

Nagel R. (2002) DarT: The embryotest with the zebrafish Danio rerio – a general model in ecotoxicology and toxicology. *ALTEX* 19: 38–48.

Oaks L.J., Gilbert M., Virani M.Z., Watson R.T., Meteyer C.U., Rideout B.A., Shivaprasad H.L., Ahmed S., Chaudry M.J.I., Arshad M., Mahmood S., Ali A. & Khan A.A. (2004). Diclofenac residues as the cause of vulture population decline in Pakistan. *Nature* 427: 630–634.

OECD (2004) *Detailed review paper on fish screening assay for the detection of endocrine active chemicals.* Organisation for Economic Co-operation and Development, Paris, France.

OECD 203 (1992a) *OECD guidlines for testing of chemicals. 203. Fish acute toxicity test.* Organisation for Economic Co-operation and development, Paris, France.

OECD 204 (1984) *OECD guidlines for testing of chemicals. 204. Fish prolonged toxicity test: 14 day study.* Organisation for Economic Co-operation and Development, Paris, France.

OECD 210 (1992b) *OECD guidlines for testing of chemicals. 210. Fish, early-life stage toxicity test.* Organisation for Economic Co-operation and Development, Paris, France.

OECD 212 (1998) *OECD guidlines for testing of chemicals. 212. Fish, early-life stage toxicity test on embryo and sac-fry stages, last updated 21st September 1998.* Organisation for Economic Co-operation and Development, Paris, France.

OECD 215 (2000) *OECD guidlines for testing of chemicals. 215. Fish, juvenile growth test.* Organisation for Economic Co-operation and Development, Paris, France.

OECD 305 (1996) *OECD guidlines for testing of chemicals. 305. Bioconcentration: Flow-through fish test, last updated 14th June 1996.* Organisation for Economic Co-operation and Development, Paris, France.

OECD 17 (2010). *Short guidance on the threshold approach for acute toxicity. ENV/JM/MONO 17. 31-May-2010.* Organization for Economic Co-operation and Development. Paris. France.

Pärt P. & Wood C.M. (2003) Cultured epithelia from fish gills. In C. Mothersill & B. Austin (Eds.), *In Vitro methods in aquatic toxicology* (pp. 35–52). United Kingdom: Springer-Chichester.

Pesonen M. & Andersson T. B. (1997) Fish primary hepatocyte culture; an important model for xenobiotic metabolism and toxicity studies. *Aquatic Toxicology* 37: 253–267.

Raimondo S., Montague B.J. & Barron M.G. (2007) Determinants of variability in acute to chronic toxicity ratios for aquatic invertebrates and fish. *Environmental Toxicology and Chemistry* 26: 2019–2023.

Russel W.M.S. & Burch R.L. (1959) *The principles of humane experimental techniques.* London: Methuen.

Sánchez, P., Llorente, M.T. & Castaño, A. (2000) Flow cytometric detection of micronuclei and cell cycle alterations in fish derived cells after exposure to three model genotoxic agents: Mitomicyn C, Vincristine and Benzo(a)pyrene. *Mutation Research* 465: 113–122.

Schirmer K. (2006) Proposal ro improve vertebrate cell cultures to establish them as substitutes for regulatory testing of chemicals and effluents using fish. *Toxicology* 224: 163–183.

Scholz S., Fisher S., Gundel U., Kuster E., Luckenbach T. &Voelker D. (2008) The zebra fish embryo model in environmental risk assessment – application beyond acute toxicity testing. *Environmental Science and Pollution Research International* 15: 394–404.

Schulte C. & Nagel R. (1994) Testing acute toxicity in the embryo of zebrafish, *Brachydanio rero*, as an alternative to acute fish test: preliminary results. *ATLA* 22: 12–19.

Segner H. (2004) Cytotoxicity assays with fish cells as an alternative to the acute lethality test with fish. *ATLA* 32: 375–382.

Segner H. & Cravedi J.-P. (2001) Metabolic activity in primary cultures of fish hepatocytes. *ATLA* 29: 251–257.

US EPA (1986) *Fish life cycle toxicity test.* EPA 540/9-86-137 United States Environment Protection Agency, Office of research and Development, Washington DC, USA

Wolf K. & Quimby M.C. (1969) Fish cell and tissue culture. In W.S. Hoar & D.J. Randall (Eds.), *Fish Physiology* (pp. 253–305). New York: Academic.

Wood C.M. & Pärt P. (1997) Cultured branchial epithelia from freshwater fish gills. *Journal of Experimental Biology* 200: 1047–1059.

Chapter 8
Chemical Risk Assessment in Toxicological Perspective

Helmut Greim

Abstract The discipline of toxicology is concerned with the health risks of human exposure to chemicals. According to the Paracelsus' paradigm toxicology is charged with describing the adverse effects of chemicals in a qualitative sense, and with evaluating them quantitatively by determining how much of a chemical is required to produce a substance specific. Taken together the intrinsic properties of an agent are described (hazard identification) and the amount of the chemical required to produce these (risk characterization) is determined. Since humans or organisms in the environment can be exposed via inhalation, skin contact or oral intake, the concentrations in the different environmental compartments, which result in human or environmental exposure, must be evaluated. Obviously risk characterization comprises the following elements:

- Hazard identification, i.e. a description of the agent's toxic potential. Dose–response, including information on the concentration above which the agent induces toxic effects to identify the no observable effect level (NOEL).
- Exposure assessment, in which the concentration of the agent in the relevant medium and time of exposure are evaluated.

Based in this information difference between the NOEL and human exposure or the risk at a given exposure is determined. Humans may be exposed to chemicals in the air, water, food, or on the skin. From the concentrations of a chemical in these different compartments the external daily exposure is estimated. The response to the chemical depends upon duration and route of exposure, the toxicokinetics of the chemical, the dose–response relationship and the susceptibility of the individual. Thus, the precise definition of the terms hazard, exposure, and risk is essential to understand toxicological evaluations (details on data requirements and procedures for risk assessment are given subsequently).

Hazard: this qualitative term represents the intrinsic toxic properties of a compound. The expression of hazard depends upon conditions of use and exposure.

H. Greim (✉)
Institut für Toxikologie und Umwelthygiene, Technische Universität München, Hohenbachernstrasse 15–17, 85350, Freising-Weihenstephan
e-mail: helmut.greim@lrz.tu-muenchen.de

J. Eriksson et al. (eds.), *Regulating Chemical Risks: European and Global Challenges*, DOI 10.1007/978-90-481-9428-5_8, © Springer Science+Business Media B.V. 2010

Dose: is determined by the concentration of the chemical and the time of exposure.

Risk: is the likelihood of an adverse effect resulting from a given exposure.

Keywords Data requirements • Dose • Hazard • Risk • Toxicology

8.1 Data Requirements for Hazard Assessment

8.1.1 Hazard Identification and Dose Response

Depending on reactivity, solubility and metabolism, the chemical or its metabolites can reach the critical external and internal target. Evaluation requires animal studies to detect irritation or corrosion on the skin, the eye, the gastrointestinal tract or the respiratory system. According to its distribution and metabolism the chemical can induce various systemic effects upon in the critical organ such as liver, kidney, brain. These organ-specific effects including the dose response of effects and the NOEL are determined by repeated dose studies in animals. In vitro and in vivo tests are applied for evaluation of genotoxicity. Sensitive methods in analytical chemistry and molecular-biological approaches including toxicokinetics and the various 'omics' have significantly improved the understanding of the modes of action of chemicals. Such information also allows reduction of animal experiments but not its complete avoidance.

Such information is obtained from acute, subchronic and chronic exposure studies in experimental animals, mostly rats and mice, via routes relevant to the use of the chemical (oral, inhalation, dermal). These include: the paradigm of Paracelsus implies that the occurrence and intensity of toxic effects are dose dependent (see Greim and Snyder 2008). This paradigm addresses the concept of threshold effects with the consequence that there no effects up to a certain dose, the NOEL. Animal or human exposure is usually defined as the dose, e.g., in mg of the chemical/kg body weight/day. This daily dose may result from oral, inhalation or dermal exposure or as a sum thereof. The external dose leads to a specific internal dose, which depends on the amount absorbed via the different routes. Absorption rates via the different routes can vary significantly, although oral and inhalation exposure usually leads to the highest internal dose. For example, about 50% of cadmium in tobacco smoke is absorbed in the lung, whereas, cadmium absorption from the gastrointestinal tract is about 10%. Ultimately, it is the dose which reaches the cellular target over a given time period that results in the toxicological response. The dose that defines the toxic potency of a chemical is the product of the interrelated external, internal, and target doses. No toxic effects will be seen if the dose is below the NOEL, whereas effects increase with increasing exposure. Using the semi-logarithmic plot of the dose–response relationship the curve is sigmoidal and varies in slope from chemical to chemical. Thus, if the curve is shallow a doubling of the dose results in

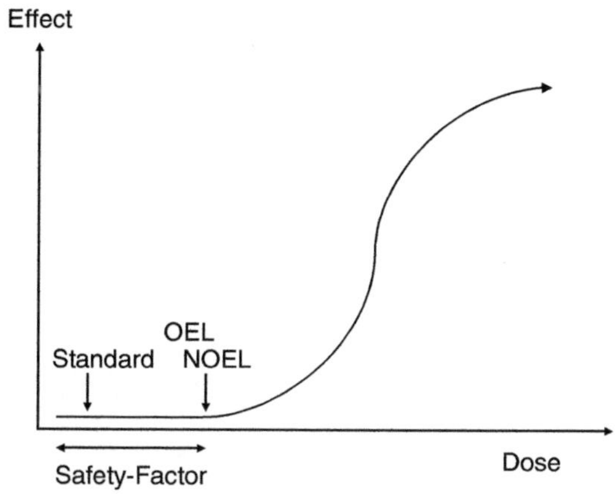

Fig. 8.1 Dose–response curve showing log dose on the X-axis and percent response (*Effect*) on the Y-axis. The figure illustrates the location of regulatory values such as the NOEL, Occupational Exposure Levels (OELs) or environmental standards such as Acceptable Daily Intake (ADI)

a small increase of effects, whereas a several-fold increase of effects occur when the slope is steep (see Fig. 8.1). The location of the curve on the abscissa is a measure of the potency of the chemical.

8.1.2 *Exposure Assessment*

Toxic effects are dose dependent. Knowledge of the extent and duration of exposure is essential. Exposure defines the amount of a chemical to which a population or individuals are exposed via inhalation, oral or dermal routes. Exposure is commonly defined by mg of the chemical/kg body weight/day. Monitoring of exposure usually requires measurement of the chemicals in the air, food, or consumer products and includes frequency, duration of exposure, concentration of substance in the product and the amount of product used per contact. Information on indoor air concentrations is of specific relevance since people stay longer indoors than outdoors and concentrations usually are higher than outdoors. Exposure in the home can be estimated by the use of appropriate modelling techniques. Children represent a special case of exposure. For example, they may be exposed to chemicals that are released from toys during mouthing or via skin contact.

This external exposure may not necessarily correlate with internal exposure. The rate of absorption through the skin, the lung or the gastrointestinal tract determines the body burden of the chemical. Measurement of the chemical, its metabolites or products of the interaction of the chemical or its metabolites with cellular macromolecules such as proteins or DNA in body fluids and tissues determines 'internal exposure'. Use of such biomarkers provides exact information on actual internal exposure (target dose) to an agent (Angerer et al. 2007; Boogaard 2007; Needham et al. 2007). When used in experimental studies in animals and humans they allow assessment of the individual and internal exposure as compared to external exposure.

In case of genotoxic carcinogens biomarkers include DNA-adducts, which are considered to be markers of exposure, and biomarkers of effects e.g. chromosomal aberrations, sister chromatid exchanges, increased frequency of micronuclei or mutations (Swenberg et al. 2008). Whereas biomarkers of exposure extrapolate down to zero, biomarkers of effect interpolate with the spontaneous or background number of mutations. Since exposure can be multiple and via different sources over a certain time several definitions are used:

Combined exposure is the total exposure to one stressor, cumulated from several sources and/or via several pathways:

- Aggregate exposure is total exposure to one stressor from several sources and/or via several pathways over time.
- Multiple exposure describes the exposure to several stressors, with possible synergetic or antagonistic effects.

Exposure assessment is plagued by many uncertainties, which often leads to overestimation of the actual exposure. Estimation of exposure is even more complicated for mixtures of chemicals.

8.2 Risk Assessment

8.2.1 The General Approach

Toxicological evaluations take different approaches for new and existing chemicals. For existing chemicals the available information is collected by a literature search and a risk assessment based on exposure data, knowledge of the dose–response relationship, and the mode of action, can be performed. In the case of newly developed drugs, pesticides or new chemicals a stepwise procedure is used starting from simple in vitro and in vivo short-term tests. Depending on the hazardous potential of the agent, studies can be extended to evaluate long-term effects by repeated dose studies, toxicokinetics and toxic mode of action.

Any evaluation needs to define the compound, its structural alerts and physical-chemical parameters like water/lipid solubility and volatility as well as the purpose of the evaluation. To screen for specific effects such as relative cytotoxicity, mutagenicity or hormonal effects simple in vitro tests may be appropriate. This allows identification of specific wanted or unwanted effects and by that selection of specific methods for more detailed evaluation.

The stepwise procedure usually starts with the determination of the LD_{50}, a short term repeated exposure test in rodents and the evaluation of genotoxicity by an in vitro bacterial test system (Ames-test) and for cytogenicity in mammalian cells. In case of indication for genotoxicity the results are verified in vivo usually by the mouse bone marrow micronucleus test. For further evaluation the compound additional tests including studies on toxicokinetics or the toxic mechanisms will follow. Such information provides information on the reactivity of the test compound, its absorption and distribution in the organism and on critical effects. This allows the decision whether the data base needs further testing by repeated dose studies in animals for 28 and 90 days, which depending on their outcome and intended use of the chemical are followed by a 6 months or life-time study to evaluate potential effects upon long-term exposure including carcinogenicity. Details can be seen in Greim and Snyder (2008) or the Technical Guidance Document of the European Chemical Bureau (TGD).

8.2.2 *The Tools for Hazard Identification*

Sufficient information on the hazardous properties requires information on effects after short and long-term exposure on the various potential end-points. These include:

- Acute, sub-chronic and chronic toxicity (oral, inhalation, dermal)
- Irritation (skin, mucous membranes, eye) and phototoxicity
- Sensitization and photosensitization
- Genotoxicity (in vitro and in vivo methods)
- Carcinogenicity (lifetime studies)
- Reproductive toxicity
- Toxicokinetics
- Mode and mechanism of action

In all cases information on the dose–response of effects is essential to identify the slope of the dose response curves, possible thresholds, the NOEL, LOEL or the maximal tolerated dose (MTD). For most of the tests guidelines are available (see OECD or European Communities).

8.2.2.1 Toxicokinetics

Toxicokinetics describe Absorption, Distribution, Metabolism and Elimination (ADME) of a chemical in humans, experimental animals or cellular systems. Of specific importance for interpretation of animal studies and for extrapolation of hazards between species is the comparative information on the exposure and the dose that reaches the critical target.

A chemical may enter the body via food, air or the skin. Upon inhalation or skin penetration the compound directly enters the circulation and distributes into the organs. When absorbed from the gastrointestinal tract the chemical enters the liver via the portal vein. The epithelial cells of the gut wall and the liver demonstrate a large capacity for metabolizing chemicals so that a compound may be extensively metabolized by this 'first pass effect' before entering the systemic circulation. Larger molecules, e.g., the glucuronosyl-conjugates can be excreted via the biliary system into the duodenum where the conjugates may be hydrolyzed so that the original compound is reabsorbed and reenters the liver. This 'enterohepatic circulation' and the first path effect, which occurs after oral exposure, may result in different effects than upon inhalation or dermal exposure, intravenous or intraperitoneal injection. Since metabolism of chemicals is species-specific evaluation of possible differences between laboratory animals and humans is essential to judge the relevance of findings from animal studies.

8.2.2.2 Omics

'Omics' data derived from gene expression microarrays or from high-throughput testing of proteins or endogenous metabolites and alterations by toxic agents become increasingly available and need to be evaluated for suitability for use in the hazard and risk assessment process. As long as the information is not related to functional changes, interpretation of such data is difficult and there is the possibility to over- or misinterpret the effects observed, although they might be useful in assessing mechanisms.

Genomics use microarray technologies to quantify genes of interest (Luhe et al. 2005). For all genes represented this technology quantifies the amount of transcript present and is therefore called transcriptomics or genomics. These two terms are commonly used as synonyms.

Metabolomics evaluate the final downstream products of the genome, which represent the total low-molecular-weight compounds (metabolites) present in cells or an organism (Dunn et al. 2005). Since these participate in metabolic reactions required for growth, maintenance, and normal function they are indicative for the response of living systems to patho-physiological stimuli or genetic modification.

Proteomics represent the sum of all proteins expressed in a tissue at a given time (Wetmore and Merrick 2004). Proteins are more complex chemically than nucleic acids and undergo extensive modification after translation, such as cleavage of signalling peptides, phosphorylation/dephosphorylation, or glycosylation. Treatment with xenobiotics not only change protein expression but also leads to drug–protein covalent binding, a protein-modification, which may affect the function of the protein.

8.2.2.3 Mode Versus Mechanism of Action

Mode of action comprises all available information on the toxic effects of a compound. Mechanistic data explain how a chemical interferes with the cellular targets and by that induces toxicity. Such information is essential to understand species specificities, species differences, sensitive populations or the interpretation of data regarding threshold or non-threshold effects. They also help to evaluate the relevance of the toxic effects to humans when the data are derived from experimental animals.

Mechanistic information is most relevant for the evaluation and classification of carcinogens. As indicated below, a carcinogenic effect, which is induced by a specific mechanism that does not involve direct genotoxicity, such as hormonal deregulation, immune suppression, cytotoxicity, the detailed search for the underlying mode of action may allow identification of a NOEL. This can also be considered for materials, such as poorly soluble fibers, dusts and particles, which induce persistent inflammatory reactions as a result of their long-term physical presence that ultimately lead to cancer (Greim and Ziegler-Skylakakis 2007).

8.2.3 Risk Assessment for Threshold Compounds

The risk assessment process differentiates between reversible and irreversible effects. From the dose–response curves of chemicals that induce reversible effects the 'no observable effects level' (NOEL) is determined. If damage persists and accumulates upon repeated exposure a NOEL cannot be determined and every exposure is related to a defined risk. This applies to effects on the highly specialized cells of the nervous system, sensitizing agents for which a NOEL is difficult to assess, and for genotoxic carcinogens. For chemicals, which induce reversible effects the NOEL of the most sensitive endpoint is compared with the human exposure to describe the Margin of Exposure (MOE). If the NOEL is derived from animal experiments a MOE of 100 or greater is considered without concern. This factor considers a tenfold difference between the sensitivity of the experimental animals and humans and another factor of 10 to take into account possible inter-individual differences among the human population. These tenfold factors allow consideration of toxicokinetic and toxicodynamic differences, and are subdivided to take into account the aspects separately (Renwick 1993, Renwick and Lazarus 1998, Dorne and Renwick 2005). A MOE of at least ten is sufficient if the NOEL is derived from human data (WHO 1987).

8.2.4 Risk Assessment for Non-threshold Genotoxic Carcinogens

The effects of genotoxic compounds are considered non-threshold. Thus, risk assessment for a given exposure is usually performed by a linear or sub-linear extrapolation from the high dose effects observed in animals to the lower human exposure. Since the outcome of the extrapolation depends on the model applied and extrapolation over different orders of magnitude is error prone, the European Food and Safety Authority (EFSA 2005) recommended to avoid this extrapolation and proposed the MOE approach. This approach uses the benchmark dose, or the T25 calculated from a carcinogenicity study and compares this with human exposure. A MOE of 10,000 and more is considered to be of minor concern. The advantage is that neither a debatable extrapolation from high to low doses needs to be performed nor are hypothetical cancer cases calculated. For details of the different approaches see, SCHER, SCCP, SCENIHR (2008).

Other concepts used are *ALARA* (exposure as low as reasonable) or *TTC* (threshold of toxicological concern) for insignificant low exposures. The ALARA principle intends to keep the exposure to substances at the lowest achievable level, usually limited by technological limitations or economic considerations. The major advantage of this approach is that only hazard identification data are needed. The disadvantage is that it does not make use of the general toxicological database and does not take into account the carcinogenic potency and the actual exposures, which may be negligibly low. It is not applicable to assess or compare risks.

The TTC principle has been developed originally for chemicals present in food such as flavouring substances. Its use has been proposed for herbal preparations, personal and household care products, and impurities in pharmaceuticals (Munro et al. 2008). Recently it has been proposed for cosmetic ingredients (Kroes et al. 2007). The TTC principle determines a human exposure value for the daily uptake of a compound, below which there would be no appreciable risk to human health. Depending on the structure of the compound, different values for the daily uptake are applied. It may avoid animal studies for compounds of very low exposure.

8.3 Classification and Labelling of Carcinogens

Classification and labelling of chemicals is hazard based (see UNECE). The potency is only taken into account when labelling compounds for effects after acute and chronic exposure. In case of carcinogens neither the potency nor the risk at a certain exposure determines classification and labelling. The systems for classification of carcinogens used by various national or international institutions were developed in the 1970s. Classification is based on qualitative criteria, and reflects essentially the weight of evidence available from animal studies and epidemiology. Classification is usually based on the certainty with which a carcinogenic potential for a chemical can be established. Generally three categories, the definitions of which slightly differ, are used:

- Human carcinogens
- Animal carcinogens, reasonably anticipated to be human carcinogens
- Not classifiable because of inadequate data

For classification mode of action and potency of a compound are either not taken into account, or at best is used as supporting arguments. The advancing knowledge of reaction mechanisms and the different potencies of carcinogens have initiated a re-evaluation of the traditional concepts.

The International Agency for Research of Cancer (IARC) and the OECD propose to use data on the carcinogenic mechanism and potency in decision-making and intend to consider information whether carcinogenicity is not likely below a certain dose. The Globally Harmonized System (GHS, see UNECE) has simplified and harmonized the classification criteria and categories of the International Agency for Research on Cancer (IARC), US Environmental Protection Agency (EPA), and the European Commission for carcinogens (Table 8.1) but does not include criteria for consideration of exposure and thus of carcinogenic risk.

Recently, the US Environmental Protection Agency (EPA 2005) proposed to apply a *weight-of-evidence-narrative* to characterize an agent's carcinogenic potential

Table 8.1 The classification systems of IARC, the European Commission and the Global Harmonized System (details see text)

IARC	Europe	GHS
Group 1	Category 1	Category 1A
Human evidence	Human evidence	Human evidence
Group 2A	Category 2	Category 1B
Limited human evidence, strong animal and mechanistic evidence	Sufficient evidence for human carcinogenesis from animal data	Animal evidence for carcinogenicity in humans
Group 2B		
Limited human evidence, less than sufficient animal evidence or strong mechanistic data		
Group 3	Category 3	Category 2
Inadequate human and animal data for classification	Inadequate data for classification	Suspected human carcinogen, inadequate data for classification
Group 4		
No indication for carcinogenicity		

and potency. However, linear extrapolation from cancer incidences in animal studies at high doses to low doses of human exposures is still required as default. The extrapolation process should utilize toxicokinetic and toxicodynamic modelling to conclude on risks if substantial data support is available. The narrative approach describes at what exposure an agent is carcinogenic by specific routes of exposure.

The MAK-committee of the German Research Foundation differentiates between non-genotoxic and genotoxic for classification (Greim and Reuter 2001). Within the European Commission this concept is applied by the Scientific Committee of Occupational Exposure Limits (Bolt and Huici-Montagud 2008).

These approaches originate from the concept that there is differentiation between mechanisms of carcinogenicity caused by non-genotoxic and genotoxic carcinogens. Thus, it is possible to identify a NOEL for non-genotoxic carcinogens, provided there is sufficient information on the primarily non-genotoxic mechanism. The American Conference of Governmental Industrial Hygienists (ACGIH 2008) uses a concept, which considers carcinogenic potency for classification since 1995.

8.4 Conclusion

Toxicology describes the intrinsic properties of an agent (hazard identification) by applying conventional and substance specific test procedures, to estimate the amount of the chemical required to produce these effects and to compare this with human exposure (risk characterization).

There is an array of testing procedures to determine the hazardous properties of a chemical. These include animal studies on acute, sub-chronic and chronic toxicity, irritation and phototoxicity, sensitization, photosensitization, genotoxicity, carcinogenicity, or toxicity to reproduction. Information on the toxicokinetics and mode of action of the toxic effects improve the relevance of the findings for man. Toxicogenomics or high-throughput testing of agents for a single end-point become increasingly available and may improve hazard identification. Of great importance is the assessment of human exposure, because risk is the likelihood of an effect at a given exposure.

References

ACGIH (2008) *Threshold Limit Values and Biological Exposure Indices*. Signature Publications

Angerer, J., Ewers, U., Wilhelm, M. (2007) Human biomonitoring: state of the art. *International Journal of Hygiene and Environmental Health* 210: 201–228.

Bolt, H.M., Huici-Montagud, A. (2008) Strategy of the scientific committee of occupational exposure limits (SCOEL) in the derivation of occupational exposure limits for carcinogens and mutagens. *Archives of Toxicology* 82(1): 61–64.

Boogaard, P.J. (2007) Human biomonitoring activities – programmes by industry. International *Journal of Hygiene and Environmental Health* 210: 259–261.

Dorne, J.L.C.M., Renwick, A.G. (2005) The refinement of uncertainty/safety factors in risk assessment by the incorporation of data on toxicokinetic variability in humans. *Toxicological Sciences* 86: 20–26.

Dunn, W.B., Bailey, N.J., Johnson H.E. (2005) *Measuring the Metabolome: Current Analytical Technologies*. Analyst, Cambridge.

EFSA (2005) Opinion of the scientific committee on a request from EFSA related to a harmonised approach for risk assessment of substances which are both genotoxic and carcinogenic. *The EFSA Journal* 282: 1–31. http://www.efsa.eu.int/science/sc_commitee/sc_opinions/1201/sc_op_ej282_gentox_summary_en1.pdf

EPA (US Environmental Protection Agency; 2005) Guidelines for carcinogen risk assessment EPA/630/P-03/001F. USEPA, Washington, DC.

European Communities, Directive 67/548/EEC Annex V Part B. http://ecb.jrc.it/testing-methods/

Greim, H., Reuter, U. (2001) Classification of carcinogenic chemicals in the work area by the German MAK Commission: current examples for the new categories. *Toxicology* 166: 11–23.

Greim, H., Snyder, R. (eds.) (2008) *Toxicology and Risk Assessment – A Comprehensive Introduction*. Wiley.

Greim, H., Ziegler-Skylakakis, K. (2007) Risk assessment for biopersistent granular particles. *Inhalation Toxicology* 19: 1–6.

IARC (International Agency for Research on Cancer). *IARC Monographs on the Evaluation of Carcinogenic Risks to Humans*. WHO, Lyon.

Kroes, R.A.G., Renwick, Feron V., Galli, C.L., Gibney, M., Greim, H., Guy, R. H., Lhuguenot, J.C., van de Sandt, J.J.M. (2007) Application of the threshold of toxicological concern (TTC) to the safety evaluation of cosmetic ingredients. *Food Chemicals Toxicology* 45: 2533–2562.

Luhe, A., Suter, L., Ruepp, S., Singer, T., Weiser, T., Albertini, S. (2005) Toxicogenomics in the pharmaceutical industry: hollow promises or real benefit? *Mutation Research* 575: 102–115.

Munro, I.C., Renwick, A.G., Danielewska-Nikiel, B. (2008) The threshold of toxicological concern (TTC) in risk assessment. *Toxicology Letters* 180: 151–156.

Needham, L.L., Calafat, A.M., Barr, D.B. (2007) Uses and issues of biomonitoring. *International Journal of Hygiene and Environmental Health* 210: 229–238.

OECD, Guidelines for the Testing of Chemicals, Section 4: Health Effects. http://lysander.sourceoecd.org/vL¼3769318/cL¼31/nw¼1/rpsv/cw/vhosts/oecdjournals/1607310x/v1n3/contp1-1.htm

Renwick, A.G. (1993) Data-derived safety factors for the evaluation of food additives and environmental contaminants. *Food Additives and Contamination* 10: 275–305.

Renwick, A.G., Lazarus, N.R. (1998) Human variability and noncancer risk assessment – an analysis of the default uncertainty factor. *Regulatory Toxicology and Pharmacology* 27: 3–20.

SCHER, SCCP, SCENIHR (2008) *Risk Assessment Methodologies and Approaches for Mutagenic and Carcinogenic Substances*. Health and Consumer Protection Directorate-General, The EU Commission.

Swenberg, J.A., Fryar-Tita, E., Jeong, Y.C., Boysen, G., Starr, T. (2008) Biomarkers in toxicology and risk assessment: informing critical dose-response relationships. *Chemical Research in Toxicology* 21: 253–225.

TGD: Technical Guidance Document in support of Commission Directive 93/67/EEC on Risk Assessment for new notified substances, Commission Regulation (EC) No 1488/94 on Risk Assessment for existing substances and Directive 98/8/EC of the European Parliament and of the Council concerning the placing of biocidal products on the market. ecb.jrc.ec.europa.eu/tgd

UNECE (United Nations Economic Commission for Europe). Globally harmonized system of classification and labelling of chemicals (GHS) http://www.unece.org/trans/danger/publi/ghs/ghs_rev02/02files_e.html

Wetmore, B.A., Merrick, B.A. (2004) Toxicoproteomics: proteomics applied to toxicology and pathology. *Toxicologic Pathology* 32: 619–642.

WHO (1987) *Principles for the Safety Assessment of Food Additives and Contaminants in Food*. Environmenal Health Criteria 70. International Programme on Chemical Safety, World Health Organisation, Geneva.

Chapter 9
Occupational Exposure Limits in Comparative Perspective: Unity and Diversity Within the European Union

Linda Schenk

Abstract This is a book about the regulation of chemical risks; this chapter specifically concerns the regulation of chemicals in the occupational setting. People are exposed to a variety of chemicals during their life; some are to our knowledge not harmful while others are. Working life may be a major contributor to a person's accumulated chemical exposure. A number of diseases have been related to the occurrence of harmful substances in the occupational setting, for instance asthma, allergies and several forms of cancer. One can conclude that the risks associated with chemicals exposure and their regulation in the work place is well worth scientific scrutiny. Occupational exposure limits (OELs) are limits of concentrations of specific substances in the air, averaged over a period of time. The rationale behind OELs is that if the dosage of a chemical is sufficiently low, no or acceptably low adverse health effects will arise. The dose–response relationship differs of course with the different inherent traits of the specific chemical. For some chemicals evidence suggests that a negative health effect only occurs above a certain level of exposure, this means that a safe level exposure is possible to achieve. For many chemicals this is not the case though, either there is not enough knowledge to derive a no effect level (NOAEL), if such one does indeed exist, or there is in fact a linear dose–response relationship without any threshold. In the latter case low-level exposure might only lead to very low individual risks but if many persons are exposed the collective exposure result in substantial population effects.

Keywords European Union • OEL • Risk • STEL • TWA

9.1 Introduction

The level of the OELs depends on the outcome of the risk assessment and risk management processes for the corresponding substances. There are also different kinds of OELs, depending on the time-frame for the exposure that is regulated.

L. Schenk (✉)
Södertörn University, SE 141 89, Huddinge, Sweden
e-mail: linda.schenk@sh.se

J. Eriksson et al. (eds.), *Regulating Chemical Risks: European and Global Challenges*, DOI 10.1007/978-90-481-9428-5_9, © Springer Science+Business Media B.V. 2010

Time weighted averages (TWAs) are usually set for an 8 h day during a 40 h week. They are intended to ensure that no adverse effects occur during the entire working life. To protect against acute effects, usually irritation, the limit is set for a shorter period of time. Short term exposure limits (STELs) usually limit a concentration for a 15 min period. Some organisations also set ceiling values for certain substances, ceiling values concern even shorter time periods than the STEL, usually 5 min. Lists of OELs were introduced as regulatory risk management tools during the twentieth century. It started with the formation of the American Conference of Governmental Industrial Hygienists (ACGIH) in 1938. The ACGIH soon became one of the most influential organisations world-wide when it comes to occupational health regulations), for instance the ACGIH TLVs were adopted by many regulatory agencies during the 1950s and 1960s (Piney 1998, Hansson 1998). Since then, national agencies have gradually developed their own OELs. Today a large number of countries have national lists of OELs, covering hundreds of different chemicals and other exposures. OELs have become the risk management tool for occupational health that is most extensively used by the regulatory authorities.

The OELs are exact numerical values which simplifies quantitative comparisons and statistical analyses. Comparisons of the final OELs can help uncover instances where risk assessment led to discordant results or principles of risk management differ. This chapter aims to describe the development of the regulation of chemicals exposure in the work place with focus is on the European Union, and what factors might affect the level of the OELs. It is important to distinguish between the setting for a regulation and its implementation. The prevalence of chemical health risks on workplaces depends not only on the chosen levels of OELs but also on other factors, including how stringently these OELs are implemented and enforced. Therefore no conclusions should be drawn about the quality of actual working conditions from the material presented in this chapter.

9.2 European Community Involvement

In 1978 the European Community announced its first Action Programme on health and safety at work, aimed at harmonising provisions and measures regarding the protection of workers' health. Previous Community involvement in occupational health and safety had been scarce and with limited influence (Walters 2002). The framework directive 80/1107/EEC was one of the most important outcomes of the 1978 Action Programme; it was the first directive to define a European legal framework for chemicals at the workplace and set out a number of preventive measures (Walters and Grodzki 2006). One of the measures prescribed was the setting of OELs (article 4(4)) (CEC 1980). This framework directive has since been replaced by the 89/391/EEC which is the framework now in effect. Indicative OELs are established through Commission directives and Binding OELs through Council directives. Adopting a Commission directive does not require a formal consultation with the European Parliament, which a Council directive does. The directive 91/322/EEC was the first to define a number of indicative OELs. These indicative OELs were proposed

Table 9.1 EU directives setting indicative and Binding OELs

Directive	Type of OEL
91/322/EEC	Indicative
00/39/EC	Indicative
98/24/EC	Binding for lead
03/18/EC	Binding for asbestos
04/37/EC	Binding for wood dusts, vinyl chloride and benzene
06/15/EC	Indicative

by the EC which to its help had an informal group of scientific experts. In 1995 this group received a formal status as the Scientific Committee on Occupational Exposure Limits (SCOEL). The SCOEL recommends health-based OELs to the EC. When they find it impossible on the basis of current knowledge to identify threshold doses below which no harm to human health can be guaranteed, the SCOEL recommends a pragmatic OEL that is deemed to carry a sufficiently low risk. The feasibility of the OELs recommended by the SCOEL is evaluated by a separate committee, the Advisory Committee for Safety, Hygiene, and Health at Work. It is an assembly of representatives from governments, employers' organisations and trade unions. Indicative OELs are established by the EC when it is concluded that there is a clear threshold dose below which there are no adverse effects on human health. The indicative exposure limits are to be taken into consideration by each member state, but the national OEL is allowed to be higher or lower than the EC indicative OEL. Binding OELs are, as the name implies, mandatory and each member state must either implement the limit set by the EC or a lower limit (Feron 2003). Up to date decisions have been made on 115 substances, resulting in 105 indicative OELs and 10 binding OELs (Table 9.1).

9.3 Aiming for Unity

Previous studies of occupational exposure limits have shown that there are large and unsystematic differences between decisions made for different chemicals with similar adverse health effects (Hansson 1997; Hansson and Rudén 2006). Case studies concerning certain areas of occupation (Haber and Maier 2002, Bigelow et al. 2004) or certain chemicals (Taylor et al. 2007, Cunningham et al. 1998) confirm that there are national differences in risk assessment and management of occupational chemical exposure.

Several steps have been taken towards a more harmonised methodology within the more general area of risk assessment of chemicals. The EU has proposed a number of 'principles for assessment of risks to man and the environment' in directive 93/67/EEC (EC 1993) as well in the Technical guidance documents in support of directive 96/67/EEC on risk assessment (EC 2003), to further common practices in risk assessment. A harmonisation of national exposure limits is to be expected, since the EU sets both binding and indicative OELs for each member state to consider in its national regulations. However as noted by Vincent (1998) a full international harmonisation of OELs is unlikely and may not even be the most efficient means to improve the working environment. The aim should be an intermediate harmonisation, with national lists of exposure limits based on national considerations but with common international criteria and methods.

In December 2006 the proposition for the new chemicals legislation within the European Community was passed by the European Parliament and the Council of the European Union. It entered into force on the first of July 2007. REACH, which is the common name of this new legislation, stands for regulation, evaluation and authorization of chemicals and is a framework on how to produce basic information about the chemicals that are on the market today. One important aspect is that greater responsibility for data-generation and risk assessment is laid on the manufacturers and importers of chemicals. How this information will affect the regulation of chemicals in the workplace is uncertain. The test strategies suggested for the substances produced or imported in volumes of 1–10 t per year will not produce enough data to determine an OEL (EC 2006; Walters and Grodzki 2006). However, REACH will help to produce initial information on a large number of substances and might according to Nielsen and Øvrebø (2008) help to keep up the pace of setting and revising OELs. For substances within the scope of REACH, excluding e.g. cosmetics and pharmaceuticals, that are produced in quantities above 10 t a chemical safety report is to be prepared. One of the requirements of this is to identify so called Derived No-Effect Levels (DNELs) for substances that have identifiable threshold effects. Within the guidance for the implementation of REACH (ECHA 2008) workers are mentioned as one subpopulation requiring a specific DNEL, and an overview of how to derive such worker-DNELs is also given in chapter R.8. For a substance without an identifiable threshold effect a Derived Minimal Effect Level (DMEL) is to be derived. A DMEL should correspond to a risk level 'which is considered to be of very low concern'. Discussions on how national regulations should relate to the DNELs have already started; the Polish case is treated by Gromiec (2008). One of the issues pointed out in this paper is that the DNELs are derived by manufacturers and importers while national OELs are developed by governmental agencies.

A major argument for the harmonisation of procedures and exposure limits is the aim to cut costs and minimise duplication of efforts. A common minimal level of occupational protection will also reduce the risk that insufficient health protection is used as a mean of competition to attract industry. As Vincent (1998) points out, harmonisation will require changes in the regulatory practices, and regulatory authorities with older and larger bureaucracies may be the ones most resistant to change. A similar conclusion is drawn by Grabbe (2001) who studied the influence of the EU on governance in post-communist countries in Central and Eastern Europe. The potential influence of the EU was very large since EU membership was a high priority for these countries. According to Grabbe (2001) a lack of stable form of governance could mean less institutional resistance. An anticipated EU membership has also been identified as a driving force for administrative reform by Lippert et al. (2001). The Commission's review of the implementation of the framework directive 89/391/EEC in the EU member states did in fact conclude that the impact of the directive was greater in member states either with less developed legislation in the field or legislation not already based on preventive principles, as the EU directives are intended to be (EC 2004).

Table 9.2 The rate of change per year and the half-life of the OELs. Based on a comparison between the OELs for substances that were present on both the initial and the last list for each country in our selection. Substances added during the period of time in question do not influence these variables (see footnote 1)

Organisation	Period	Rate of change	Half-life (years)
ACGIH[a]	1995–2005	0.985	45
Estonia	2001–2007	0.993	98
EU	1991–2006	0.990	64
Finland	1993–2005	0.980	34
Germany	1995–2005	0.974	26
Poland	1998–2005	0.9996	1569
Sweden	1996–2005	0.995	139
United Kingdom	1995–2005	0.986	48

[a]American Conference of Governmental and Industrial Hygienists.

Another aspect of OELs is that they tend to decrease gradually over time as they are revised. The half-life[1] of OELs displayed in Table 9.2 is an example of that. This has also been shown in other several studies, e.g. Hansson (1998) which includes a review of the Swedish OELs from 1969 to 1992. Greenberg (2004) made a review of the documentation of British asbestos exposure limits from 1898 to 2000 and Markowitz and Rosner (1995) reviewed the TLV for silica from 1935 to 1990. Both these studies show that the OELs are lowered as more information on adverse effects becomes available and the protection of worker's health is given higher priority. This aspect points at the influence of each organisation's time-frame for the update procedures.

9.4 National Diversity

Walters and Grodzki (2006) conclude that there are strong similarities between EU member states in their systems for setting OELs. They refer to the influence of the ACGIH and also, to a lesser extent, to the Nordic countries, Germany and the EU. The Nordic countries have a history of being active in taking precautionary measures in the occupational setting. Taylor et al. (2007) compared how the European OEL for lead has been implemented in 14 EU member states. Their results show that the OELs for lead set nationally are mostly the same as the EU binding OEL for lead; in five cases the exposure limits were lower. The biological limit values (allowable concentrations of lead in the blood of the employees) for lead exhibited a larger variation between countries.

[1]When OELs are revised, decreases are much more common than increases. The yearly decrease rate of an OEL for a particular substance can be calculated as (b/a) (1/n), b being the OEL n years after the year that a was the OEL for that same substance. The decrease rate of a list is the geometric mean of the decrease rates of all its substances. The decrease rate can also be expressed as the half-life of an exposure limit. If the OEL was 200 ppm in 1990 and 100 ppm in 2005 the exposure limit has been halved in 15 years, which corresponds to an average yearly decrease of 4.5% (1–0.5(1/15)). It should be noted that the OELs in actual fact are changed in a step-wise fashion, whilst half-life is based on a linear model.

The data presented in this chapter are collected from the standard-setting documents of seven different countries, the private organisation ACGIH and the European Union. The seven countries are (year of collected regulations in brackets): Estonia (2001, 2007), Finland (1993, 2002, 2005), France (2005), Germany (1995, 2000, 2005), Poland (1998, 2002, 2005), Sweden (1996, 2000, 2005) and UK (1995, 2000, 2005). The ACGIH TLVs have been very influential world-wide and thus they are included in some comparisons even though the ACGIH is a private organisation seated in the USA (ACGIH 1995, 2005). The OELs have been investigated in respect to three different measures: (1) what chemicals have been selected, i.e. coverage, (2) the average level[2] of exposure limits for all chemicals, and (3) the similarity[3] between the OELs of different EU member states and the OELs recommended by the European Commission.

It has been showed that concerning coverage the national OELs of the non-European countries were more similar to each other in respect to coverage of substances than the European countries were (Schenk et al. 2008a). A plausible interpretation could be that the non-European countries cluster around the ACGIH. Many of the investigated authorities did refer to the ACGIH TLV and documentation in their own national regulations. The EU is a more recent actor and the substances assigned an EU OEL were all previously regulated with OELs in several European countries. Thus the EU is not a pioneering agency concerning coverage of substances. Another prominent reason is the fact that a relatively low number of substances have been assigned a binding or indicative OEL, thus the EU can for statistical reasons not exert a corresponding clustering force as the ACGIH.

Table 9.3 displays the number of individual exposure limits, of the European countries Finland has the highest number of exposure limits (760) and Germany the lowest number (325); the ACGIH has 763 exposure limits on its list. The average number of substances on the national lists is approximately 500. Three countries have recently implemented major changes in the composition of their national OEL lists (Schenk et al. 2008b). The number of OELs on the Polish list has been substantially increased while the number of OELs on the German and United Kingdom lists has decreased. The Hazardous substances committee of Germany has undertaken a review of its OELs, leading to the withdrawal of a great number of exposure limits, suspected not to be truly health based, up to 2005 (Castleman 2006).

---

[2]The average level of the lists of OELs was calculated by the geometric means method. The list of OELs set by the European Commission is used as a comparison list to standardize all national lists, resulting in a new list of ratios for each nation. The geometric mean of these ratios is what in this chapter is called *the average level* of OELs. The EU level equals 1; a geometric mean above 1 means that the OELs on average are higher than the EU OELs, and vice versa. For further elaboration on the method, including the choice of geometric over arithmetic means refer to Schenk et al (2008a).

[3]The geometric similarity is a measure of the distance of the national OELs from the EU OELs. In assessing a particular list the ratio for each substance, between its value on the list in question and the EU list, is used. Ratios above 1 are inverted while ratios below 1 are kept, resulting in a new list of similarity ratios. The geometric mean of the *similarity ratios* is then calculated. It can only assume positive values below or equal to 1. A geometric mean of 1 corresponds to complete similarity and as the geometric mean decreases so does the similarity of the exposure limits to the EU OELs.

Table 9.3 The number of regulated substances on each list of OELs

Country/ organisation	Total no of OELs
ACGIH	763
Estonia	443
EU	105
Finland	760
France	556
Germany	325
Poland	541
Sweden	436
United Kingdom	414

Table 9.4 The geometric mean of ratios and the geometric similarity of the most recent national lists, using the EU list as a comparison list. The Geometric mean is the average level of the OELs, the average of the EU list is 1. The more similar a list is to the EU list the closer to 1 is the geometric similarity measure (see also footnotes 2 and 3)

Country/organisation	No[a]	Geometric mean	Geometric similarity
ACGIH (2005)	95	1.158	0.650
Estonia (2007)	95	0.986	0.986
Finland[b] (2005)	101	0.816	0.805
France (2005)	97	0.948	0.728
Germany (2005)	70	0.964	0.746
Poland (2005)	100	0.809	0.665
Sweden (2005)	91	0.959	0.604
United Kingdom (2005)	81	1.058	0.769

[a]Number of substances both on the individual list and the comparison list.
[b]The Finnish OELs are said to be harmful concentrations, i.e. not health protecting, safe concentrations as is the intention of the other countries' OELs.

In the United Kingdom the Occupational Exposure Standards and Maximum Exposure Limits were replaced in 2005 by Workplace Exposure Limits and 'principles of good practice'. The Occupational Exposure Standards and Maximum Exposure Limits were transferred to the new Workplace Exposure Limit system but those exposure limits for which there was insufficient evidence that they protect human health, were withdrawn (Walters and Grodzki 2006).

The data presented in Table 9.4 are comparisons of the different list to the European Union list. The comparisons concern all substances with an EU OEL except substances without Chemical abstracts number (CAS) and asbestos, the latter due to it being regulated as number of fibres rather than a concentration. After these restrictions 102 substances are left. No national list includes all of the substances on the EU list, although Finland has 101 of 102 possible matches. Most European countries have about 90 substances in common with the EU list.

No list has an average level identical to the EU's. All of the included countries except the United Kingdom have a geometric mean of ratios below one. The ACGIH average level is also above one. This means that most countries determine exposure limits that on average are lower than the EU OELs. The Polish list has the lowest level;

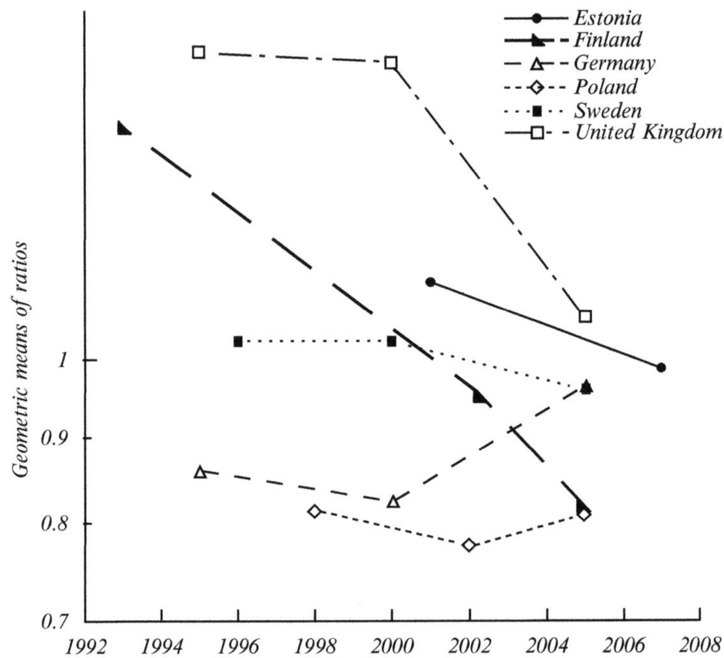

Fig. 9.1 Overall level of exposure limits

its average level of OELs is 19% lower than the EU list. The ACGIH OELs, for those substances that are also on the EU list, are on average higher than both the EU and United Kingdom. That the level of the OELs differs is only expected since each organisation defines what level of protection the OELs are aimed at. A simple way to classify the level that is aimed at is to divide the OELs between whether economical and technical factors are essential when determining them or if the standard setter claims to base the limits on health considerations only. Interesting though is the fact that the two organisations which claim only human health is considered, the ACGIH and the EU (for indicative OELs), are those that have the highest average levels. It could be interpreted as the other regulators are determining the lowest feasible levels and that the health effects of chemicals in the work place are no longer a concern. Unfortunately this is not a plausible explanation. It should also be noted that the Finnish OELs are not aimed at constituting a safe level; the values listed are air-borne concentrations at which harmful effects are shown. These limit values are not enforced but supposed to be used as guidance when assessing work place risks. There is a discrepancy between the claims of the different standard setters and the actual levels of their OELs.

As can be seen in Fig. 9.1 the overall level of exposure limits has decreased in four of the studied countries during the past 10 years. Germany is the most obvious exception as the average level of the exposure limits increased from 2000 to 2005. This increase coincides with the removal of several substances as was described earlier in this chapter. Poland's OELs have not changed noticeably from 1998 to

2005 (from 0.815 to 0.811). The level in 2002 was lower though (0.777) which could mean that substances added from 2002 to 2005 obtained values that were not as much lower than the EU OELs than the previous Polish OELs. That the geometric means increase is probably not an effect of OELs being revised and assigned a higher level, in Table 9.2 one can see that the rate of change is below one, which means that the OELs generally have been lowered when revised. There are several driving forces for this development, among them increasing knowledge of dangers, better technology available and lowered acceptance of occupational health risks.

The final column in Table 9.4 displays the geometric similarity measure for the current lists of OELs. This similarity measure can only range between 0 and 1. The closer the national exposure limits are to EU exposure limits, the closer is the geometric similarity to one. As can be seen in Table 9.4 the Estonian exposure limits are the ones most similar to the EU OELs. Least similar to the EU OELs are the Swedish exposure limits. The geometric similarity of the national OELs tends to increase over time. This means that the national exposure limits are approaching the EU level.

Up to date the Polish list has showed the highest level of assimilation of added substances on OEL lists. 27 of the 154 substances (18%) added in the Polish regulation have been added after the same substances have been regulated in an EU directive. These substances have a generally lower level that the other congruent EU substances while the similarity measure shows that the new substances actually are closer to the EU levels than previously set exposure limits. The high number of substances added in the recent years, the large proportion of them also being recently added by the EU, and the fact that the geometric similarity is increasing, support the hypothesis that Poland is in a process of harmonising with the EU OELs. The Estonian list also displays a large similarity with the EU list, especially concerning the level of OELs, which is almost identical for the substances (Table 9.4).

9.5 Scrutinising Diversity

Earlier in this chapter it has been shown that there are mechanisms in force that could lead to more uniform OELs among the countries of the European Union. And the numerical comparisons performed indicate that the levels of the OELs are converging. But still there are large differences for a number of substances. Now the focus will be turned from the general comparison of entire lists to the regulation of individual substances. Again the 102 substances from the EU list are used. Table 9.5 lists eight substances for which the OELs vary by more than a factor of ten. For the 102 substances seven OELs had the same level, i.e. no variation, on all the lists on which they were included. It should be noted that the lowest OEL is in force in only one country for all the substances in Table 9.5. On the other hand the highest level is in force in several countries for three of the substances, indicating that it is more common for the regulators to agree on the less restrictive levels. Also the EU has set an OEL corresponding to the highest level for these three substances.

Table 9.5 Substances with OELs differing by a factor of ten or larger. Only substances that are regulated by the European Union are included. All OELs displayed are TWAs

Substance	CAS	No OELs	Highest OEL	Country	Lowest OEL	Country	Ratio
p-Dichlorobenzene	106-46-7	6	25 ppm	GB	0.75 ppm	FR	33.3
Fluorine	7782-41-4	6	1.58 mg/m^3	EU, EE, FR, GB	0.05 mg/m^3	PL	31.6
Ethyl Chloride	75-00-3	6	1000 ppm	FR	50 ppm	GB	20.0
2-Butoxyethanol	111-76-2	7	25 ppm	UK	2 ppm	FR	12.5
Chlorobenzene	108-90-7	6	47 mg/m^3	PL	4.6 mg/m^3	GB	10.2
Allyl alcohol	107-18-6	6	2 ppm	EE, EU, GB, SE	0.2 ppm	FR	10.0
2-Butoxyethyl acetate	112-07-2	6	20 ppm	EE, EU, FI, DE, GB	2 ppm	FR	10.0
n-Butylacrylate	141-32-2	7	10 ppm	SE	1 ppm	GB	10.0

Country codes according to ISO 3166-1:
DE – Germany
EE – Estonia
EU – European Union
FI – Finland
FR – France
GB – UK
PL – Poland
SE – Sweden

Table 9.6 The OELs and the concluded critical effects for p-dichlorobenzene (106-46-7)

Agency	OEL	Year of limit	Year of RA	Critical effect
ACGIH	10 ppm	1993	2001	Eye irritation
EU	20 ppm	2000	1994	Liver and kidney toxicity
Estonia	20 ppm	2007	[a]	–
Finland[b]	20 ppm	–	1998	Irritation
France	0.75 ppm	–	2003	Carcinogenicity
Germany	No MAK	2001	2001	Carcinogenicity
Sweden[b]	10 ppm	2000	1998	Irritation
United Kingdom	25 ppm	–	[c]	–

[a]Documentation not available.
[b]Nordic Group of Experts document.
[c]The HSE referred to the ACGIH documentation.

Table 9.7 Overview of key references concerning p-dichlorobenzene.

Key studies	EU	NEG	ACGIH	Germany	France
Hollingswoth et al. (1956)	+++	+++	+	−+−	−
Dow Chemical Co. (1978)	−	−	+++	−	−
Riley et al. (1980)	++	−	−	−	−
ICI (1980)	−	−	−	+++	−
NTP (1987)	−	+	−	+++	−
JBRC (1995)	−	−	−	+++	+++
US EPA (1996)	−	−	−	+++	−

+++ Referred to as a key reference.
++ Referred to as important support to key reference.
+ Reviewed but not used as key reference.
−+− Reviewed and criticised.
− Not reviewed.
Shaded areas indicate that the studies were not available to the risk assessor at the time of data collection.
ACGIH – American Conference of Governmental and Industrial Hygienists.
NEG – Nordic Expert Group.
References to key studies are listed in full in the reference list of this chapter.

That no of the lowest limits are EU OELs is not surprising considering the high average level presented earlier in this chapter.

The largest variability is found for the substance p-dichlorobenzene, which is used in for instance moth balls and space deodorants. Details collected from the documentations of the different OELs for p-dichlorobenzene are presented in Tables 9.6 and 9.7. The ACGIH is included due to its influence, and also the UK health and safety executive refers to the ACGIH documentation for the UK OEL.

Table 9.6 lists statements of critical effect or main adverse effect of p-dichlorobenzene that the OEL is supposed to protect against together with the OELs and year of risk assessment. The large variability of the OELs to some degree can be explained by the difference in what critical effect that has been stated. The documentation also was searched for what literature was reviewed and specifically key references, these can be found in Table 9.7. Key references are those that supply the risk assessor with a point of departure for an OEL, it could be a no effect level, effect level or benchmark dose derived from animal or human data. This analysis shows that the scientific opinion on the hazardous properties has changed markedly over the years. Going back to 1971, the then current ACGIH document reports of a number of effects in exposed animals. Among these were liver necrosis, irritation and CNS symptoms. In humans the handling of products containing p-dichlorobenzene in some instances was believed to have lead to subjective CNS symptoms and clouding of the eyes. The conclusion was that 75 ppm should be sufficient to protect workers from these adverse effects (ACGIH 1971). In 1991 the German DFG lowered their OEL for p-dichlorobenzene from 75 to 50 ppm, due to observations of increased liver weight in animals. Tumours seen in animals were of unknown human relevance, but concluded to not be caused by genotoxicity (DFG 1992). The Swedish criteria group collaborated with the US NIOSH on collecting the documentation for p-dichlorobenzene in 1992. The conclusion in this document was that a genotoxic mechanism could not be excluded, but CNS effects and liver toxicity were considered to be of main importance for an OEL (Hellman 1992). The EU Scientific expert group produced their documentation in 1994. In this document the carcinogenicity noted in animals is concluded to be due to a rat specific pathway and not genotoxicity (SEG 1994). In 1998 the Nordic expert group concluded that p-dichlorobenzne is not genotoxic and that the critical effect is irritation (Elovaara 1998). The German DFG did a new review of data on p-dichlorobenzene in 2003. This time the DFG determined that a genotoxic mechanism could not be excluded and removed their threshold OEL (DFG 2003). The same year also France reviewed the toxicity of the substances and lowered their OEL to 0.75 ppm due to carcinogenicity (Group d'experts sur la santé 2003). In the two latter instances particularly one study (JBRC 1995) seems to have contributed with important information on the carcinogenic properties of p-dichlorobenzene. It should be noted though that it is a report from a research centre and not published in a peer-reviewed journal. The German DFG refers to it as being a brief summary report to the Japan Ministry of Labour. Only four of the documentations listed in Table 9.7 are recent enough to have had potentially access to this study. The ACGIH (2005) and the Nordic expert group did not review this report and also the critical effect they conclude on is not carcinogenicity, but liver and kidney toxicity and irritation, respectively.

As this example with p-dichlorobenzene shows the different levels of OELs are connected to differing scientific opinions on the harmful effects of the substance. The selection of what data to review obviously is important for the outcome of the risk assessment. It can also be concluded to not only being dependent on the time-related availability of the data. The potential for a selection bias of what literature to review in risk assessments has been shown by Rudén for the risk assessments concerning the carcinogenic properties of trichloroethylene (Rudén 2001). Although a case-study of acryl amide (Rudén 2004) shows that a selection bias by no means is an inevitable consequence. Another aspect of which the conclusions from a risk assessment are heavily dependent on is the evaluation of the reviewed data. In Table 9.7 there is an example of one study being very differently evaluated by the risk assessors. While the European Union scientific expert group and the Nordic group of experts used Hollingsworth (1956) as a key reference, the ACGIH reviewed it without specific comments and the German DFG even criticised its reliability.

Another interesting aspect of differences in the documentation is the use of uncertainty or safety factors. These are commonly used in food safety and environmental risk assessment, but customarily not used when deriving occupational exposure limits. A study of the margins of safety used in OELs comprising 14 substances and 45 OELs and the documentation for these only found four instances where explicit safety or uncertainty factors had been used (Schenk, 2010). Two of these instances concerned p-dichlorobenzene making this substance rather unique. The documentations in question are from the EU and France. The magnitude of the uncertainty factors differs between them; the EU applied a factor of 10 and France a factor of hundred. But this is again explainable by the different severity of the concluded critical effects (Table 9.6).

9.6 Unity and Diversity

The list of OEL started as a very uniform occurrence. The industrialised countries almost without exception started their own work with the lists of OELs by copying the ACGIH TLVs. As time passed and national agencies introduced their own risk assessment procedures there has been an increasing diversity and today both coverage and level of OELs vary considerably amongst countries. In this chapter is has been shown that the average level of exposure limits is approximately 25% higher for United Kingdom than Poland. No evidence has been found of this variation being explainable by differences in legal status or by deviations in the principles for risk assessment and risk management explicitly stated, such as the intended level of health protection. Nevertheless, the ACGIH still has a noticeable impact on national regulations and new forces for unifying the OELs have arisen under the aim of harmonisation. EU regulations can be expected to have a significant effect on the coverage of substances on national lists, considering that a national risk assessment and management process is mandatory for the substances that are assigned an indicative OEL. But up to now most substances given indicative OELs had national exposure limits already before the directive in question. The actual effect of the EU on the coverage is thus not clear, possibly that countries that develop new occupational health and safety regulations are more influenced by the EU standards than countries with already institutionalised practices.

The currently high level of the EU OELs compared to the studied European countries can be perceived as somewhat surprising. A general harmonisation process could be expected to lead to exposure limits at intermediate level. Instead harmonisation seems to take the form of adjustment upwards that may reverse the previous trend of OELs becoming lower over time. This development could be cause for concern if it leads to a lower margin of safety being accepted. To estimate the size and nature of a possible such effect, again further study of toxicological documentation for each individual substance is needed.

There is no demand on the individual countries to implement the exact value of the EU indicative OELs. But a calculation of the similarity to the EU levels for substances added nationally after being assigned an EU OEL showed that Finland and UK have assimilated the exact same levels as the EU OELs while Germany, Poland and Sweden have not (Schenk et al. 2008b). However, it has been showed that the national exposure limits have become more similar to the EU OELs since the mid-1990s. It cannot be determined whether this is an effect of harmonisation without further scrutiny of the each country's motives for the individual OELs. One has to bear in mind that the national response to a harmonising incentive, like an EU directive, depends on the current national circumstances. Harmonisation is usually conceived as a conscious process. As can be seen in Fig. 9.1 the average level of the OELs varied more among countries only 10 years ago than it does today. Thus the process in progress during the past 10 years could well be a result of an aspired harmonisation.

Many benefits come from harmonisation, as it joins several perspectives into one process. Among those benefits are reduced costs for each participant and that low demands on occupational health will not become a mean of competition to attract industries. Also, international collaboration would be able to improve the speed at which new substances are added and old limits are revised, an objective that is well worth to aspire. But, since most European countries have an average level of exposure limits that is lower than that of the EU, harmonisation could lead to regulations offering less protection for human health. That is, if the indicative OELs of the EU are simply assimilated without adjustment, it would lead to an increase in the average level since a majority of the previously set national exposure limits seem to have resulted in lower exposure limits than the present EU indicative OELs. Harmonisation can also have another negative effect. Important changes in OELs are often introduced by pioneering agencies suggesting advancements of safety demands and methodology. One not very desirable effect of harmonisation could be that these front-runners will become scarce in the future.

It is generally accepted that risk decision processes should be transparent. An important question is how the transparency of the decision-making process for occupational health and safety will be affected by an increasing centralisation of decision-making to the EU. The transparency of an OEL is to a large degree depending on the accessibility of its documentation. Accessibility means in this case being both available and clear concerning data selection, evaluation and what extrapolations are made in order to derive the level of the OEL. The focus for the harmonisation should not be on acquiring identical levels of OELs in different countries. Rather it should aim to develop methodology of the performance of risk assessments and especially on resulting in consequent and transparent documentation.

References

ACGIH – American Conference of Governmental Industrial Hygienists (1971) Documentation of the threshold limit values for substances in workroom air, Third Edition 1971. ACGIH, Cincinnati.

ACGIH – American Conference of Governmental Industrial Hygienists (1995) Threshold limit values for chemical substances and biological exposure indices. ACGIH, Cincinnati.

ACGIH – American Conference of Governmental Industrial Hygienists (2005) Documentation of the TLVs and BEIs with other worldwide occupational exposure values CD-ROM. ACGIH, Cincinnati.

Bigelow P., Moore D., Yassi A. (2004) Assessing the health implications for healthcare workers of regulatory changes eliminating locally developed occupational exposure limits in favor of TLVs: An evidence-based bipartite approach. *International Journal of Occupational and Environmental Health* 10(4): 433–444.

Castleman B.I. (2006) Legacy of corporate influence on threshold limit values and European response. *American Journal of Industrial Medicine* 49: 307–309.

Cunningham E.A., Todd J.J., Jablonski W. (1998) Was there sufficient justification for the 10-fold increase in the TLV for silica fume? A critical review. *American Journal of Industrial Medicine* 33(3): 212–223.

CEC (1980) Council Directive 80/1107/EEC of 27 November 1980 on the protection of workers from the risks related to exposure to chemical, physical and biological agents at work.

DFG – Deutsche Forschungsgemeinschaft (1992) Occupational toxicants volume 4. VCH Verlagsgesellschaft mbH, Weinheim, Germany.

DFG – Deutsche Forschungsgemeinschaft (2003) Occupational toxicants. Volume 20. Wiley VCH GmbH&Co. KGaA,, Weinheim, Germany.

Dow Chemical Co. (1978) Preliminary Study into the Environmental Fate of PARADOW Blocks, May 17, 1973. TSCA 8(d) Submission 8DHQ-0978-0299. Washington (DC):1978. U.S. Environmental Protection Agency.

EC (1993) Commission Directive 93/67/EEC of 20 July 1993 laying down the principles for assessment of risks to man and the environment of substances notified in accordance with Council Directive 67/548/EEC.

EC (2003) Technical guidance document in support of the commission directive 93/67/EEC in risk assessment for new notified substances and the commission regulation (EC) 1488/94 on risk assessment for existing substances. http://www.ecb.jrc.it/tgdoc. Cited March 2007.

EC (2004) Communication from the Commission to the European Parliament, the Council, the European Economic and Social Committee and the Committee of Regions – on the practical implementation of the provisions of the health and safety at Work Directives 89/391 (Framework), 89/654 (Workplaces), 89/655 (Work Equipment), 89/656 (Personal Protective Equipment), 90/269 (Manual handling of Loads) and 90/270 (display Screen Equipment). COM (2004) 62 Final. Brussels.

EC (2006) Regulation (EC) No 1907/2006 of the European Parliament and of the Council of 18 December 2006 concerning the Registration, Evaluation, Authorisation and Restriction of Chemicals (REACH), establishing a European Chemicals Agency, amending Directive 1999/45/EC and repealing Council Regulation (EEC) No 793/93 and Commission Regulation (EC) No 1488/94 as well as Council Directive 76/769/EEC and Commission Directives 91/155/EEC, 93/67/EEC, 93/105/EC and 2000/21/EC.

ECHA – European Chemicals Agency (2008) Guidance on the implementation of REACH. Guidance on information requirements and chemical safety assessment Chapter R.8: Characterisation of dose [concentration]-response for human health.

Elovaara E. (1998) The Nordic expert group for criteria documentation of health risks from chemicals 122. Dichlorobenzenes. *Arbete och hälsa* 1998: 4.

Estonia – Estonian ministry of Social Affairs (2001) Limit values for chemical hazards in the working environment. Regulation No. 293 of the Government of the Republic of 18 September 2001. RT I [State Gazette] 2001, 77, 460.

Estonia–Estonian ministry of Social Affairs (2007) Limit values for chemical hazards in the working environment. Regulation No. 293 of the Government of the Republic of 23 October 2007. RT I [State Gazette] 2007, 55, 369.

Feron V.J. (2003) Setting occupational exposure limits in the European Union. In: *Occupational Exposure Limits – Approaches and Criteria*. Proceedings from a niva course held in Uppsala, Sweden, 24–28 September 2001. Ed. Johanson G. Arbete och hälsa, 2003: 17, Stockholm, Sweden.

Finland – Arbetsministeriet (1993) *HTP-värden 1993*. Säkerhetsmeddelande Nr 23. Arbetsministeriet Delegationen för arbetarskydd beträffande kemiska ämnen, Tammerfors, Finland.

Finland – Social – och hälsovårdsministeriet (2002) *HTP-värden 2002*. Social och hälsovårdsministeriet, Tammerfors, Finland.

Finland - Social – och hälsovårdsministeriet (2005) *HTP-värden 2005*. Social och hälsovårdsministeriet, Helsinki, Finland

France – Insititut National de Recherche et de Sécurité (2005) *Note documentaire: Valeurs Limites d'exposition professionelle aux agents chimiques en France*. ND2098. Insititut National de Recherche et de Sécurité, France.

Germany – Deutsche Forschungsgemeinschaft (DFG) (1995) *MAK- und BAT-Werte-Liste 1995*. Senatskommision zur Prüfung gesundheitsschadlicher Arbeitsstoffe. Mitteilung 31.VHC Verlagsgesellschaft, Weinheim, Germany.

Germany – Deutsche Forschungsgemeinschaft (DFG) (2000) *List of MAK and BAT Values 2000*. Commission for the investigation of Health Hazards of Chemical Compounds in the Work Area. Report No. 36. Wiley-VHC, Weinheim, Germany.

Germany – Deutsche Forschungsgemeinschaft (DFG) (2005) *List of MAK and BAT Values 2005*. Commission for the investigation of Health Hazards of Chemical Compounds in the Work Area. Report No. 41. Wiley-VHC, Weinheim, Germany.

Grabbe H. (2001) How does Europeanization affect CEE governance? Conditionality, diffusion and diversity. *Journal of European Public Policy* 8(6): 1013–1031.

Greenberg, M. (2004) The British approach to asbestos standard setting: 1898–2000. *American Journal of Industrial Medicine* 46(5): 534–541.

Gromiec J. (2008) Problems concerning the integration of 'derived-no-effect-levels' (DNELS) into occupational safety and health regulations. In Polish. *Medycyna Pracy* 59(1): 65–73.

Group d'experts sur la santé (2003) Fiche de synthèse: 1,4-dichlorobenzene. October 2003. Acquired through e-mail 2007-11-09 INRS.

Haber L.T., Maier A. (2002) Scientific criteria used for the development of occupational exposure limits for metals and other mining related chemicals. *Regulatory Toxicology and Pharmacology* 36: 262–279.

Hansson S.O. (1997) Critical Effects and Exposure Limits. *Risk Analysis* 17(2): 227–236.

Hansson S.O. (1998) *Setting the Limit. Occupational Health Standards and the Limits of Science*. Oxford University Press, New York.

Hansson S.O., Rudén C. (2006) Evaluating the risk decision process. *Toxicology* 218: 100–111.

Hellman B. (1992) Basis for an occupational health standard: Chlorobenzene. *Arbete och hälsa* 1992: 31.

Hollingsworth R.L., Rowe V.K., Oyen F., Hoyle H.R., Spencer H.C. (1956) Toxicity of paradichlorobenzene; determinations on experimental animals and human subjects. *AMA Archives of Industrial Health* 14: 138–147.

ICI – Imperial Chemical Industries (1980) Paradichlorobenzene: long term inhalation study in the rat. Central Toxicology Laboratory: 1980. CTL study no. PRO130 No. CTL/P/447. Unpublished report.

JBRC – Japan Bioassay Research Center (1995) Toxicology and carcinogenesis studies of 1,4-dichlorobenzene in F344/DuCrj rats and Crj:BDF1 mice (two years inhalation studies). Report to the Ministry of Labour in Japan. November 1995.

Lippert B., Umbach G., Wessels W. (2001) Europeanization of CEE executives: EU membership negotiations as a shaping power. *Journal of European Public Policy* 8(6): 980–1012.

Markowitz, G., Rosner, D. (1995) The limits of thresholds – silica and the politics of science, 1935 to 1990. *American Journal of Public Health* 85(2): 253–262.

Nielsen G.D., Øvrebø S. (2008) Background, approaches and recent trends for setting health-based occupational exposure limits: a minireview. *Regulatory Toxicology and Pharmacology* 51(3): 253–269.

NTP – US National Toxicology Program (1987) Technical report on toxicology and carcinogenesis studies of 1,4-dichlorobenzene in F344/N rats and B6C3F1 mice (gavage studies). Research Triangle Park (NC): National Toxicology Program NT;1987. NIH Publication No. 87–2575.

Piney M. (1998) Exposure limits and health risks arguments for a paradigm shift. In: Bal, R. and Halfmann, W. (eds) *The Politics of Chemical Risk. Scenarios for a Regulatory Future*. Kluwer, Dordrecht.

Poland – The Minister of Labour and Social Policy (1998) The ordinance on the maximum admissible concentrations and intensities of harmful to health agents in the working environment. Journal of Laws 1998, No 79, item 513, Poland.

Poland – The Minister of Labour and Social Policy (2002) The Ordinance on the maximum admissible concentrations and intensities of harmful to health agents in the working environment. Dziennik Ustaw, No 217, item 1833, Poland.

Poland – The Minister of Labour and Social Policy (2005) The Ordinance on the maximum admissible concentrations and intensities of harmful to health agents in the working environment, Dziennik Ustaw 2005, No. 212, item 1769, Poland.

Riley R.A., Chart I.S., Doss A., Gore C.W., Patton D., Weight T. M. (1980) Paradichlorobenzene long term inhalation study in the rat. Macclesfield (UK): ICI Central Toxicology Laboratory. Unpublished report.

Rudén C. (2001) The use and evaluation of primary data in 29 trichloroethylene carcinogen risk assessments. *Regulatory Toxicology and Pharmacology* 34: 3–16.

Rudén C. (2004) Acrylamide and cancer risk – expert risk assessments and the public debate. *Food and Chemical Toxicology* 42: 335–349.

Schenk L., Hansson S.O., Rudén C., Gilek M. (2008a) Occupational exposure limits: a comparative study. *Regulatory Toxicology and Pharmacology* 50: 261–270.

Schenk L., Hansson S.O., Rudén C., Gilek M. (2008b) Are occupational exposure limits becoming more alike within the European Union? *Journal of Applied Toxicology* 28: 858–866.

Schenk L. (2010) Comparison of Data Used for Setting Occupational Exposure Limits. *International Journal of Occupational and Environmental Health* 16: 249–262.

SEG – Scientific Expert Group (1994) Recommendation from Scientific Expert Group on Occupational Exposure Limits for 1,4-Dichlorobenzene. SEG/SUM/65, 1994.

Sweden – Arbetarskyddsstyrelsens Författningssamling (1996) Hygieniska gränsvärden och åtgärder mot luftföroreningar, AFS 1996: 2.

Sweden – Arbetarskyddsstyrelsens Författningssamling (2000) Hygieniska gränsvärden och åtgärder mot luftföroreningar, AFS 2000: 3.

Sweden – Arbetarskyddsstyrelsens Författningssamling (2005) Occupational exposure limit values and measures against air contaminants, AFS 2005: 17.

UK – Health and Safety Executive (1995) Occupational Exposure Limits 1995. HSE Books, UK.

UK – Health and Safety Executive (2000) Occupational Exposure Limits 2000. HSE Books, UK.

UK – Health and Safety Executive (2005) List of approved workplace exposure limits. http://www.hse.gov.uk/coshh/table1.pdf. Cited March 2007.

US EPA – US Environmental Protection Agency (1996) One year study of p-dichlorobenzene administered orally via capsule to beagle dogs. Monsanto Company, Environmental Health Laboratory (EHL). No. ML-94-210. Project No. EHL 94093. Unpublished report.

Vincent J.H. (1998) International exposure standards: a review and commentary. *American Industrial Hygiene Association Journal* 59: 729–742.

Taylor A., Angerer J., Arnaud J., Claeys F., Kristiansen J., Mazarraza O., Menditto A., Patriarca M., Pineau A., Valkonen S., Weycamp C. (2007) Differences in national legislation for the implementation of lead regulations included in the European directive for the protection of the health and safety of workers with occupational exposure to chemical agents (98/24/EC). *International Archives of Occupational and Environmental Health* 80: 254–264.

Walters D. (2002) The framework directive. In: Walters, D., and Lang, P. (eds) *Regulating Health and Safety Management in the European Union*. A study of the dynamics of change. College of Europe Publications, Brussels.

Walters D., Grodzki K. (2006) *Beyond Limits? Dealing with Chemical Risks at Work in Europe*. Elsevier, Oxford.

Chapter 10
Scientific Uncertainty and Science-Policy Interactions in the Risk Assessment of Hazardous Chemicals

Christina Rudén and Michael Gilek

Abstract In this chapter relatively recent European Commission risk assessment reports for three potential PBT/vPvB chemicals are used as examples to illustrate scientific uncertainty in the risk assessment process, and how science and policy interact when such uncertainty is handled. The studied risk assessment reports are for pentabromodiphenylether (Penta), octabromodiphenylether (Octa), and decabromodiphenylether (Deca) and the analyses focus on the scientific basis for assessing the risk of potential PBT and vPvB properties as described in these documents. The purpose of this effort is to contribute to a discussion aiming at clarifying the nature of science-policy interactions, and improving the transparency of the risk assessment process.

Keywords Hazard • REACH • Risk assessment • SVHC • Uncertainty

10.1 Identifying Substances of High Concern

In the European chemicals legislation, REACH, persistent, bioaccumulating and toxic chemicals (PBT), as well as chemicals that are very persistent and very bioaccumulating (vPvB) have been recognized as 'Substances of Very High Concern' (SVHC). According to REACH, substances with these properties should be prioritized for risk management decisions aiming at reducing exposures. Other properties of high concern include carcinogenicity, mutagenicity, reproductive toxicity and endocrine disruption.

C. Rudén (✉)
Royal Institute of Technology (KTH),
Teknikringen 78B, SE 100 44, Stockholm, Sweden
e-mail: cr@abe.kth.se

M. Gilek
Södertörn University, SE 141 89, Huddinge, Sweden
e-mail: michael.gilek@sh.se

J. Eriksson et al. (eds.), *Regulating Chemical Risks: European and Global Challenges*,
DOI 10.1007/978-90-481-9428-5_10, © Springer Science+Business Media B.V. 2010

The motivations for prioritizing substances that are persistent and bioaccumulating are that, if continuously emitted, the environmental concentrations will increase over time and the substances will accumulate in biota and magnify in top predators including humans. The long-term exposures that will follow, in combination with the long life-cycle of humans and several marine species, increases the concern of adverse effects that cannot be detected at an early stage. This is problematic since cessation of emissions will only slowly result in a reduction in chemical concentration (European Commission 2003a).

Data that enables a first assessment of a chemical's degradability are required in REACH for prioritized substances in the lowest tonnage band (1–10 tonnes per year). For substances that belong to this category the standard Ready biodegradability test may be required. For all substances with a production volume over 10 tonnes, data from standard tests for abiotic degradation (hydrolysis) should also be presented. For substances in the higher volume categories (>100 tonnes per year) simulation testing in water, soil and sediment may furthermore be required on a case-by-case basis, as well as the identification of breakdown products and data on to what extent the substances have the potential for bioconcentration in fish.

Specific criteria for categorizing PBT and vPvB chemicals have been defined and are incorporated in Annex XIII of the REACH legislation. The present PBT/vPvB criteria are based on standard laboratory test methods, in particular long-term toxicity testing, data on half-lives as estimated by simulation testing in various compartments, and bioconcentration tests in fish. Hence, for these criteria to be applicable, testing according to the requirements for high volume chemicals (i.e. substances with a yearly production volume that exceeds 100 tonnes per year and manufacturer) is needed. For substances below this tonnage band the REACH test requirements are insufficient for the PBT/vPvB criteria to be applicable.

The REACH PBT/vPvB criteria are currently under review by the European Commission and in this process several challenges in identifying chemicals with these properties have been recognized. Several sources of uncertainties in the risk assessment of them have also been identified (ECHA 2008). One aspect that has been discussed is the general preference for standard test methods over non-standard ones. In this context a standard method refers to a method that is specified in internationally accepted testing guidelines such as the OECD guidelines, and non-standard methods refer to tests performed according to any other research method of sufficient scientific quality. Within the regulatory framework standard tests are usually accepted without restrictions. The major advantages of using standard test methods are that the results are directly comparable across substances and that the data they generate will be accepted across jurisdictions. The major disadvantage is that the standard methods do not always represent the most sensitive and relevant testing approach given the type of substance under investigation and the endpoint hypothesized. Therefore, it is widely agreed that results from non-standard tests may contribute relevant information. Data from non-standard methods are usually taken into account in risk assessment, and their reliability and relevance are evaluated on a case-by-case basis (European Commission 2003a). However, their use for hazard classification, e.g. according to the PBT/vPvB criteria, is hampered since the criteria directly refer to standard tests (ECHA 2008).

To increase flexibility and relevance in regulatory hazard and risk assessment, it has been argued that a more open 'weight-of-evidence' approach should be introduced that takes all relevant information into account, such as evidence of long range transport and measured concentrations in the environment (ECHA 2008; Reineke 2008; Breitholtz et al. 2006). However, there are still different opinions within science as well as among decision-makers about (i) what constitutes relevant information, (ii) when to consider data generated by novel test methods sufficiently reliable, and (iii) how to operationalize a weight-of-evidence approach. Overall this results in diverging views on several key questions concerning the data needs for regulatory risk management decision-making. For example, what and how much data are needed to warn the users of the chemical of a suspected effect? What and how much data are needed to take actions to reduce the highest exposures? What and how much data are needed to ban the use of a chemical? The established criteria for classification and warning labelling give some guidance (Council Directive 67/548 EEC), but in general there is little consensus about how these questions should be answered.

10.2 Uncertainty in Risk Assessments of Potential PBT/vPvB Substances

The European Commission risk assessment reports for the diphenylethers: Penta (European Commission 2001), Octa (European Commission 2003b), and Deca (European Commission 2002 and 2004) were used to identify examples of scientific uncertainties in the risk assessment process for potential PBT/vPvB substances. A systematic search for indicators of scientific uncertainty in these documents was performed, including how these uncertainties are described and handled in the risk characterization and in the conclusions section, and taking into consideration the use of and weight given to non-standard data in the final conclusions.

The three analysed risk assessments are contemporary and have all been performed by the same European Union member states (the UK and France except for the risk assessment for Penta that the UK was the sole rapporteur for). They furthermore cover three substances that are chemically closely related, representing three brominated diphenyl ethers, used as flame retardants, with different degrees of bromination.

In the next sections the identified uncertainties are briefly summarized. First uncertainties in exposure assessment are presented and discussed, second we turn to the uncertainties in the hazard assessment part of the risk assessment, third uncertainties pertaining to the overall conclusions are analysed. Finally we describe how the risk assessors have proposed to deal with these uncertainties.

10.2.1 Uncertainties Identified in the Exposure Assessment

With respect to exposure assessment the risk assessors report uncertainties about the magnitude of the actual emissions for all three substances as well as in the measured environmental concentrations, trends in concentrations, and the kinetics of the substances in various biota and environmental matrices.

In all three documents concern is raised about the validity and relevance of models used to predict exposures, such as EUSES, which rely on the substance's lipophilicity as estimated by its octanol-water partitioning coefficient (Kow) and other chemical characteristics. The reason for this is that these models are generally not applicable to substances with a very high lipophilicity. Furthermore, the determination of some main physico-chemical properties such as lipophilicity, water solubility and vapour pressure is also stated as sources of uncertainty for substances with very high lipophilicity.

For Octa and Deca, lack of detailed knowledge about the mechanisms and rates of degradation (debromination) as well as the bioaccumulative and toxic properties of degradation products are stated as sources of significant uncertainty.

To handle these uncertainties the risk assessors use several strategies. They propose conservative default values for emissions, worst-case estimates and realistic worst-case assumptions for amounts used and emitted, and best estimate/expert judgment for estimating bioconcentration factors. They furthermore recommend method development (improved chemical analyses), further testing (degradation pathways), and more chemical analyses (environmental monitoring).

10.2.2 Uncertainties Identified in the Hazard Assessment

With respect to the hazard identification part of the risk assessments, the main uncertainties discussed in these documents relate to the human health relevance of the observed developmental neurotoxicity in rodents. These are uncertainties that have been highlighted in all three documents. Likewise, the predicted no effect concentration for contaminated sediments is considered uncertain in all three risk assessments. The predicted no-effect concentration (NOEC) for water is considered uncertain for Octa and Deca, and the predicted no-effect concentration for the terrestrial compartment is identified as uncertain for Octa.

Concerns about using data from sub-chronic toxicity tests (30 and 90-day studies, respectively) for bioaccumulating substances are described as a source of uncertainty in the Penta and the Octa documents.

Thus, in summary, most of the identified sources of uncertainty relating to the hazard assessment concern the relevance of different methods for hazard identification. The relevance of sub-chronic testing is considered particularly uncertain for persistent and highly lipophilic substances for which low-level, long-term exposures are predicted.

To handle these uncertainties the risk assessors use a range of different estimates for the predicted no effect concentrations (using a diversity of methods and assumptions) to provide a better understanding of the magnitude of this uncertainty.

The relevance of the observed rodent liver effect is evaluated using the common default assumption in risk assessment that an effect seen in experimental animals is considered relevant to human health risks in the absence of evidence to the contrary

(European Commission 2003a). However, the risk assessors seem to be more reluctant to use this default assumption to assess the relevance of the observed indications of developmental neurotoxicity in rodents. For this endpoint additional data are requested.

Additional information is also requested from toxicity testing in birds (Octa and Deca), although it is also questioned whether such testing would actually eliminate uncertainties considering the inherent difficulties in achieving sufficiently high exposures in the available tests.

10.2.3 Uncertainties Identified in the Overall PBT Conclusions

The risk assessors conclude that Penta is 'highly persistent, bioaccumulative and of particular note, has been detected, albeit in relatively low levels, in human breast milk, the levels increasing with time.' The uncertainties mentioned in relation to this conclusion are mainly connected with the toxicity/hazard assessment. In particular, it is stated that there are significant uncertainties in the risk characterization (i.e. the calculated margin-of-safety for health risks (MOS[1]). Especially the margin-of-safety for breastfed babies is identified as uncertain and potentially insufficient.

The risk assessors conclude that Octa can be considered persistent in the environment but that the available laboratory data indicates a low potential for bioaccumulation. Furthermore, it is concluded that Octa shows no toxicity towards aquatic organisms up to the limit of water solubility, and that effects in other organisms are only observed at relatively high concentrations, based on standard laboratory tests. Nevertheless, the risk assessors note that environmental monitoring of concentrations in biota indicate that Octa, as well as HexaBDE, and HeptaBDE are present at low concentrations in fish, marine mammals and predatory birds' eggs.

The risk assessors note that the occurrence of Octa (and also Deca) 'appear to contradict the conventional wisdom that molecules such as decabromodiphenyl ether and to a lesser extent octa- and heptabromodiphenyl ether are too large to pass through biological membranes and should not accumulate in organisms.'

Finally, it is concluded that the Octa levels found in wildlife are 'below those that are predicted to cause effects on fish-eating species using the PEC/PNEC[2] approach. However, the sample sizes are small, and the trend in these levels is unknown. It is also possible that higher concentrations could be found in other organisms.'

---

[1]The human health risk characterization is typically carried out by comparing the No-Observed-Adverse-Effect-Level (NOAEL) to the human exposure level. The ratio is called Margin of Safety. If human exposure is estimated to exceed the NOAEL, the substance is considered to be 'of concern'. If the exposure estimate is less than the NOAEL, the appropriate 'margin of safety' is assessed case-by-case (European Commission 2003a).

[2]The environmental risk characterization is typically carried out by comparing the predicted no effect concentration (PNEC) to the predicted environmental concentration (PEC). A PEC/PNEC ratio above 1 indicates that the substance poses a potential risk to the environment (European Commission 2003a).

The uncertainties identified in connection with these conclusions include questions about the adequacy of the method used to calculate a MOS for bioaccumulative substances like Octa, as well as a similar methodological uncertainty linked to the current PEC/PNEC approach for secondary poisoning in terms of both the PEC and the PNEC (which could lead to an underestimation of risk). The risk assessment report also identifies a general uncertainty connected with the strength of the scientific basis for drawing conclusions on the current and future environmental risks of Octa.

Similar to what is the case for Octa, the risk assessors conclude that Deca can be considered persistent in the environment but that the available laboratory data indicates a low potential for bioaccumulation. Furthermore, the conclusion of the risk assessment is that Deca shows no toxicity towards aquatic organisms at or below the water solubility of the substance, and that standard laboratory tests only indicate effects in other organisms at relatively high concentrations. But, again, the risk assessors note that Deca has been found in low concentrations in fish, marine mammals and predatory birds' eggs, and that these observations are unexpected given the common assumption that molecules the size of Deca are too large to pass through biological membranes.

Regarding human health risks the Deca risk assessors note that scientific information on bioconcentration in human adipose tissues and subsequent elimination via for example breast milk is insufficient, but that other brominated diphenyl ethers (i.e. Hexa, Penta and Tetra) are excreted with breast milk. Based on the low rate of oral absorption in rats and the low bioaccumulation potential, the Deca risk assessors state that they 'might anticipate a rather low excretion of this compound in the breast milk.'

The uncertainties identified in connection with these conclusions relate to the suitability of the current risk assessment approach for secondary poisoning, bioaccumulation, and whether or not debromination of the higher brominated molecules gives rise to other more toxic and bioaccumulative BDEs (such as Penta) is occurring at a significant rate in the environment. The risk assessors conclude that this combination of uncertainties raises concerns about the possibility of long-term environmental effects that cannot easily be predicted and that a strict PEC/PNEC approach may not be appropriate for this substance.

Concerning the results of the risk assessments, all three documents state that the margin-of-safety for health risks is significantly uncertain. For Octa and Deca uncertainty in the PEC/PNEC ratio is also stressed, and it is proposed that the traditional PEC/PNEC approach may not be appropriate for these highly lipophilic and largely water insoluble substances.

To handle these uncertainties different strategies are proposed. For Penta it is argued that risk management should be considered without delay because of the bioaccumulation potential and risks to breastfeeding babies. For Octa it is recommended that further information should be gathered in order to refine the risk assessment, but also that risk management options should be considered even in the absence of adequate scientific knowledge (a number of technical experts from EU member states that immediate risk reduction measures are warranted). Finally, for Deca it is recommended that further information should be gathered in order to refine the risk assessment, and at the same time 'the need to investigate risk management options should be considered'. Again this conclusion was questioned by a number of technical experts who considered immediate risk reduction measures warranted.

The Octa and Deca risk assessors also indicate the need to develop more suitable methods for risk assessment of bioaccumulating substances.

So, concerning the overall risk assessment conclusions the documents differ. For Penta it is concluded that 'risk management should be considered without delay'. For Octa 'risk management options should be considered also in the absence of adequate scientific knowledge', and finally, for Deca it is concluded that 'Consideration should be given [...] about the need to investigate risk management options now in the absence of adequate scientific knowledge'. However, the overall conclusions for Octa and Deca were not uncontested since the risk assessment documents also states that a number of experts considered the available information for Octa and Deca sufficient to warrant risk reduction measures directly.

10.3 Discussion

Within risk management scientists, experts, stakeholders and decision-makers interact to address fundamental questions such as what type and how much data should be required before a particular risk is judged as intolerable and when and to what extent measures to reduce exposures are required. It is evident that these questions include both scientific and policy-related aspects. One specific illustration of these science-policy interactions can be seen in the discussions and various opinions expressed among experts, stakeholders and decision-makers in connection with the current review of the REACH criteria for PBT/vPvB substances where e.g. environmental NGOs such as WWF (Reineke 2008) as well as the European Chemicals Agency (ECHA 2008) argue that the criteria need to be adjusted so that also non-standard scientific information such as monitoring data should be considered if relevant and available.

When attempting to manage the environmental and health risks posed by chemicals, the main challenge is usually a substantial lack of scientific data. Even for the most well investigated substances uncertainty usually remains concerning the actual magnitude of risks. This scientific uncertainty need to be taken into account in risk management (Renn 2008). We have in this contribution addressed this important issue by examining how scientific uncertainties are described and handled in three recent European Commission risk assessments of penta-, octa-, and deca-brominated dipehylether. This scrutiny of the risk assessment process identified scientific uncertainties that relate to lack of test data, but also to lack of knowledge about the actual reliability and relevance of different endpoints and test methods used to generate knowledge. Clearly, the relevance and reliability of measurements and methods are sources of uncertainties that will not easily be reduced by just performing additional standard tests.

In the regulatory setting standard tests are generally preferred to non-standard ones. The main reason for this is that standard test procedures make results comparable across substances, and contribute to ensure a high scientific quality of the test and the reporting of data. However, by using a non-standard approach the test procedures can be adjusted to what is most relevant for the substance at hand and the hypothesized effects. For assessment of PBT/vPvB properties of substances already in commerce, relevant non-standard data sources also include information from environmental monitoring of contaminant concentrations and long-range transport. Field data reflects the emissions, fate, and behaviour of a chemical in the environment. Therefore, it represents a highly relevant type of information. On the other hand, using various types of field data in chemical risk assessments may introduce higher variability and uncertainty since the reliability (reproducibility) of field measurements may vary. However, data with high relevance that are of sufficient quality should not easily be dismissed in the risk assessment. On the other hand, it is important to recognize that risk assessment of chemicals cannot as a rule rely on monitoring data, since this would preclude a proactive assessment and management of hazardous chemicals. Instead, in the long run we agree with the conclusion of the analysed risk assessment documents that more suitable and environmentally relevant standardized methods need to be developed for some regulatory purposes but also that data requirements need to be strengthened so that the bioaccumulation, toxicity and degradation can be estimated for all industrial chemicals. In the short perspective, however, there are several potential PBT/vPvB chemicals in use today with measurable concentrations in various environmental compartments. It would seem inappropriate not to use all available information of sufficient quality to assess the environmental and health risks of potentially hazardous chemicals. Furthermore, we believe that there are good possibilities to develop a more general framework (or guidance) for how to weigh and integrate various lines of evidence by drawing on weight-of-evidence methodologies developed for site-specific risk assessments (e.g. Long and Chapman 1985; Suter et al. 2000; Burton et al. 2002).

There are important differences between scientists and regulators in how they can address scientific uncertainties and what sort of considerations and trade-offs are connected with their handling of identified uncertainties. Usually, in science the focus is on avoiding false positives.[3] (A false positive means that you conclude that a specific chemical is hazardous even though it is in fact not). Identified uncertainties in science can be handled by performing additional testing before a final conclusion can be made on the properties of a chemical. A decision-maker on the other hand will always have to treat every substance *as if* its properties were known, regardless of scientific uncertainties. If she, for instance decides to take no action to reduce exposures, this would correspond to a standpoint that the substance has

[3]This traditional scientific focus on purely minimising false positives has been criticized as being inadequate for applied sciences such as toxicology since costs of false negatives (i.e. concluding that a hazardous chemical is safe) are larger than in non-applied sciences. For applied sciences it has been argued that it is scientifically justifiable to shift the burden of proof somewhat towards reducing false negatives (i.e. adopting a more precautionary approach) (Peterman and M'Gonigle 1992).

tolerable properties given current exposures. On the other hand, if the decision-maker concludes that risk reduction measures should be taken, based on current knowledge, then this decision corresponds to the standpoint that the substance has properties that makes current exposures intolerable. A decision-maker can thus never act as if the properties of a substance are unknown.

Looking at the three studied risk assessment documents a traditional scientific approach towards handling uncertainties (i.e. by proposing additional studies) can be observed in response to several of the identified sources of uncertainty. In two of the examined risk assessment documents (i.e. Octa and Deca) the technical experts from several European Union member countries disagree with the conclusions made in the documents on how the identified uncertainties should be treated. Such fundamental disagreement among technical experts on the overall conclusion obviously reflects underlying uncertainty in the risk assessments and should always be transparently described. A thorough uncertainty analysis that is not just focused on the particular details of the risk assessment but also on the scope of possible conclusions helps decision-makers to estimate the degree of precaution that is involved in different decision alternatives.

Furthermore, from a regulatory vantage point, it can be mentioned that several risk reduction measures have been taken for these chemicals both within the European Union and in international conventions. The use of Penta, Octa and Deca in electrical and electric equipment is, for example, restricted in EU legislation through the RoHS Directive (2002/95/EC). Penta is also included (and Octa is suggested for inclusion) in the Stockholm Convention on persistent organic contaminants. Thus, decision-makers appear to have treated the available and often uncertain information on Penta, Octa and Deca as if their properties, use and emission were known to pose intolerable risks (at least to the extent reflected by the above described risk reduction measures).

There is no generally accepted and systematically applied approach for assessing and handling uncertainties in the risk assessment process. It relies heavily on case-by-case expert judgement. This introduces flexibility in the system but also makes it less predictable since the results will to some extent depend on what experts are performing the risk assessment and their particular training, experience and expertise.

10.4 Conclusions and Recommendations

How much and what type of data are needed before a substance should be classified as a PBT or vPvB chemical is described in the REACH criteria. However, as has been shown above there are cases where these criteria do not cover all the information available. A strict application of the criteria could, in such cases, lead to either over- or underestimation of actual risk.

Awaiting full scientific proof of an effect before taking actions to reduce emissions is not a proactive strategy and may be a particularly problematic approach for persistent chemicals since concentrations in biota will increase over time and decline only slowly after emissions have ceased. The aim should therefore be that risk reduction measures are considered before substances with these properties are emitted to the environment. On the other hand, far-reaching actions to reduce emissions based on very limited data might be associated with unnecessary monetary and social costs, if the preliminary data is later proven wrong. Consequently, there is a need to strike a balance in how much data is required for different risk management decisions, and to acknowledge that whatever criteria are used, the risk evaluation involves an interplay between science and policy.

If the three substances used as illustrations in this chapter are reasonably representative of risk assessments for persistent and bioaccumulating substances, then these examples indicate a need to improve the scientific basis of PBT and vPvB risk assessments so that industrial chemicals with these properties can be identified proactively, before they can be measured in human tissues and other top-predators. An example of a new approach focusing on the molecules' inherent reactivity has recently been proposed (Green and Bergman 2005). There is also a need for establishing a more systematic framework for assessing and handling uncertainties in the risk assessment of PBT/vPvB substances. This approach should include consideration of both uncertainties connected with the reliability and relevance of individual data and methods and those relating to the overall conclusions. In this respect the development of a systematic and widely accepted approach for utilising relevant non-standard data in chemical risk assessments (as well as in PBT/vPvB assessments within REACH) deserves special attention. Basically, these methodological improvements will mean that risk assessments to a larger extent than today would incorporate the best available scientific knowledge.

References

Breitholtz M., Eriksson J., Green N., Rudén C. (2006) 'Testing and Risk Assessment of Persistent and Bioaccumulating Chemical Substances – improvements within REACH?' *Human and Ecological Risk Assessment* 12(4):782–805.

Burton G.A., Chapman P.M., Smith E.P. (2002) Weight-of-evidence approaches for assessing ecosystem impairment. *Human and Ecological Risk Assessment* 8:1657–1673.

Council Directive (EEC) 67/548 of 27 June 1967 on the approximation of laws, regulations and administrative provisions relating to the classification, packaging and labeling of dangerous substances. *Official Journal* 196: 1–5.

ECHA (2008) Comments on Document CA/56/2008. Review of REACH Annex XIII. European Chemicals Agency. Helsinki, Finland December 12 2008 http://chemicalwatch.com/downloads/ECHA%20paper%20on%20Annex%20XIII.pdf. Accessed April 21, 2009.

European Commission (2001) European Union Risk Assessment ReportDiphenyl ether, Pentabromo derivative CAS No: 32536-52-0EINECS No: 251-087-9 1st Priority List Volume: 5.

European Commission (2002) European Union Risk Assessment ReportDiphenyl ether, Decabromo derivative CAS No: 32536-52-0EINECS No: 251-087-9 1st Priority List Volume: 17.

European Commission (2003a) Technical guidance document in support of commission directive 93/67/EEC on risk assessment for new notified substances and commission regulation (EC) No 1488/94 on risk assessment for exiting substances. Luxembourg.

European Commission (2003b) European Union Risk Assessment ReportDiphenyl ether, Octabromo derivative CAS No: 32536-52-0EINECS No: 251-087-9 1st Priority List Volume: 16.

European Commission (2004) Update of the risk assessment of bis(pentabromophenyl) ether (decabromodiphenyl ether) CAS Number: 1163-19-5 EINECS Number: 214-604-9 Final Environmental Draft of May 2004.

Green N., Bergman Å. (2005) Chemical reactivity as a tool for estimating persistance. *Environmental Science and Technology* 39: 480A–486A.

Long E.R., Chapman P.M. (1985) A sediment quality triad: measures of sediment contamination, toxicity, and infaunal community composition in Puget Sound. *Marine Pollution Bulletin* 16:405–415.

Peterman, R.M., M'Gonigle, M. (1992) Statistical power analysis and the precautionary principle. *Marine Pollution Bulletin* 24(5): 231–234.

Reineke, N. (2008) REACH must allow use of 'real world' PBT evidence. Guest column. Chemical Watch European Business Briefing.http://www.chemicalwatch.com/. Accessed April 21, 2009.

Renn O. (2008) *Risk Governance: Coping with Uncertainty in a Complex World*. London: Earthscan.

Suter G.W., Efroymson R.A., Sample B.E., Jones D.S. (eds) (2000) *Ecological Risk Assessment for Contaminated Site*. Boca Raton, FL: Lewis Publishers.

Chapter 11
Assessing Chemical Risks: Evaluating Products Rather than Substances, and the Case of Anti-fouling Paints

Britta Eklund and Jenny Karlsson

Abstract In this chapter, the case of anti-fouling paints will be used as an example on how risks might remain undiscovered due to how the regulations are used and interpreted. The example will be used to illustrate that it is important to continuously develop the legislation to avoid unwanted effects on both human and environmental health. It is also of utmost importance that the interpretation of the new legislation, especially in the beginning, which will be guidance for future decisions, is interpreted in a sound way that will lead to a sustainable use of chemicals.

Keywords Anti-fouling • Biocidal • Chemicals • Products • Risk assessment

11.1 Introduction

Clearly, the use of chemicals has given society enormous benefits, but also substantial problems since many of them pose a threat to both the environment and the human health. Chemicals are present in mostly everything used in every-day life, including cleaning materials, clothing, cosmetics, furniture, paints, etc. During the last 50 years, the world's yearly chemical production has increased dramatically from less than 10 million tonnes to over 400 million tonnes (KemI 2008). In 2007, there were more than 72,000 chemical products registered in the Swedish Chemicals Agency's product register, with more than 13,000 different chemical substances included (statistics, www.kemi.se). The majority of the chemical products are used within the industries and the households are estimated to have access to around 20% of all chemical products that are handled in the society (KemI 2008).

B. Eklund (✉) and J. Karlsson
Department of Applied Environmental Science, Stockholm University,
SE 106 81, Stockholm, Sweden
e-mail: britta.eklund@itm.su.se; jenny.karlsson@itm.su.se

J. Eriksson et al. (eds.), *Regulating Chemical Risks: European and Global Challenges*,
DOI 10.1007/978-90-481-9428-5_11,

An important source of diffusive spread of environmentally harmful chemicals are the chemical products, which are included in a large amount of goods such as building material, cars, toys and clothes (KemI 2008). The large amount of goods is difficult to survey and the diffusive spread of harmful substances from both products and goods is an increasing problem (KemI 2008). Consumer products could therefore constitute an important source to the diffusive spread of environmentally harmful substances.

Substances added to products with intended toxic effect are regulated by specific legislations. Pesticides used within agriculture are regulated according to the Plant Protection Products Directive while other biocides are regulated by the Biocidal Products Directive and encompass both products and individual chemicals (Council Directive 91/414/EEC; Council Directive 98/8/EC). There are also other chemical substances that are regulated by specific legislations such as pharmaceuticals, veterinary medicinal products, cosmetics, and food additives (e.g. Council Directive 76/768/EEC; Council Directive 89/107/EEC, Council Directive 2001/82/EC; Council Directive 2001/83/EC).

Until the adoption of the new chemical legislation, REACH, within the European Union in June 2007 there was no requirement of identification, assessment and evaluation of all other chemicals used in products. REACH will in a step-wise transition period be fully enforced by 2018. One important improvement with the new legislation is that the importers and producers now will be responsible to collect data and evaluate toxicological and ecotoxicological effects of the chemical content of their products before putting them on the market. Another important improvement is that both existing and new chemicals will be evaluated.

Due to the regulations and how these are interpreted some chemicals and products will fall between two stools. There are several borderline cases between the Biocidal Products Directive and other directives e.g. cosmetics, medicinal products for humans and veterinary medicinal products (Doc-Biocides-2002/01 Version 08.01.2008; Doc-Biocides-2002/03-rev1 24.05.2004). One example is shampoos that control dandruff (Doc-Biocides-2002/03-rev1 24.05.2004). These can be regarded as both cosmetic and biocidal products. However, from a legal point of view they cannot be cosmetic and biocidal products at the same time. Also between REACH and the other legislations problems may arise. An example is anti-fouling paints which commonly are used on boat hulls to reduce attachments of organisms. The anti-fouling paints can either be working by chemical or physical means and due to this fall under different regulations. In the first case, when a paint contains one or more substances intended to have a toxic effect on fouling organisms, it is considered a biocidal product and thus regulated according to the Biocidal Product Directive. In the second case, when a paint consist of only basic chemicals and do not contain any substances intended to have an anti-fouling effect it is considered as basic substances and accordingly regulated by REACH. But what happens when a paint intended to function by physical means, still leaks toxic substances?

11.2 Background: Anti-fouling Paints

11.2.1 History of Anti-fouling Paints Globally

Traditionally anti-fouling paints are complex mixtures to which toxic substances have been added with the purpose to prevent attachment of organisms, e.g. barnacles and algae, to surfaces submerged in the sea. Fouling organisms is a huge problem since the attached organisms cause increased drag, which significantly reduces the speed as well as impairs the manoeuvrability of the boat and thus increases fuel consumption. The problems with fouling organisms are not a new concern as it has been a problem for boat-owners in all times (WHOI 1952). Already the Phoenicians were said to use copper sheets on their boats to avoid fouling (WHOI 1952; Lunn 1974). Mankind has been inventive and solved the problems, e.g. by using anti-fouling paints containing different kinds of biocides e.g. tributyl tin (TBT), irgarol, diuron and copper. The anti-fouling paints have during the last 30 years received increased attention as high concentrations of several of the biocides have been measured in marinas all over the world (e.g. Dahl and Blanck 1996; Voulvoulis et al. 2000; Haglund et al. 2001; Viglino et al. 2004; Cornelissen et al. 2008; Eklund et al. 2008) as well as in shipping lanes (e.g. Strand et al. 2003) and unwanted effects on aquatic organisms have been observed (Alzieu et al. 1986; Bryan et al. 1986; Strand and Jacobsen 2002; Konstantinou and Albanis 2004; Magnusson et al. 2005). An example is the use of TBT in paints, which started in the 1960s (Walker et al. 2001; Yebra et al. 2004). Since it was very efficient its use increased rapidly around the world. However, later organic tin compounds were associated with the decline in oyster population and imposex in gastropods (Alzieu et al. 1986; Bryan et al. 1986). These endocrine effects lead to restrictions of TBT in many countries for use on smaller boats < 25 m (Council Directive 89/677/EEC; USEPA, 1987; Chau et al. 1997). The reason for focusing on restrictions for this group was that pleasure boats mainly sail in the near shore waters, which are the most important recruiting areas for a number of organisms. In 2001, the International Maritime Organisation (IMO 2001) adopted the International Convention on the Control of Harmful Anti-fouling Systems on Ships (the AFS Convention), which states a global ban of the use of TBT in ship paints. The convention has been ratified by 25 countries, encompassing >25% of the world tonnage, and entered into force in September 2008. Europe is leading the way and since 2003 TBT is prohibited on all European flagged ships (with the exception of the navy). Further, since January 2008 a ban exists in EU, which forbids ships to make port in European harbours if they are painted with anti-fouling paints containing organic tin compounds unless it is sealed so they cannot leak into the water (EG 782/2003).

Different copper compounds mainly replaced TBT as the active component in anti-fouling products (Dahl and Blanck 1996; Yebra et al. 2004). However, several algal species showed tolerance to copper and to achieve protection against these tolerant species, a number of so-called booster biocides, e.g. zinc pyrithione, irgarol and diuron, were introduced together with copper to enhance the antifouling effect (Voulvoulis et al. 2002). The two most commonly used booster biocides are irgarol 1051 and diuron (Konstantinou and Albanis 2004). Consequently, elevated concentrations of both irgarol and diuron have been detected all over the world in areas with intense boat traffic, particularly in marinas and harbours (Dahl and Blanck 1996; Voulvoulis et al. 2000; Haglund et al. 2001; Thomas et al. 2001, 2002; Okamura et al. 2003). Also, laboratory studies with irgarol and diuron have shown negative effects on aquatic organisms at very low concentrations (Okamura et al. 2000; Férnandez-Alba et al. 2002; Karlsson et al. 2006). Because of the high concentrations in coastal waters and negative impact on the aquatic life some countries, e.g. Sweden, UK and Denmark, have restricted or banned the use of these two anti-fouling agents (KemI 1992; Thomas et al. 2002; Danish EPA 2008).

11.2.2 History of Regulations in Sweden

Ever since the ban of TBT containing paint for use on pleasure boats in 1989 (Council Directive 89/677/EEC), the regulations in Sweden have gradually been strengthened as several of the active substances have shown to constitute a risk to both human health and the aquatic environment. The approval of existing anti-fouling substances has been surveyed twice and after risk/benefit analysis new restrictions was adopted in 1992 (KemI 1992) and 1998 (KemI 1998a, b).

The toxicity of many compounds, especially metals, are strongly dependant on salinity and since Sweden has a long coastline, ranging from limnic to fully marine conditions, the restrictions were divided in three different areas i.e. West coast (Norwegian border – Trelleborg), East Baltic Sea (Trelleborg- Örskär) and Bothnian Sea-Bothnian Bay. The main reason for this is that the problems with fouling organisms are reduced in lower salinities (KemI 1992, 1998a, b). Further, as the risks are considered higher in the near coastal areas, firmer restrictions were introduced for anti-fouling paints for use on smaller boats < 12 m, i.e. most pleasure boats, than for ships, which mainly operate on the open oceans.

In the 1992 decision, the benefits were not considered to outweigh the risk in order to allow any chemically acting anti-fouling paints for use in fresh waters or in the northern part of the Baltic Sea, i.e. the Gulf of Bothnia, and on boats weighing less than 200 kg, which easily can be hauled ashore. Due to its carcinogenic effects, diuron was no longer approved as active substance in any anti-fouling paints for use on any boats. Further, isothiazoline was not approved for use on any boats shorter than 25 m. Copper was approved for pleasure boats only if the leakage rate was less than 75 and 150 μg Cu/cm^2 for the first 14 days for use on the east coast and west coast, respectively (KemI 1992).

In the next risk/benefit analysis performed in 1998 the approval and use of anti-fouling paints was further restricted much due to new information on the risks of copper in less saline waters (e.g. Andersson and Kautsky 1996; KemI 1998a, b). It was decided that none of the copper leaking anti-fouling paints would be allowed for use on the east coast for boats shorter than 12 m. Such paints were only allowed on the west coasts and only if the leakage rate was below 200 μg/cm^2 during the first 14 days and less than 350 μg/cm^2 up to 30 days. To let the market adapt to the new regulation a transition period was permitted during which the salesmen could sell out their stock. After this period the public was allowed to use their personal stock of old paint. In the new regulation, all boats shorter than 12 m were covered by the same rules, regardless if they were used as pleasure boats or for commercial activities. For boats longer than 12 m, copper with maximal steady state leakage rate of 55 μg/cm^2/day was allowed. Higher rates were allowed on vessels mainly sailing on the oceans (KemI 1998a, b).

11.2.3 The Situation of Today on Anti-fouling Paints in Sweden

The intention of the authorities, with the restriction of chemical leaking paints was to reduce the release of toxic substances to the sensitive productive coastal areas (Kemi 1998a). The Baltic Sea is considered an environmentally sensitive area, which later was acknowledged internationally by the International Maritime Organisation (IMO 2005) and classified as a Particularly Sensitive Sea Area (PSSA) in 2004 (IMO 2005). This was based on its unique environment, with low salinity and low biodiversity. Also, the Baltic Sea is in an evolutionary perspective a young sea and the organisms live in a very stressful environment, as they have not had time to fully adapt to the brackish water conditions (Kautsky and Tedengren 1992; Rydén et al. 2003).

Today in Sweden, the active substances used in anti-fouling preparations for use on pleasure boats on the west coast are mainly different kinds of copper compounds alone or in combination with the herbicide irgarol. For larger boats (>12 m), anti-fouling paints containing the biocides zinc pyrithione and isothiazoline are also allowed for use.

Instead of chemically leaking paints, especially for use in the Baltic Sea, the aim was to direct the protection of hulls to physical means. As a result of the new regulation, a number of new paints entered the market. These paints claimed to be friendlier to the environment as they were said not to contain any biocides and function by mere physical means. The most common types are polishing paints and silicone based paints. The polishing paints have a surface layer, which is continuously peeling off and with this any fouling organisms. The silicone based paints have a surface structure, which makes it difficult for the fouling organisms to attach tightly and thus they will easily be washed off when the boat speeds through the water.

11.3 The Biocidal Products Directive, REACH and Pitfalls

On the EU level, biocidal products are mainly regulated by the Biocidal Products Directive (98/8/EG), which came into force in 2000. A biocidal substance or product is defined as: 'Active substances and preparations containing one or more active substances, put up in the form in which they are supplied to the user, intended to destroy, deter, render harmless, prevent the action of, or otherwise exert a controlling effect on any harmful organism by chemical or biological means.'

Therefore, according to the Biocidal Products Directive, it is only when a substance deliberately is added to a product in order to have a biocidal effect, that an approval from the national authorities is required. The aim of the Biocidal Products Directive is to harmonize the market for biocidal products within Europe. This means that a substance used in a certain product approved in one country can apply in any other European country and by mutual recognition, the product will almost automatically be approved for use also in this new country. Thus, a full assessment of the active substance in a certain product will only be carried out in the country responsible for this substance. This evaluation process of different products is in progress and in all, 23 different product types will be evaluated and classified within the Biocidal Products Directive. To begin with, wood preservatives for use in building and construction materials and rodenticides are evaluated. The next product group is anti-fouling products. All manufacturers of these types of products have to register the active compounds that they intend to use in their products along with substantial documentation of the biocidal action (see mainly annex IIA, IIB in the Biocidal Product Directive, 98/8/EG). For these mentioned product groups the time limit for the registration period has ended and now the competent authorities in the European countries are in the process of evaluating the delivered documentation. The responsibility for the evaluation of different registered active substances has been divided among the member countries. For instance France is responsible for copper which has been registered as active substance both for use as wood preservative and in anti-fouling products. Even if the time limit for documentation of wood preservatives was 28th of March 2006 at the latest no decision has yet been taken for any of the substances and products which are in the process.

When an active substance has been approved for use in a certain product the manufacturer can apply for marketing of his product in any other European country and claim mutual recognition of the product. The aim is that this should be more like a formal registration. However, in theory, each country has the possibility to claim special consideration as to why not to apply the mutual recognition. This could be for instance, that the Baltic Sea is considered as an area of special concern or that the low temperature in a Nordic country slows down the degradation rate of a substance and therefore might pose a greater risk than in a warmer country. As the process has not yet come so far this clause has never been tested. Copper in anti-fouling paint might be the first case to test the possibility for a member state to have more strict rules than the other member states. Today Sweden has special and exclusive regulations compared to other European countries concerning the use of copper in anti-fouling paints for use on smaller boats sailing in the Baltic Sea. If France is approving copper in anti-fouling paints, the producers of such paints will probably apply for mutual recognition also in Sweden. When this happens Sweden has a chance to claim special consideration in order to keep the more strict regulation of the nation. The question is how simple this trial is and if a nation actually is able to apply more strict rules than the rest of the European countries.

All chemicals within EU, which do not fall under any of the former mentioned legislations or are exempted by special paragraphs in REACH (EC 1907/2006), are since 1st of June 2007 regulated by REACH. The information requirements and the time limit for registration vary according to the tonnage in which the substance is manufactured and are set at 1–10, 10–100 and >100 t. The producer is required to register and deliver a technical dossier describing the inherent properties of the substance to the European Chemicals Agency (ECHA) if produced in amounts over 1 t per year. For chemicals produced in more than 10 t per year, chemical safety assessment and exposure scenarios are required for substances classified as PBT (Persistent, Bio-accumulative, Toxic) and vPvB (very Persistent and very Bio-accumulative).

The basis for which regulation that will be applicable is that it is up to the producer to decide whether any substances with a biocidal purpose have been added and whether an approval is needed according to the Biocidal Products Directive. In the case when no biocides are added, REACH will regulate the assessment of the chemicals involved and the amount produced will determine the requirements for testing. In the Biocidal Products Directive on the other hand, there is no lower limit i.e. all chemicals added as biocides shall be tested no matter what amount they are produced. The different requirements on documentation in the Biocidal Products Directive compared to in REACH may have the consequence that some chemicals that are toxic or have other unwanted properties, but only produced in lower amounts can be used in certain products without any evaluation. This will be illustrated with the experiments on anti-fouling paints claimed to function by physical means.

11.4 Experimental Results of Toxicity from Physically Working Paints

The toxicity of ten paints for use on pleasure boats on the east coast of Sweden was investigated. Nine of the paints were so-called physically working paints, while one (EcoMar 2000) was an, at that time, approved chemically working paint (www.kemi.se). The toxicity of the paints was compared to a banned paint based on leakage of the toxic agents copper and irgarol. Leakage waters from the different paints were produced by placing a painted piece in artificial seawater during 14 days (Karlsson and Eklund 2004; Karlsson et al. 2006). The different leakage waters were tested on two common brackish water organisms, the red macro alga *Ceramium tenuicorne* and the harpactacoid copepod *Nitocra spinipes*. On the alga a growth inhibition test (Eklund 2005 ISO 2010) was performed and on the crustacean a mortality test (SIS 1991) was carried out.

The results are compiled in Table 11.1 and show the EC/LC50 values (the concentration where the growth of the alga was inhibited 50% or where 50% of the animals were dead) of the two test methods. The results showed that leakage water from several of the so-called physical working paints were toxic to the two test

Table 11.1 EC50 and LC50 values (% leakage water) for C. tenuicorne and N. spinipes exposed to leakage waters (2 weeks of leakage) of anti-fouling paint. In brackets, 95% confidence intervals (CI) are presented

	Ceramic	Nitocra	
	EC_{50} (%)	LC_{50} (%)	
Paint	(95% CI)	(95% CI)	Function
Lefant H2000[a] (Lotérc AB)	2.2 (1.2–3.5)	57 (49–68)	Polishing
Mille light[a] (Hemple Färg AB)	2.2 (1.6–3.2)	55 (31–150)	Polishing
Micro Eco[a] (International Färg AB)	0.60 (0.55–0.65)	1.1 (0.91–1.4)	Polishing
SSC-44[a] (US Gloss Europe AB)	0.91 (0.17–0.22)	9.0 (7.6–11)	Waxy with scale structure
Lefant SPF[b] (Lotérc AB)	0.56 (0.46–0.69)	49 (31–96)	Polishing
Cruiser Eco[b] (International Färg AB)	0.35 (0.27–0.46)	65 (54–80)	Polishing
SafeBoatskin[b] (Sailway)	>100	>80	Polymer wax
Aurora VS721[b] (Thulica AB)	>100	>80	Wax
EcoMar2000[b, c] (Thulica AB)	>100	>80	Capsaicin (pepper extract)
Vc17mEco[b] (International Färg AB)	>100	>80	Teflon
Cruiser Superior[a,d] (International Färg AB)	>100	>80	Copper and irgarol

[a]Karlsson and Eklund (2004) New biocide-free anti-fouling paints are toxic. Marine Pollution Bulletin 49: 456–464.
[b]Karlsson et al. (2006). A practical ranking system to compare toxicity of anti-fouling paints. Marine Pollution Bulletin 52: 456–464.
[c]Approved by the Swedish Chemicals Agency (www.kemi.se.)
[d]Reference paint containg copper and irgarol.

organisms and in many cases as toxic as the reference paint, which contained copper and irgarol. Two of the paints were even more toxic to the crustacean than the reference paint, where only 1.1% and 9% of the leakage waters killed half of the animals compared to 30% leakage water from the copper and irgarol leaking reference paint (Karlsson and Eklund 2004; Karlsson et al. 2006). The paints that exhibited toxic effects were generally polishing paints. Positively, the results indicated that three of the physically working paints, SafeBoatskin, Aurora VS721 and Vc17m Eco, did not show toxic effects to the two organisms. Even 100% leakage water did not affect the alga and neither did the highest test concentration of 80% affect the crustacean. These paints were based on teflon and wax. Also, the leakage water from the approved chemically working paint, EcoMar2000, did not affect the crustacean and alga, negatively.

11.4.1 How Can Products Only Containing Basic Chemicals Be Toxic?

The experimental study with boat paints clearly shows that leakage waters from several of the so-called physical working paints contained something that negatively affected the two test organisms. Some paints were even more toxic than the reference paint, which contained the biocides copper and irgarol. This was unexpected since physically working paints were supposed not to leak toxic substances. So how is it possible that a product only containing basic chemicals still is toxic?

11.4.1.1 Effects of Regulations

The main explanation is that since the producers have not deliberately added anything to the product with the purpose of having an inhibiting effect on living organisms, the product is not a biocidal product. In this case the requirements regarding, e.g. documentation are very different from when a biocide has been added. When a biocide has been added the producer is obliged to provide the authorities with substantial documentation before the product can be approved and put on the market. However, when the producer claims that the chemicals added to a product have no chemical or biological action no approval is needed and in this case it is sufficient to register the product with the national chemical authorities. To the Swedish products register, only substances included over 5% by weight must be specified unless they are classified as dangerous to human health or the environment. Exempted are substances already classified as toxic, mutagenic, carcinogenic, allergenic, harmful to reproduction, bio-accumulative and persistent, for which there are no lower limit and information on all such included substances should be specified (KIFS 2008:2).

Some of the physically working paints on the market in 2001 were tested in ecotoxicological tests and several were found toxic even if they did not contain any biocide and only consisted of basic chemicals. Thus, this is an example of a mixture/product consisting of basic chemicals but still produce a toxic leachate. Consequently, if these physically working anti-fouling paints had not been tested the toxicity would not have been discovered. The same has been shown by Pettersson et al. (2000) who found toxicity of varying degree to Daphnia from different detergents and softener products. This risk is apparent also with a number of other mixtures and products commonly used besides paints, e.g. cleaning products for households, cars and boats, defrosting chemicals, dispersion agents, glues. The example with boat paints demonstrates a risk that products can be distributed on the market, resulting in release and discharge of toxic substances in the environment without the awareness of both the society and the authorities. In theory, since the REACH regulation exempt registration and any documentation of production of substances produced in lower quantities than 1 t per year and legal unity it could open up for the production of very potent substances. One of the most potent substances known today is tributyltin (TBT). According to the proposed quality standards by the Common Implementation Strategy for the Water Frame Directive (2005) the safe level for the pelagic community is 0.0002 μg TBT/L. This means that one ton of a substance with comparable potency could be used in up to 5,000,000 L and then just be below what is considered as safe level. Hence, a producer with no sense of responsibility could produce very toxic substances in quantities below 1 t and not be obliged to register the substance. The consequences could be a repetition of the stories with e.g. TBT, DDT and PCB as a worst case scenario.

11.4.1.2 Toxic Chemicals Added for Other Reasons Than Biocidal

Apart from possible biocides, anti-fouling paints normally also contain, e.g. solvents, preservatives, fixing agents. The paints exhibiting toxic effects in the study were the polishing paints and at least three of the polishing paints (Micron Eco, Cruiser Eco and Lefant SPF) contain zinc oxide according to declarations on the cans and their safety data sheet. This substance has recently been classified as harmful to the environment and may therefore be one explanation to the observed toxicity. The question is whether this should be regarded as a biocide in future formulations of anti-fouling paints?

Currently in Sweden, preservatives in chemical products may be imported and handled without authorization from the Swedish Chemicals Agency even though an authorization of these kinds of chemicals will be required in the future (KemI 2005, KIFS 2008:3 Chap.4, §4). Therefore, preservatives in paints are not prohibited today and it cannot be excluded that these may have a biocidal effect after application of the paints. According to the safety data sheet found on the website of the producer, Micron Eco contains the biocide zinc pyrithione. In this case, the intention of the addition of zinc pyrithione could be as a preservative and not as an anti-fouling agent. Some of the other paints have content lists including, e.g. solvents and fixing agents, whereas the contents in other paints are unknown to the consumer. Thus, the observed toxicity could be due to other substances added to the paint.

Similarly, special agents are added in smaller amounts for specific reasons to a number of products. In tires for example, special substances are added to protect the degradation of the rubber and other are acting to increase the elasticity (Wik and Dave 2005, 2006). Similarly, special agents are added to products made of plastic to increase wanted specific properties of the end-product (Lithner et al. 2009). The results show that within each tested product group leachates of varying toxicity have been detected, some very toxic and other not toxic.

On the initiatives of independent researchers from universities, ecotoxicological tests have been performed on the whole product in a similar way as what has been done with the anti-fouling paint. The results show that within each group several products have been detected, which produces a toxic leachate that is diffusively spread in the environment. These examples show that it is possible to discriminate between toxic and non-toxic products.

11.4.1.3 Synergistic Effects

Another explanation to the observed toxicity in some of the paints could be that the included chemicals exhibit synergistic effects and thereby together are more dangerous than the additive effect of the single chemicals. Impurities of the included chemicals could also be part of an explanation.

We argue that it is of great importance to pay extra attention to products likely to be distributed in the environment, e.g. boat paints, discharges from tires, cleaning agents for boats and cars, anti-frost compounds, and that all such products, regardless if they contain biocides or not, should be tested in the form they actually are used, i.e. the whole product. In this way consideration is taken to all possible additive or synergistic effects, which are not considered if only the single chemicals are evaluated.

11.5 Classification of Substances and Labelling of Products

Single substances are classified according to their harmfulness. This classification forms the basis for the risk phrases that the substance should be labelled with. Before July 2002 only single substances needed to be labelled with an appropriate risk symbol. After this date also products should be labelled (Council Directive 67/548/EEC). The primary aim of the labelling is to inform the consumer on the risks with the product. However, the labelling depends on which substances that are included and in what concentrations they are used and consequently not on the whole combined product. This means for example, that all paints containing zinc oxide, which today is classified as 'dangerous to the environment' needs to be labelled with the 'dead fish and dead tree' symbol (KemI 2005). Since most physical working paints contain zinc oxide this is most likely the reason for the risk symbol. This labelling has resulted in unclearness among the consumers about the physically working paints. On one hand the product is claimed to function by physical means and with that understood by laymen not to leak toxic substances, and on the other hand it is labelled with the 'dead fish and dead tree' symbol, which means that the product contains substances that are dangerous to the environment. Since the physical working paints at first were promoted as environmental friendly it does not rhyme well with the label. The same situation may arise with other products as well. We claim that a labelling that is referring to the whole product would enhance the risk communication to the common man and the society.

11.6 Conclusion

The example with physical working paints show that these may be as toxic as the paints functioning by leaking added biocides and that a risk exists on unexpected spread of toxic substances. This may be the case for many other compounds and products, e.g. plastics, tires, detergents. This risk can be minimized if:

- All products likely to be spread in the environment are classified and risk assessed.
- The leakage from the whole product is the base for classification instead of only single substances.

A classification of whole products would minimize the risk of not discovering possible additive and synergistic effect of chemical mixtures. A labelling of the risk of the whole product would be less confusing and increase the clarity of the risk communication to consumers.

References

Alzieu, C.L., Sanjuan, J., Deltreil, J.P., Borel, M. (1986) Tin contamination in Arcachon Bay: effects on oyster shell anomalies. *Marine Pollution Bulletin* 17: 494–498.

Andersson, S., Kautsky, L. (1996) Copper effects on reproductive stages of Baltic Sea Fucus vesiculosis. *Marine Biology* 125: 171–176.

Bryan, G.W., Gibbs, P.E., Hummerstone, L.G., Burt, G.R. (1986) The decline of the gastropod Nucella lapillus around south-west England: evidence for the effect of tributyltin from anti-fouling paints. *Journal of the Marine Biological Association of the UK* 66: 611–640.

Chau, Y.K., Maguire, R.G., Brown, M., Yang, F., Hummerstone, L.G. (1997) Occurrence of organoton compounds in the Canadian aquatic environment five years after the regulation of anti-fouling uses of tributyltin. *Water Quality Research Journal of Canada* 32: 453–521.

Common Implementation Strategy for the Water Frame Directive (2005) Environmental Quality Standards (EQS). Substance Data Sheet. Priority Substance No 30, Tributyltin compounds (TBT-ion) Cas-No 688-73-3 (36643-28-4).

Cornelissen, G., Pettersen, A., Nesse, E., Eel, E., Helland, A., Breedveld, G.D. (2008) The contribution of urban runoff to organic contaminant levels in harbour sediments near the Norwegian cities. *Marine Pollution Bulletin* 56: 565–573.

Council Directive 67/548/EEC of 27 June 1967 on the approximation of laws, regulations and administrative provisions relating to the classification, packaging and labelling of dangerous substances.

Council Directive 76/768/EEC of 27 July 1976 on the approximation of the laws of the Member States relating to cosmetic products.

Council Directive 89/107/EEC of 21 December 1988 on the approximation of the laws of the Member States concerning food additives authorized for use in foodstuffs intended for human consumption.

Council Directive 89/677/EEC of 21 December 1989 amending for the eighth time Directive 76/769/EEC on the approximation of the laws, regulations and administrative provisions of the member states relating to restrictions on the marketing and use of certain dangerous substances and preparations. 31989L0677R(01) and (02).

Council Directive 91/414/EEC of 15 July 1991 concerning the placing of plant protection products on the market.

Council Directive 98/8/EC of the European Parliament and of the council of 16 February 1998 concerning the placing of biocidal products on the market.

Council Directive EU 2001/83/EC of the European Parliament and of the Council on the Community code relating to medicinal products for human use.

Dahl, B., Blanck, H. (1996) Toxic effects of the anti-fouling agent 1051 on Periphyton communities in coastal water microcosms. *Marine Pollution Bulletin* 32: 342–350.

Danish EPA. http://www.mst.dk/Bekaempelsesmidler/Biocider/For+fritidssejlere/Regler/01070101.htm Cited in June 2008.

Directive 2001/82/EC of the European Parliament and of the Council on the Community code relating to veterinary medicinal products.

Doc-Biocides-2002/01 Version 08.01.2008, Guidance document agreed between the Commission services and the competent authorities of the Member States for the Biocidal Products Directive 98/8/EC and for the Medicinal Products for Human Use Directive 2001/83/EC and the Veterinary Medicinal Products Directive 2001/82/EC Borderline between Directive 98/8/EC concerning the placing on the market of biocidal products, Directive 2001/83/EC concerning medicinal products for human use and Directive 2001/82/EC concerning veterinary medicinal products.

Doc-Biocides-2002/03-rev1 24.05.2004, Guidance document agreed between the Commission services and the competent authorities of Member States for the biocidal products Directive 98/8/EC and for the cosmetic products Directive 76/768/EEC Borderline between Directive 98/8/EC concerning the placing on the market of biocidal products and Directive 76/768/EEC concerning cosmetics products.

EC 1907/2006 of the European Parliament and of the council of 18 December 2006 concerning the Registration, Evaluation, Authorisation and Restriction of Chemicals (REACH), establishing a European Chemicals Agency, amending Directive 1999/45/EC and repealing Council Regulation (EEC) No 793/93 and Commission Regulation (EC) No 1488/94 as well as Council Directive76/769/EEC and Commission Directives 91/155/EEC, 93/67/EEC, 93/105/EC and 2000/21/EC.

EG 782/2003 of the European Parliament and of the Council of 14 April 2003 on the prohibition of organotin compounds on ships. 32003R0782.

Eklund, B. (2005) Development of growth inhibition test with the marine and brackish water red alga Ceramium tenuicorne. *Marine Pollution Bulletin* 50: 921–930.

Eklund, B., Elfström, M., Borg, H. (2008) TBT originates from pleasure boats in Sweden in spite of firm restrictions. *Open Environmental Journal* 2: 124–132.

Férnandez-Alba, A.R., Hernando, M.D., Piedra, L., Chisti, Y. (2002) Toxicity evaluation of single and mixed anti-fouling biocides measured with acute toxicity bioassays. *Analytica Chimica Acta* 456: 303–312.

Haglund, K., Pettersson, A., Peterson., M., Kylin, H., Lord, S.C. and Dollenmeier, P. (2001) Seasonal distribution of the anti-fouling compound Irgarol 1051 outside a marina in the Stockholm archipelago. *Bulletin of Environmental Contamination and Toxicology* 66: 50–58.

IMO (2001) *International Convention on the Control of Harmful Anti-Fouling Systems on Ships*, IMO (the AFS Convention). London 18 October 2001.

IMO (2005) Revised guidelines for the identification and designation of Particularly Sensitive Sea Areas (PSSAs). IMO resolution A. 982(24).

International Standardisation Organisation (ISO) (2010) Water Quality—Growth inhibition test with the brackish water and marine macroalga *Ceramium tenuicorne*. ISO 10710.

Karlsson, J., Eklund, B. (2004) New biocide-free anti-fouling paints are toxic. *Marine Pollution Bulletin* 49: 456–464.

Karlsson, J., Breitholtz, M., Eklund, B. (2006) A practical ranking system to compare toxicity of anti-fouling paints. *Marine Pollution Bulletin* 52: 456–464.

Kautsky, N., Tedengren, M. (1992) Ecophysiological strategies in Baltic Sea invertebrates. In: Björnestad E., Hagerman L. and Jensen K. (eds.) *Proceedings of the 12th Baltic Marine Biologists Symposium*. Olsen and Olsen, Fredensborg.

KemI (1992) Antifouling Products – Pleasure boats, commercial vessels, nets, fish cages and other underwater equipment. Eds. Debourg, C., Johnson, A., Lye, C., Törnqvist, L. and Unger, C. (In English KemI Report No 2/93). Stockholm.

KemI (1998a). *Biocidprocessen, Antifoulingprodukter. Fritidsbåtar*. Eds.Eriksson, U., Lindgren, P., Olsson, B. and Unger, C. Kemikalieinspektionen PM 1998. Stockholm (In Swedish).

KemI (1998b) *Biocidprocessen, Antifoulingprodukter för yrkesmässigt bruk, Fartyg längre än 12 meter*. Eds. Lindgren, P., Olsson, B. and Unger, C. (In Swedish).

KemI (2005) *Båtbottenfärger till fritidsbåtar. Ett inspektionsprojekt*. KemI PM 3/03. Stockholm (In Swedish).

KemI (2008) *Underlag till den andra fördjupade utvärderingen av miljökvalitetsmålet Giftfri miljö*. Stockholm (In Swedish).

KIFS (2008:2) *The Chemical Products and Biotechnical Organisms Regulations*. Last amendment is KIFS 2008: 5.

KIFS (2008:3) Kemikalieinspektionens föreskrifter (KIFS 2008:3) om bekämpningsmedel. Stockholm.

Konstantinou, I. K., Albanis, T.A. (2004) Worlwide occurrence and effects of anti-fouling paint booster biocides in the aquatic environment: a review. *Environment International* 30: 235–248.

Lithner, D., Damberg, J., Dave, G. Larsson, Å. (2009) Leachates from plastic consumer products – screening for toxicity with Daphnia magna. *Chemosphere* 74: 1195–1200.

Lunn, I. (1974) *Antifouling: A Brief Introduction to the Origins and Developments of the Marine Antifouling Industry*. BCA, Thame.

Magnusson M, Borgegren A, Granmo Å (2005) *Investigation of imposex on the Swedish west coast using Nassarius nitidus as a bioindicator species 2005*. Report to the Swedish Environmental Protection Agency 2005, Stockholm, Sweden.

Okamura, H., Aoyama, I., Liu, D., Maguire, R.J., Pacepavicius, G.J., Lau, Y.L. (2000) Fate and ecotoxicity of the new anti-fouling compound Irgarol 1051 in the aquatic environment. *Water Research* 34: 3523–3530.

Okamura, H., Aoyama, I., Ono, Y., Nishida, T. (2003) Anti-fouling herbicides in the coastal waters of western Japan. *Marine Pollution Bulletin* 47: 59–67.

Pettersson, A., Adamsson, M., Dave, G. (2000) Toxicity and detoxification of Swedish detergents and softener products. *Chemosphere* 41: 1611–1620.

Rydén, L., Migula, P., Andersson, M. (2003) *Environmental Science*. A Baltic University Publication, Sweden.

SIS (1991) *Determination of acute lethal toxicity of chemical substances and effluents to Nitocra spinipes Boeck- Static procedure*. Swedish standard SS 02 81 06, Stockholm, Sweden.

Strand, J., Jacobsen, J.A. (2002) Imposex in the sublittoral neogastropods common whelk (Buccinium undatum) and red whelk (Neptunea antiqua) in Danish water. *Marine Ecology Progress Series* 244: 171–177.

Strand, J., Jacobsen, J.A., Pedersen, B., Granmo, Å. (2003) Butyltin compounds in sediment and mollusks from the shipping strait between Denmark and Sweden. *Environmental Pollution* 124: 7–15.

Thomas, K.V., Fileman, T.W., Readman, J.W., Waldock, M.J. (2001) Anti-fouling paint booster biocides in the UK coastal environment and potential risks of biological effects. *Marine Pollution Bulletin* 42: 677–688.

Thomas, K.V., McHugh, M., Waldock, M. (2002) Antifouling paint booster biocides in UK coastal waters: inputs, occurrence and environmental fate. *The Science of the Total Environment* 293: 117–127.

USEPA, US Congress (1987) The organising antifouling paint control act of 1988 (33 USC 2401). Public law 1988: 100–333.

Viglino, L., Pelletier, E., St-Louis, R. (2004) Highly persistent butyltins in northern marine sediments: a long-term threat for the Saguenay Fjord (Canada). Environmental Toxicology and Chemistry 23: 2673–2681.

Voulvoulis, N., Scrimshaw, M.D., Lester, J.N. (2000) Occurrence of four biocides utilized in antifouling paints, as alternatives to organotin compounds, in waters and sediments of a commercial estuary in the UK. *Marine Pollution Bulletin* 40: 938–946.

Voulvoulis, N., Scrimshaw, M.D., Lester, J.N. (2002) Comparative environmental assessment of biocides used in anti-fouling paints. *Chemosphere* 47: 789–795.

Walker, C.H., Hopkin, S.P., Sibly, R.M., Peakall, D.B. (2001) *Principles of Ecotoxicology*, 2nd edition. Taylor & Francis, New York.

Wik, A., Dave, G. (2005) Environmental labelling of car tires- toxicity to Daphnia magna can be used as a screening method. *Chemosphere* 58: 645–651.

Wik, A., Dave, G.(2006) Acute toxicity of leachates of tire wear material to Daphnia magna – variability and toxic components. *Chemosphere* 64: 1777–1784.

Woods Hole Oceanographic Institution (WHOI) (1952) Marine fouling and its prevention. US Naval Institute, Annapolis, Maryland. http://hdl.handle.net/1912/191

Yebra DM, Kiil S, Dam-Johansen K. (2004) Antifouling technology – past, present and future steps towards efficient and environmentally friendly antifouling coatings. *Progress in Organic Coatings* 50: 75–104.

Part II
Chemical Regulation: Politics, Policy and Management

Chapter 12
Global Trends in Chemicals Management

Gunnar Bengtsson

Abstract International cooperation in chemical safety has existed for more than a 100 years. Over the past half century it has grown considerably, and there are now upwards of one hundred international agreements, programs and initiatives concerning chemical safety. Within a few decades, almost all chemicals of significance will have been assessed for their hazards, and a global system will ensure that chemical substances and preparations are classified and labelled in a uniform way with respect to their hazards. This chapter discusses the challenges and trends in global chemical management with focus on three major issues: (1) the difficulties for both developed and developing countries to live up to the many international agreements, particularly with respect to reducing risks, (2) the difficulties for non-governmental stakeholders to keep abreast of the enormously complex information and to contribute to chemical safety development, and (3) how to best cope with the risks from the many substances that emerge from various sources with volumes doubling within a generation and reaching the environment and man via a multitude of pathways. Addressing these issues requires new policy developments. In this respect, the chapter examines the potential for the global chemical safety framework to adopt some of the regulatory and economic instruments in use in global radiation safety.

Keywords Chemicals • Development • GHS • Global • International cooperation

12.1 Drivers for International Cooperation in Chemical Safety

Toxic chemicals were used thousands of years ago in medicine, for suicide, and to poison adversaries in warfare and in struggles for power. Regulatory control of chemicals in most countries began with a focus on specific chemicals known to be

G. Bengtsson (✉)
Luruddsvägen 32, Ekero SE-17854, Sweden
e-mail: gunnarbengt@telia.com

J. Eriksson et al. (eds.), *Regulating Chemical Risks: European and Global Challenges*,
DOI 10.1007/978-90-481-9428-5_12, © Springer Science+Business Media B.V. 2010

toxic in the context of criminal law. Over the past two centuries, as countries have developed, so has legislation to control hazardous chemicals in pharmaceuticals, occupational safety, food quality and clean air and water. This approach has, during the past half century, been broadened to consider most chemicals as potentially hazardous (Lönngren 1992). In the following, several factors that drive global cooperation in chemical safety are discussed.

12.1.1 Developed Countries Lead Legislation and Its Implementation

Today, extensive chemical safety legislation and reasonably effective implementation exist in two of the blocks of countries that dominate international negotiations: the JUSCANNZ block (Japan, USA, Switzerland Canada, Australia, Norway and New Zealand) and the 27 countries of the European Union (EU). These are referred to here, for the sake of abbreviation, as developed[1] countries, although this is an oversimplification used in order to facilitate description of the trends in chemical safety. There is less extensive legislation in the G77 block of developing countries (Group77 + China: the Group of 77 developing countries plus China, which in reality comprises about 130 developing countries), and its implementation is often inadequate. These countries together with countries with economies in transition are referred to here, for the sake of abbreviation, as developing countries. In general, the legislation in these countries is advancing along the route travelled by the countries of the first two blocks, starting with some of the most hazardous substances. Much of the global work on chemical safety deals with bridging the gap between the countries that have extensive chemical safety legislation and those that do not.

12.1.2 Developing Countries Suffer the Worst Effects of Chemicals on Health

The health impacts of chemicals are significant, though difficult to quantify. Some estimates are available (WHO 2002) indicating that about 5% of the global burden of disease can be attributed to environmental chemicals exposures. The distribution of effects across the globe is the result of the combined effect of the volume of chemicals use and the effectiveness of chemical safety measures. In the countries of the OECD (Organisation for Economic Co-operation and Development) in general

[1]The designations 'developed' and 'developing' are intended as broad descriptions and do not necessarily express a judgement about the stage reached by a particular country or region in the development process. Reference to developing countries is generally intended to include countries with economies in transition as well.

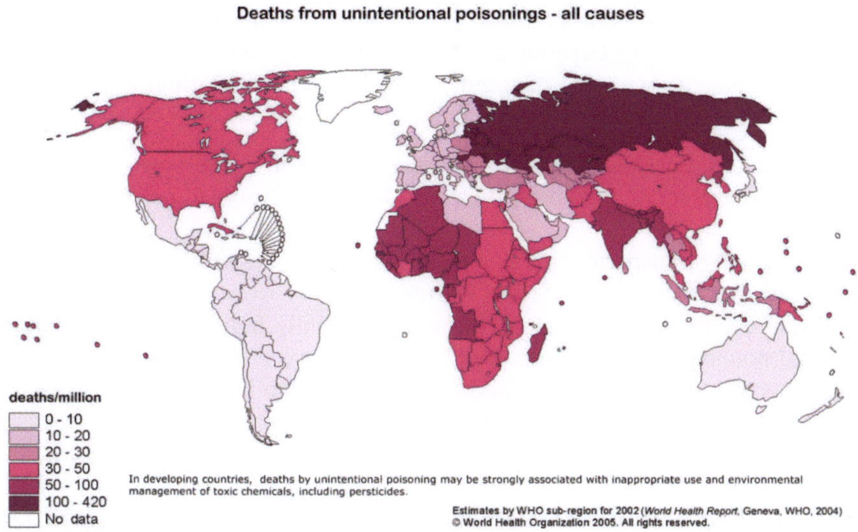

Fig. 12.1 Differences in annual death rates due to unintentional poisonings. Data are for many countries based on scant information. Still, it appears that the highest rates occur in some countries in Western Africa, north-east Asia and south-east Asia. (Reproduced with permission from WHO 2005)

the use of chemicals per person is more than tenfold compared to non-OECD countries, but the risks are compensated for by more well developed chemical safety systems. Fig 12.1 illustrates differences in the rate of unintentional poisoning across the globe. Unintentional poisonings were in 2002 estimated to kill 355,000 people globally each year (WHO 2005), and two-thirds of these deaths occurred in developing countries. Such poisonings are strongly associated with excessive exposure to, and inappropriate use of, toxic chemicals. In general, chemical safety in developing countries leaves much to be desired. Accidents are abundant, although statistics are often of poor quality. An unfortunate milestone in this regard is provided by the accident in Bhopal, India in 1984 where a chemical factory accidentally released methyl isocyanate into the air, killing thousands of people.

12.1.3 Chemicals Cross National Borders

There are several reasons why international cooperation is sought for chemical safety. The first international instrument for restricting chemicals use may have been the St. Petersburg Declaration from 1868 dealing with the use of fulminating substances, a type of explosive. The Brussels International Formulary dealing with pharmaceuticals was concluded in 1906. The First Hague Convention on Exercising Control over Opium was signed in 1912. These examples show that similar to the national level, warfare and pharmaceuticals were of early interest with regard to international control. Occupational safety, food quality and clean air and water followed as areas of interest (Lönngren 1992). Reasons for instituting international control of chemicals risks include that hazardous chemicals may be

- Manufactured across the globe, and increasingly in developing countries with less stringent safety measures, by a workforce that is sometimes transnationally recruited
- Intentionally transported across national borders in the form of chemical preparations, manufactured goods and wastes
- Released into the environment and unintentionally transported across international boundaries through air, water and migratory species and endanger living organisms, contaminate food and water and lead to health concerns that have been noted even in pristine Arctic ecosystems

12.1.4 Production and Use Move Towards Developing Countries

The chemicals industry is very diverse, producing thousands of substances that are used by other industries and that are present in countless consumer products. Approximately 600 million tonnes are produced annually, in addition to some 4,000 million tonnes of petroleum products. An earlier study (OECD 2001) predicted that the industry would continue to expand over the next 20 years, with faster growth rates in non-OECD countries. The demand for chemicals would, over that period, more than double in the non-OECD countries, while it would increase by about half that rate in the OECD countries. Still, come 2020 the OECD countries would account for twothirds of the world demand. These trends have been essentially confirmed by later studies (CEFIC 2008).

Trade across regions would continue to be large, almost 30% of the total production, and even more is traded across national borders. The trends for chemical safety in the OECD study (2001) were predicted to be

- Greater focus on safety over the life cycle of chemicals
- Increasing involvement of all stakeholders, with industry taking more responsibility for generating and assessing health and environmental data and other stakeholders involved in oversight
- Increasing outreach to non-OECD countries to help them build up their chemical safety in order to cope with the rapid expansion of their chemicals industries

12.1.5 A Multitude of Chemicals May Harm Health and the Environment

Early concerns for risks from chemicals started from very obvious toxic effects on humans. Carcinogenic effects were recognised quite early, and later less immediate effects came in, such as hereditary disease and neurological effects.

Today, exposures are known to be widespread and the chemicals known to be predominantly harmful. The number of hazardous chemicals runs into the tens of thousands, albeit many with low production volumes. For instance, in a major database (Prevent 2009) more than 21,000 of about 32,000 substances are classified as hazardous to health or to the environment. Of all substances, some 2,200 are classified as or suspected to be carcinogens, almost 1,600 as hazardous to reproduction, about 750 as mutagens and some 7,400 as hazardous to the environment. Of 85 million tonnes of chemical preparations in Sweden in 2006, 70 million tonnes were classified as hazardous to health and 27 million tonnes as hazardous to the environment (KemI 2009). The largest globally produced volumes are for petroleum products, and these are typically toxic, carcinogenic and hazardous to the environment.

Much larger material flows are due to mining and other activities, which mobilise more than 50,000 million tonnes per year (OECD 2008). While these mainly give rise to considerations of sustainability and resource efficiency (Fig. 12.2), their contaminants may entail considerable chemical safety concerns. For instance, the coal mining of 7,000 million tonnes annually mobilises associated contaminants of some 40,000 tonnes of lead, 20,000 tonnes of arsenic and 600 tonnes of cadmium, based on estimates of contaminants in internationally traded coal (CSIRO 2009).

More than 100 substances in urine or blood are subject to a large-scale monitoring program (CDC 2005), and more than 100 phenolic chlorinated or brominated substances have been detected in human blood in Sweden (Hovander et al. 2002).

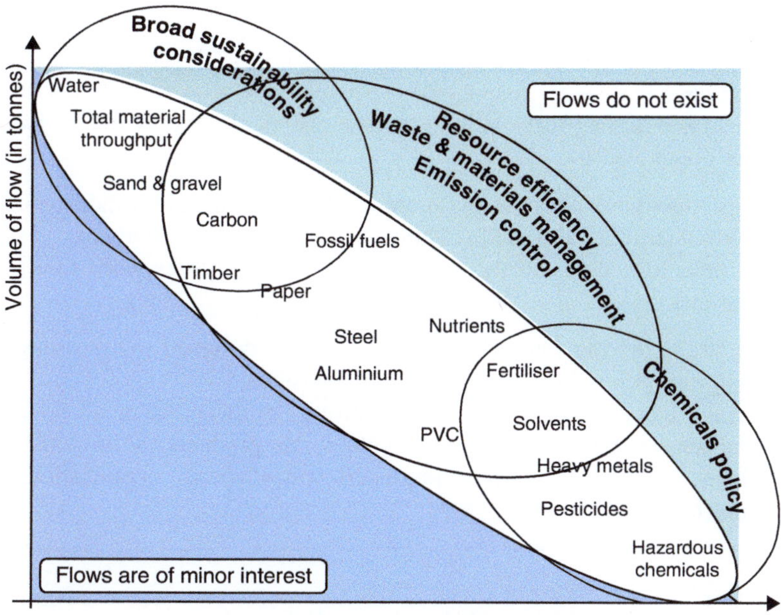

Fig. 12.2 Schematic representation of material flows, environmental impact and policy uses. (Reproduced from Measuring material flows and resource productivity Synthesis report, p. 9, © OECD 2008 (OECD 2008) with permission from OECD)

Exposure of newborns has attracted particular attention, and almost 300 chemicals have been detected in the umbilical cord blood in newborn babies (Houlihan et al. 2005) and hundreds of substances in breast milk (Massart et al. 2008).

These exposures may sometimes have long-term effects, not least on children. For instance, reduced birth weights or birth lengths have been associated with environmental levels of bisphenol A (NTP-CERHR 2008), PFOA (Fei et al. 2008), phthalates (Latini et al. 2003) and others (IFCS 2003a), and effects on the central nervous system that may lead to effects on intelligence or behaviour (IFCS 2003a) have been associated with environmental levels of lead, mercury and PCBs.

Effects have also been observed on the environment. Rachel Carson's Silent Spring in the 1960s drew attention to effects on wildlife. Our Stolen Future (Colborn et al. 1996) created an interest in disruption of the endocrine system by chemicals, even in animals. Many cases of effects on natural animal populations can now be inferred from laboratory experiments, although confirmation in the environment is extremely difficult. For instance, organochlorine concentrations in adult male bottlenose dolphins are approaching the levels associated with adverse effects found in marine mammals (Carballo et al. 2008), and exposures of tadpoles to a mixture of nine pesticides at environmentally occurring levels lead to developmental effects in most frogs, while none were observed when the pesticides were applied one at a time (Hayes et al. 2006).

12.1.6 Reducing Differences May Help Industry, Trade and Health

The gap in chemical safety management between countries in different stages of development is a driving force in international chemical safety work. Reducing the differences could lead to

- Better production conditions and more level playing fields for industry
- Smoother international trade in chemicals and manufactured goods
- Lower risks to health and the environment in countries exposed to chemicals from far away

These are strong driving forces for special measures designed to assist developing countries in improving their chemical safety measures.

A special driver has recently come from strong European chemicals policies that include substances and waste relating to electronic products (Selin 2009). These have influences on policy makers, producers and advocacy organisations across many countries.

12.1.7 Chemical Safety Helps Overall Development

Chemical safety is only one aspect of development. Achieving chemical safety may assist in fighting poverty, protecting vulnerable groups and advancing public health and human security. Global agreements on chemical safety have reflected commitments to respect human rights and fundamental freedoms, to understand and respect ecosystem integrity and to address the gap between the current reality and global ambitions.

The overall development effort of the global community is codified in the Millennium Development Goals (United Nations 2008), adopted by the United Nations in 2001. The eight Millennium Development Goals have been adopted by the international community as a framework for the development activities of over 190 countries in ten regions; they have been articulated in more than 20 targets and more than 60 indicators. Donor countries and organisations are also pursuing the Millennium Development Goals in the development of chemical safety.

12.1.7.1 The Poor Are at Greatest Risk

Among the Millennium Development Goals, Goal 1 is to 'Eradicate extreme poverty and hunger'. It has a target to 'Halve, between 1990 and 2015, the proportion of people whose income is less than $1 a day'.

A notion of chemical safety related to Goal 1 has been presented by UNDP (2009):

> The poor are at higher risk of exposure to toxic and hazardous chemicals because of their occupations, living locations, and lack of knowledge about chemicals. Sound chemicals management can improve their living environment, and consequently their health, and can help increase their revenue (e.g. proper use of pesticides can boost crop yields and protect the productivity of freshwater and marine fisheries).

A detailed assessment has been made in a report to the World Bank (2002).

A concrete example relates to the environmental levels of e.g. PCB, dioxin, pesticides or mercury, or air contamination including small particles. These have been associated with impaired perception, intelligence and mobility in children, entailing impaired earning ability and enhanced social inequity. Several case studies have indicated a loss of around 5 IQ points in a population due to hazardous chemicals, which might mean about a 10% loss of income or about the same loss of worker productivity (Trasande et al. 2005, IFCS 2003a).

12.1.7.2 Chemical Safety Has Links to Development Beyond Poverty Aspects

Other links are that sound management of chemicals can:

- Through awareness raising help reduce the occurrence of chemical related accidents.
- Through women's empowerment help protect women and their families.
- Combined with better nutrition, improve children's working and living conditions, decrease their sensitivity to chemicals and reduce child mortality.
- Lower women's risk of contamination, improve maternal health and, thus, the health of future generations.

- Minimise the side effects of malarial medications (prophylactics) and other chemical products (for example treated mosquito bed nets) that prevent millions of deaths worldwide; almost a million people still die each year of malaria (WHO 2008).
- Prevent and/or minimise the entry of harmful chemicals into the environment and reduce the need for difficult and costly environmental remediation.

Work on elaborating links between chemical safety and development in a practically useful toolkit is in an advanced stage (UNDP 2009).

12.1.8 A Broad Range of Stakeholders Are Involved

The value of stakeholder involvement is generally recognised in donor policies and is slowly seeping into national preparations for chemicals management. In almost all countries, industry and academia can contribute significantly. Industry is responsible for making available to stakeholders data and information on health and environmental effects and is often interested in clean production programs. A distinction should be made between the chemicals producing industry and the using industry. The producing industry is often farther away from the end customer and has a stronger interest in defending the chemicals it produces, as witnessed by the very strong lobbying against the European REACH initiative (DiGangi 2003).

The chemicals using companies, such as car and furniture manufacturers, are often much closer to the market than the chemicals producing companies. Movements in the market towards environmental protection are often clearly reflected in the policies of the chemicals using companies. Because of the enormous variety of producing companies, they have great difficulties of influencing global chemical safety work but increasingly put pressure on their chemicals suppliers for better chemical safety.

The strength and competence of environmental, consumer and labour organisations varies, but in developed countries it is generally recognised that their input should be sought and may often be substantial. Environmental organisations have a role to play in information and awareness raising but always have difficulties of funding, particularly for work at the global level. The government of Sweden has systematically funded environmental organisations to raise awareness and to be an independent critical voice. In particular, in the chemical safety sector, a Secretariat has received government funding for participation in international discussion. It operates on behalf of environmental organisations (Chemsec 2009).

12.1.9 Main Contentions: Protecting Industry vs. Funding for Developing Safety

The production of chemicals has, up to now, been dominated by developed countries, although a change is underway (see Section 12.1.4). Developing countries have had to deal with their own lower volume chemical safety issues but also with exports of waste from developed countries. In international negotiations, it often happens that some chemicals producing countries oppose stricter management (see, for instance, Eckley et al. 2006) and put pressure on developing countries to join them. Developing countries are pressed by their needs for general development and often fear imposition of excessively strict standards by the developed countries. Therefore, a major issue brought up in international negotiations concerns funding for the implementation of safety measures in developing countries. A proposal at the very heart of the controversy concerns a direct tax on the global chemicals industries; this is put in perspective in Section 12.3.3.

12.2 Global Development of Chemical Safety

The drivers discussed above have resulted in international agreements for efficient management of hazardous chemicals. There are now around one hundred international agreements, programs or initiatives dealing with chemical safety at the international level. They are so many that several programs have been instituted just to coordinate international work.

12.2.1 Excessively Comprehensive Cooperation?

12.2.1.1 Almost 100 International Agreements and Programs

An extensive review of international agreements (Buccini 2004) discusses 22 global and 27 regional agreements on chemical safety, as well as 39 international programs and initiatives going beyond direct support for the agreements.

The list is not exhaustive; for instance, it does not contain:

- The 1972 Convention on the Prevention of Marine Pollution by Dumping of Wastes and Other Matter
- The 1998 Convention on Access to Information, Public Participation In Decision-Making And Access To Justice In Environmental Matters
- The 2001 International Convention on Civil Liability for Bunker Oil Pollution Damage

Furthermore, some ten new agreements have been made after the list was compiled (Mitchell 2009). The most significant of these are the global Strategic Approach to International Chemicals Management (SAICM) and the European Union Regulation concerning the Registration, Evaluation, Authorisation and Restriction of Chemicals (REACH), both adopted in 2006.

According to the review by Buccini 2004, most agreements, programs and initiatives deal with risk management for hazardous chemicals (a discussion of the different types of instruments used is found in Section 12.2.1.3). In addition, several deal with monitoring and evaluation and some with problem identification and risk assessment, particularly among the programs and initiatives.

12.2.1.2 Policy from the Highest International Level

Chemical safety gained significantly increased global visibility with the advent of the chemical safety Chapter 19 of Agenda 21 that emerged from the 1992 Rio meeting of the United Nations Conference on Environment and Development (UNCED).

This chapter addressed six substantive areas:

(a) Assessment of risks
(b) Harmonisation of classification and labelling
(c) Information exchange
(d) Risk reduction
(e) Strengthening of national capacities
(f) Prevention of illegal international traffic

In addition, a short subsection deals with the enhancement of cooperation related to several programme areas.

Progress in these areas after 10 years was reviewed at the World Summit on Sustainable Development in Johannesburg. Its Plan of Implementation (WSSD 2002) set the goal that by the year 2020 chemicals should be produced and used in ways that minimise significant adverse impacts on the environment and on human health. This has been the portal goal that has since been at the forefront of global chemical safety efforts, not least SAICM (Section 12.2.4.1).

12.2.1.3 Policy Instruments: Binding and Voluntary

A broad range of instruments for environmental management has been reviewed by Sterner (2002). These can be characterised as

- Command-and-control regulations
- Provision of information, e.g. classification and labelling, green labelling and emission registers
- Economic incentives, e.g. taxes, fees, permit trading and green procurement
- Construction of institutions for the allocation of rights that are fundamental for any market mechanism, e.g. bodies that make rules for liability and courts

In the choice of policy instruments, efficiency is important, but so are aspects of distribution, information, politics and implementation. Where monitoring and access to technology and credits are particularly difficult, it is crucial to consider policies that avoid antagonism and encourage cooperation and involvement (Sterner 2002).

In line with the latter, instruments at the global level tend in practice to be voluntary for the ratifying governments. This has been the case, for instance, with the Globally Harmonized System of Classification and Labelling of Chemicals (GHS, Section 12.2.3.2) and SAICM (Section 12.2.4.1). However, even with the formally binding Basel, Rotterdam (Section 12.2.3.3) and Stockholm (Section 12.2.3.4) Conventions, the mechanisms discussed for treatment of parties in non-compliance are softly worded and generally concern arbitration in different forms. The non-compliance mechanisms of the Rotterdam and Stockholm Conventions have still not been agreed to, 10 and 8 years, respectively, after being signed, an indication that many states remain highly protective of their sovereignty. This is a general experience from environmental agreements (Selin 2009).

The international chemical safety agreements deal to a large extent with command-and-control regulations to be established by the ratifying states, even when the agreements themselves are voluntary. For instance, the voluntary GHS has already been transformed into binding regulations in a number of countries, and in some countries emission registers are required by law in line with the Aarhus convention Protocol on Pollutant Release and Transfer Registers (UNECE 2003).

Information instruments have been widely used in global chemical safety. For instance, the generation of hazard information has been a main issue since Agenda 21. The GHS is very much about providing information for staff working in protection and for end users, and part of this involves the information instrument of Safety Data Sheets. The Pollutant Release and Transfer Registers/emission inventories (Section 12.2.3.3) are to large extent information tools that may help citizens dialogue with enterprises on releases.

Economic incentives have not been used much in chemical safety (OECD 2009e). At the national level some countries use fees, and there are a few examples of taxes. The fees for registration of pesticides have worked as deterrents to the marketing of many pesticides. It remains to be seen whether the REACH registration in the European Union will have a similar effect for low production volume substances.

Facilitation of market mechanisms has not been used to any great extent. An example is provided in the National Implementation Plan of China for the Stockholm Convention (GEF 2007b), which contains activities to 'determine the principles and mechanism for responsibility sharing among stakeholders for different types of activities, e.g. non-profitable and profitable activities'. These might include suggestions and recommendations to remove barriers to market oriented operations in response to reducing emissions from combustion and managing wastes.

12.2.2 International Coordination Is Extensive

With upwards of 100 international agreements, programs and initiatives, there is an obvious risk for overlaps and duplications. Therefore, many mechanisms of international coordination have emerged. They include eight initiatives that are briefly described in the following in approximate order of appearance that mirrors the trends described above.

12.2.2.1 Organisation for Economic Cooperation and Development (OECD)

OECD brings together 31 countries committed to democracy and the market economy; the number is slowly growing. The most important part of OECD chemical safety work has been carried out by the special chemicals groups that have appeared under different names since 1971. At present, the work is presented under Chemical safety. The first important tasks dealt with mercury, cadmium and polychlorinated biphenyls (PCBs), in addition to coordination and information exchange. They were later followed by issues in accreditation and good laboratory practice, chemicals testing and hazard assessment. Today, the work of the OECD covers twelve headings, from accidents and biocides to nanomaterials and pesticides (OECD 2009a).

12.2.2.2 International Program on Chemical Safety (IPCS)

The IPCS has since 1980 been a joint undertaking by the United Nations Environment Program (UNEP), the International Labour Organisation (ILO) and the World Health Organisation (WHO). The initial work dealt with assessment of hazards and risks from general chemicals, food additives and contaminants, and pesticide residues. From the very onset, the IPCS had technical cooperation with member states, in particular developing countries, as an important objective. Today, the IPCS still produces assessments of hazards and risks from chemicals, including in food, with methodology development being an important element (IPCS 2009). Poisoning Prevention and Management is a major activity, lending support to almost one hundred poison centres around the world. The WHO Chemical Alert and Response Team identifies, alerts, tracks and, when appropriate, coordinates a response to chemical incidents and emergencies on a global basis.

12.2.2.3 The International Council of Chemical Associations (ICCA)

The ICCA is the world-wide voice of the chemical industry, representing chemical manufacturers and producers all over the world. It is the main channel of communication between the industry and various international organisations that are concerned with health, environment and trade relations (ICCA 2009). The program Responsible Care® has since 1985 committed the worldwide chemical industry to continual improvement in all aspects of health, safety and environmental performance and to open communication about its activities and achievements. Since 1998, ICCA, in co-operation with the OECD and its member countries, has produced harmonised, internationally agreed upon data and initial hazard assessments for high production volume substances representing more than 90% of the global chemicals production. The ICCA also has programs on product stewardship (management of health, safety and environmental aspects of a product throughout its total life cycle, working in cooperation with upstream and downstream users) and on long term research.

12.2.2.4 Intergovernmental Forum on Chemical Safety (IFCS)

The great need for international coordination in chemical safety was discussed in the beginning of the 1990s as the United Nations Conference on Environment and Development was being prepared. Eventually, the discussions led to the chemicals Chapter 19 of Agenda 21. The IFCS was formed in 1994 to promote the implementation of Chapter 19 as an over-arching mechanism to develop and promote strategies and partnerships among national governments, intergovernmental organisations and non-governmental organisations (IFCS 2009). In 1994, it produced Priorities for Action in global chemical safety, dealing with the six areas of Chapter 19.

In 2002, these priorities were updated and revised. The IFCS also developed indicators of progress towards achieving the Priorities and compiled the outcomes. Before the next revision of the Priorities, a new instrument took over the formulation and follow-up of objectives: the Strategic Approach to International Chemicals Management, SAICM. At the first conference in 2009 to follow up SAICM achievements, participants decided to leave it to the IFCS to decide on its own future, thus rejecting at that time a proposal to make the IFCS a follow-up mechanism for the SAICM process. For lack of funding the operation of the IFCS was in July 2009 suspended for the foreseeable future.

12.2.2.5 Inter-Organisation Programme for the Sound Management of Chemicals (IOMC)

The IFCS was an organisation independent of the UN system, with unique contributions on the part of developing countries and non-governmental organisations. The governmental organisations saw a need to promote coordination at about the same time as IFCS was formed, and in 1995 formed the IOMC. This has defined itself as the pre-eminent mechanism for initiating, facilitating and coordinating international action to achieve the World Summit on Sustainable Development 2020 goal of sound management of chemicals. There are seven Participating Organisations and two observer organisations of the IOMC; of the latter, the World Bank has decided to become a full member.

The IOMC works by strengthening international cooperation in the field of chemicals, increasing the effectiveness of the programmes of the nine organisations and promoting coordination of policies and activities pursued jointly or separately (compare its resource guide IOMC 2009). The coordinated views, programs and studies are represented by IOMC in the governing bodies of international organisations and other fora.

12.2.2.6 The International POPs Elimination Network (IPEN)

Public interest non-governmental organisations have generally found it difficult to work at the global level, both for economic and language reasons. The advent of

Internet greatly facilitated working in large networks. The first strong global activity in the chemical safety field was the IPEN, launched in 1998. It is a global network of more than 600 public interest non-governmental organisations working together for the elimination of persistent organic pollutants (POPs) on an expedited yet socially equitable basis (IPEN 2009).

IPEN focuses on mobilizing resources for NGO activities in developing countries and countries with economies in transition. IPEN has established eight regional hubs working in the regional languages to promote the implementation of IPEN's international projects. At present, three projects are in operation, dealing with the elimination of POPs, egg sampling for analysis of several POPs substances and awareness raising and engagement for SAICM.

12.2.2.7 Strategic Approach to International Chemicals Management (SAICM)

The IFCS had managed to promote a rapid global agreement on actions against Persistent Organic Pollutants (POPs) and to establish the framework for the voluntary global harmonisation of classification and labelling. However, the follow-up of the other Priorities for Action indicated that there was little commitment on the part of many countries, even when it came to providing IFCS with progress information. There were calls for stronger instruments with associated financial resources that would permit developing countries to give higher priority to chemical safety. In 2006, SAICM was agreed upon. This was a new international framework that replaced the Priorities for Action of the IFCS and that was supplied with a special fund to facilitate implementation. Work to define processes, including indicators for follow-up, is still in progress. Important steps were taken at the second International Conference on Chemicals Management (ICCM-2) in 2009, including initiating action on electronic waste, which had until then tended to fall through the cracks between the organisations. SAICM is described more in detail in Section 12.2.4.1.

12.2.2.8 Coordination of Basel, Rotterdam and Stockholm Conventions

With upwards of 100 international initiatives in chemical safety, countries are struggling to keep track of and implement the large number of agreements. They have in SAICM called for significant coordination. The functions of the International Conference on Chemicals Management include promoting the implementation of existing international instruments and programmes and coherence among chemicals management instruments at the international level. Three key international instruments have been reviewed in this regard for ICCM-2, namely, the Basel Convention on the Control of Transboundary Movements of Hazardous Wastes and their Disposal, the Rotterdam Convention on the Prior Informed Consent Procedure for Certain Hazardous Chemicals and Pesticides in International Trade and the

Stockholm Convention on Persistent Organic Pollutants (ICCM 2009). These formally independent treaties cover partially different life-cycle issues. Strong coordination of these with SAICM has been proposed, and a special joint Extra Conference of the Parties for the tree conventions was held in February 2010.

12.2.3 Domination by Developed Countries

From Agenda 21 in 1992 until the adoption of SAICM in 2006, the concerted collaboration on chemical safety globally was structured according to the six substantive areas of Chapter 19 of Agenda 21. Implementation was dominated by work of developed countries as follows.

12.2.3.1 Assessment of Risks

IFCS in 2003 stated that facilitation of global consistency and global collaboration in data generation and accessibility would have several advantages:

- Improved safe use of chemicals with respect to human health and the environment, including increased transparency
- Minimised use of laboratory animals for testing
- Economy of testing and assessment
- Reduced barriers to trade

These have underpinned international attempts at coordination. A difficulty has been that certain companies bear the burden of producing the data and have a commercial interest in sharing this data only if they are compensated. At the same time, there is a strong public interest in essential health, safety and environmental information being accessible.

Initially, assessments of hazards and risks were made substance by substance according to the most pressing needs. More systematic programs were operated by the IPCS and a few OECD member countries, particularly for pesticides. Several attempts were made to establish registers or websites linking the different sources of information, but they never came into broad use. Over time, balances between interests have been struck to enable coordinated large-scale programs for hazard assessment. Today, there is considerable coordination in the generation and dissemination of data for thousands of general chemicals (OECD 2009b). The process is significantly aided by more than 150 detailed and internationally agreed upon testing methods used by government, industry and independent laboratories to assess the safety of chemical products (OECD 2009c) and by common principles developed mainly under IPCS Section 12.2.2.2) for harmonised approaches for performing and reporting health and environmental risk assessments. Important tools for minimising costs and use of laboratory animals for testing have been developed in the form of computer models for the properties of chemical substances and elaborate testing strategies.

For pesticides, there has been considerable technical work performed to enable coordination. The OECD countries have adopted a vision (OECD 2009d) that by the end of 2014 governments will routinely accept 'dossiers' prepared by stakeholders in the OECD format, will routinely exchange 'monographs' (containing reviews of the data submitted) and will use OECD 'monographs' as a basis for independent risk assessments and regulatory decisions for new and existing pesticides.

12.2.3.2 Harmonisation of Classification and Labelling

A very important chemical safety measure is informing users about the hazards of chemicals through symbols and phrases on the packaging labels and through additional comprehensive safety information. The skull and crossbones symbol is a well-known early example applied to poisons.

There has been growing pressure to harmonise the hazard information for several reasons

- An increasing use of many different chemicals
- An increasing international trade; and
- An increasing knowledge of many types of hazards to health and the environment

A globally harmonised system of classification and labelling was conceived around 1950 in discussions of the Chemicals Industries Committee of the ILO. Via a tortuous path, it was eventually followed up in 1992 with a decision on an action area of Chapter 19 in Agenda 21. The first operative version of the Globally Harmonized System of Classification and Labelling of Chemicals (GHS) was agreed upon 10 years later in 2002. The work had been coordinated and managed under IOMC, with separate technical focal points for completing the work concerning hazard communication and classification of health and environmental hazards as well as physical hazards.

In the Priorities for Action of the IFCS in the year 2000, the target year 2008 was set for implementation of a harmonised system, and this was endorsed by the World Summit on Sustainable Development in 2002. The system adopted in late 2002 was revised in 2005 and 2006 (UNECE 2007). Responsibility rests with the United Nations Economic and Social Council Sub-Committee of Experts on the GHS under the United Nations Economic Commission for Europe (UNECE). The target of full GHS implementation by 2008 has not been reached, but implementation is well underway. Major actors have set deadlines, such as the United States (transportation 2010) and the European Union (2010 for classification of substances and 2015 for mixtures).

The text is about 560 pages long with much technical detail. It describes physical, health and environmental hazards. For each type of hazard (for instance explosivity, carcinogenicity and hazards to the aquatic environment) there is a definition, criteria for classification (for instance as suspected human carcinogen) (Fig. 12.3) and hazard communication instructions with a designated:

Fig. 12.3 Health hazard symbol

- Symbol, for instance the symbol for Health hazard as shown to the right – there are nine symbols
- Signal word, for instance Warning – there are three levels (Warning, Danger, No signal word)
- Hazard statement, for instance Suspected of causing cancer – there are about 70 hazard statements

In addition, there is guidance for those who classify substances or mixtures of substances.

Recent work has involved classification for environmental effects with respect to environmental fate and toxicity to aquatic organisms. Discussions are ongoing concerning toxicity to terrestrial organisms. Classification and labelling are cornerstones of chemical safety, and today tens of thousands of substances are classified as hazardous.

The GHS also prescribes information in 16 headings to be used for describing hazards of chemical products in a Safety Data Sheet (SDS). They deal with inherent properties as well as safety measures for e.g. fire fighting, disposal, transport and accidents. The use of such data sheets was recommended in the Priorities for Action.

Several databases deal with classifications and with safety data sheets, as a simple search on the Internet will reveal. However, there is no internationally coordinated database. An extensive listing of safety data sheet sites is available (MSDSonline 2009). A major challenge is to make all of this information available in the languages of the users.

12.2.3.3 Information Exchange

Access to information is a fundamental right in a democracy, enabling informed decisions by citizens. It has been given much weight in chemical safety ever since Agenda 21 in 1992. The Priorities for Action 2000 called for national arrangements for exchange of information on chemicals with recognition of the language issue. At that time, gaining access to Internet was also high on the agenda. A program was instituted for that purpose (CIEN 2009) and has provided Internet access, documents, databases, a website building tool and workshops to more than 40 countries in Africa, Central America and Mexico.

IFCS in 2000 also recognised the role of information exchange in relation to toxic chemicals in the Rotterdam Convention on the Prior Informed Consent Procedure for Certain Hazardous Chemicals and Pesticides in International Trade. It encouraged its implementation. The Convention went into force in 2004. It requires information prior to export of some 30 chemicals that are listed, most of these with very little circulation. Additions of 'live' chemicals have been very controversial. At the Conference of the Parties in 2008, agreement was reached to add tributyl tin compounds but not concerning chrysotile asbestos and endosulfane. The IFCS in 2000 also recognised the importance of providing all relevant parties with safety information consistent with the safety data sheets.

Many attempts have been made to set up comprehensive registers, databanks or portals with data on chemicals. They have in general failed because of the enormity of the task. There are some 30,000 chemical substances in use in hundreds of thousands of chemical preparations and millions of manufactured products. Information needs to be available in many languages. A compilation within the IFCS (2003b), in need of updating, listed some 25 major databases on hazards, exposures and risks, almost exclusively in English. The IFCS has given particular attention to the development of emission registers, which it has sorted under Risk reduction.

12.2.3.4 Risk Reduction

In the best of worlds, risks of an undertaking can be assessed and balanced against the costs and benefits resulting in an agreed risk management option. When it comes to most chemicals, information is neither available nor accessible, and decisions on risk management have to be taken under considerable uncertainty. As guidance in this situation, the Rio Declaration from 1992 contains Principle 15. 'In order to protect the environment, the precautionary approach shall be widely applied by States according to their capabilities. Where there are threats of serious or irreversible damage, lack of full scientific certainty shall not be used as a reason for postponing cost-effective measures to prevent environmental degradation.' The interpretation of precaution has been the subject of considerable discussion, but chemicals negotiations have in the end relied on Principle 15.

Implementation of risk reduction actions agreed in the Priorities for Action in 2000 is dealt with in Table 12.1. In addition, risk reduction initiatives on other chemicals of major concern were to be put on the future agenda. This was a placeholder for the controversial topics of metals, such as cadmium, lead and mercury, and organic substances, such as polybrominated compounds. It took many years, but in 2009 the Governing Council of UNEP embarked on a plan that was to end in a treaty to strongly reduce mercury use (UNEP 2009a). Several new organic substances are considered under the Stockholm Convention, and the need for global action related to lead and cadmium is currently under discussion.

In the 17 years from UNCED to ICCM-2, progress has been made in many areas. Still, it should be recognised that in more complex cases, risk reduction takes decades. For example, some of the substances called PCBs were initially detected in

Table 12.1 The risk reduction actions of Priorities for Action in 2000 and corresponding actions taken until 2009

Priority 2000	Actions until 2009
Countries should establish ecologically sound and integrated strategies for the management of pests (IPM) and vectors for communicable disease.	Evaluation 2004 demonstrated widespread and lasting developmental impacts. FAO is now mainstreaming IPM principles into all of its work on crop production. The World Bank requires that all pest management activities it finances are within the context of an IPM approach.
Countries should establish relevant action plans with respect to identification, neutralisation and safe disposal of obsolete stocks of pesticides and other chemicals.	An Internet based pesticides stock management system is in operation. FAO facilitated programs are operated in some ten developing countries. Under the Stockholm Convention most countries are planning actions for relevant substances.
Special attention should be paid to persistent and bio-accumulating toxic chemicals. Work on a global convention on POPs should continue with a view to its entry into force as soon as possible, preferably by 2004.	The Stockholm Convention (UNEP 2009a) entered into force in 2004, restricting 12 persistent and bio-accumulating toxic chemicals. An additional 9 substances were included in 2009. In coming decades, the major issue under the convention will be the reduction of emissions of dioxins and furans from combustion.
Major industrial accidents must be prevented. National systems for emergency preparedness and response should be developed in all countries.	ILO Industrial accident convention 174 has been ratified by 13 countries. UNEPs accident program Apell has been applied in 30 countries. A revised version of the Guiding Principles of the OECD was published in 2003.
Poisoning of pesticide users in developing countries and countries with economies in transition must be prevented. Initial input was requested on the extent of the problem of acutely toxic pesticides as well as guidance for sound risk management and reduction.	WHO Guidelines on Situation analysis for public health pesticide management was published in 2005. A feasibility demonstration project was planned in 2008 for the WHO Initiative on the Impact of Pesticides on Health: Preventing intentional and unintentional deaths from pesticide poisoning.
Countries should play an active role in the observance of the revised International Code of Conduct on the Distribution and Use of Pesticides.	The Code was revised in 2002, and about a dozen associated guides are available. CropLife International and the leading companies of the plant science industry have agreed to abide by the Code.
Poison centres providing toxicological information and advice with clinical and analytical toxicological facilities should be established and strengthened.	Almost 100 countries have one or more poison centres.
Pollutant Release and Transfer Registers (PRTRs)/emission inventories should be established in additional countries.	About a dozen countries have PRTRs. A global portal centre and a data centre are available.
Governments and industry should consider granting the public's right-to-know in relation to chemical constituents of consumer products.	Chemicals in consumer products have been addressed in workshops and in the Stockholm Convention. A UNEP working group will report in 2011 on increased availability of and access to information on chemicals in products in the supply chain and throughout their life cycles.

environmental samples in 1966, first banned in Sweden in 1972 and globally restricted by the Stockholm Convention in 2001, with efforts for sound management of waste required no later than 2028. Clean-up of contaminated sites is a major risk reduction challenge that has so far barely been addressed in international cooperation.

12.2.3.5 Strengthening of National Capacities

There has been an increasing urgency in strengthening national capabilities and capacities for management of chemicals in developing countries and countries with economies in transition. When the chemicals part of the Rio Conference was being prepared at a London meeting in 1991, 43 developing countries and 28 others were represented. Fifteen years later when SAICM was adopted, there were more than twice as many developing countries but about the same number of others. This may reflect a strongly increased interest in chemical safety on the part of developing countries.

The Priorities for Action contained the same main elements for strengthening national capacities as the SAICM Global Plan of Action, even though SAICM contained many more details.

12.2.3.6 Prevention of Illegal International Traffic

Prevention of illegal international traffic in toxic and dangerous products was made a priority in Chapter 19 of Agenda 21, as it was for waste in Chapter 20; for waste the Basel Convention on the Control of Transboundary Movements of Hazardous Wastes and Their Disposal was signed as early as in 1989. For the first years after Agenda 21, the issue was put on hold, since it was felt that chemical safety legislation had to be in place first. The issue was given more weight over the years, and in the SAICM Global Plan of Action it was once again a separate area for action. In 2007, UNEPs Governing Council called in general terms for the implementation of existing international instruments in the area but up to 2010 little concerted action had been taken. This long delay reflects the complexity of the issue and illustrates how over 2 decades it has been so difficult to achieve concrete and specific actions and why the many calls for specific working groups have gone unheeded.

12.2.4 Developing Countries Start with the Most Hazardous Chemicals

When the Global plan of action of SAICM was adopted in 2006, many of the priorities from Agenda 21 of interest to developed countries had been met or their targets were well on the way towards being fulfilled.

A new instrument, such as SAICM (2006), was thus neither of major interest to the developed countries nor did most of them see the need for a continuation of the IFCS. Funding for international chemical safety was limited, and there were increasing difficulties funding these undertakings in addition to the basic work of the OECD and the three conventions of Basel, Rotterdam and Stockholm. SAICM, therefore, tended to be focused on the developing countries, and a new funding mechanism was devised for it with focus on developing countries. The funding for the IFCS dwindled, and some developed countries pulled out of IFCS activities. Non-governmental organisations and developing countries felt that they had a better platform in the IFCS than in the more formal SAICM mechanism and advocated its continuation. A new role for the IFCS was to be discussed at the first follow-up conference of SAICM in 2009. This was the setting for the implementation of SAICM from the point of view of the developed countries.

At the same time, the developing countries were struggling to cope with extensive intoxications from pesticides, lead and mercury poisoning, legacies of stockpiled obsolete pesticides and polychlorinated biphenyls and other issues that had to a large extent already been taken care of in the developed countries.

12.2.4.1 The SAICM Contents and Implementation

In Dubai in 2006 countries agreed to SAICM, which is distinguished by its:

- Comprehensive scope
- Ambitious '2020' goal for sound chemicals management
- Multi-stakeholder and multi-sectoral character
- Endorsement at the highest political levels
- Emphasis on chemical safety as a sustainability issue
- Provision for resource mobilisation
- Formal endorsement or recognition by the governing bodies of key intergovernmental organisations

SAICM comprises:

- The Dubai Declaration on International Chemicals Management, expressing high-level political commitment to SAICM.
- An Overarching Policy Strategy, which sets out its scope, needs, objectives, financial considerations, underlying principles and approaches, and implementation and review arrangements. Objectives are grouped under five themes, which significantly overlap with the six areas of the Priorities for Action: risk reduction, knowledge and information, governance, capacity-building and technical cooperation, and illegal international traffic.
- A Global Plan of Action that serves as a working tool and guidance document to support implementation of SAICM and other relevant international instruments and initiatives.

A new funding mechanism called the Quick Start Programme was set up. Its objective is to 'support initial enabling capacity building and implementation activities in developing countries, least developed countries, small island developing States and countries with economies in transition.'

The strategic priorities of the Quick Start Programme are:

- Development or updating of national chemical profiles and the identification of capacity needs for sound chemicals management
- Development and strengthening of national chemicals management institutions, plans, programmes and activities to implement the Strategic Approach, building upon work conducted to implement international chemicals-related agreements and initiatives
- Undertaking analysis, interagency coordination and public participation activities directed at enabling the implementation of the Strategic Approach by integrating – i.e., mainstreaming – the sound management of chemicals in national strategies and thereby informing development assistance cooperation priorities.

These compare well with the elements from the Priorities for action: National profiles, national action plans and incorporating chemical safety issues in national development plans. In addition, the Priorities had an element of access to information on capacity building involving the development of an Information Exchange Network on Capacity Building for the Sound Management of Chemicals. This element was transferred to the Secretariat of SAICM (SAICM 2010).

The Global Plan of Action also contained a list of 36 work areas with 273 activities dealing with essentially all aspects of chemical safety. An enormous program of this kind can only be partially dealt with, even by very advanced countries, and initially it may have confused more than helped developing countries setting up the first parts of a chemical safety program. Extensive work on indicators might help to focus efforts. However, the proposal decided on at ICCM-2 suffers partly from the same disease, going far beyond the more limited information that was not able to be collected by the IFCS and leaving many difficulties of interpretation if one is to get away from the 30-page draft surveys.

In the following section, only the three more realistic and limited priorities of the Quick Start Programme mentioned above will be discussed.

12.2.4.2 National Profiles

A national profile is a document assessing and diagnosing a country's existing infrastructure for the sound management of chemicals. By 1999, national profiles had been prepared by 61 countries. These were often first drafts, sometimes called micro- or mini-profiles. In the beginning of 2009, the National Profile homepage reported 114 profiles, many of which had been revised at least once (UNITAR and ECB 2009), and an additional 25 countries had Profiles under preparation.

Implementation of national profiles was, thus, well established as the starting point for further development of chemical safety.

12.2.4.3 National Action Plans

The production of national action plans for chemical safety had a slow start in the 1990s. The poor statistics available indicated that 46 national plans were produced or under preparation by 1999. The drafting was strongly boosted with the start of implementation of the Stockholm Convention after its entry in force in 2004. Through the Global Environment Facility, which is the financial mechanism of the Convention, funding for drawing up national implementation plans was awarded quickly and with little bureaucracy. In the beginning of 2009, 89 countries had submitted such plans and an additional 73 countries had committed themselves to do so according to the Convention. The plans that were submitted were often extensive documents based on National profiles.

The Stockholm Convention plans are likely to be the core of the work on national action plans that began after SAICM was adopted in 2006. Preparatory work has been started in about 60 countries, with funding mainly from the Quick Start Program. The preconditions are, thus, significantly more favourable than those that prevailed after Agenda 21 in 1992.

12.2.4.4 Chemical Safety and National Development Priorities

Donors of developmental aid have emphasised the need to link chemical safety to overall development. This is also reflected in the priorities of the SAICM Quick Start Program.

While it may be very important to attend to the links between chemical safety and national development, this is seldom done in a systematic way. National chemical safety planners tend to look at their activity in isolation and not to identify synergies with other areas of chemical safety. For instance, actions such as legislation, national committees or data bases may be proposed separately for PCBs, dioxins, contaminated sites, chemical hazard information and monitoring, while some of these could be combined.

Looking at their most proximate interests in a similar way, national sector planners and politicians tend to look at one sector at a time and miss the fact that chemical safety issues exist over a broad range of sectors, having a combined impact far beyond the impact within any one sector. For instance, implementation of conventions may be viewed separately for each convention, while there may be synergies in some degree of coordination. They may also fail to assess the full picture, with chemicals promoting as well as counteracting national development goals.

SAICM has an emphasis on combining the perspectives of chemical safety and national development. Hopefully, this emphasis in combination together with the funding instrument of the Quick Start Program could help in dealing with this combination. Similarly, and with much more economic clout, the Global Environment Facility (GEF) has modified its focal area strategies to include:

- Strategic Program 1: Integrating Sound Chemicals Management in GEF Projects, and
- Strategic Program 2: Articulating the Chemicals-related Interventions Supported by the GEF Within Countries' Frameworks for Chemicals Management (GEF 2007a).

One might say that Program 1 reflects the perspectives of the development planners and Program 2 those of the chemical safety planners.

12.2.4.5 How To Build Capacity

Chemicals production may in a few decades be dominated by today's developing countries, and even now these represent a larger number of countries and a greater population than developed countries. Therefore, global chemical safety will over the next decades likely be dominated by capacity building and other actions in developing countries. The three elements discussed above – national profiles, national action plans and alignment with national development priorities – will be cornerstones in this work. If this work is to be successful, a number of other considerations should also be taken into account and applied whenever appropriate (IFCS 2003c). Projects should, for instance:

- Consider long-term coaching and support and make use of twinning arrangements.
- Have an ownership/demand driven approach with clear definition of who the owners of a project and the stakeholders that have an interest in a successful outcome are.
- Build on successful previous or existing bilateral projects ('A successful project is also the donor's success').
- Use clear indicators and validation data to measure progress and success of projects.
- Seek opportunities to link to direct positive economic results and to promote active participation of trade, industry and chemicals consumers.

An internationally agreed guidance document on strategy for capacity building has been prepared by the IOMC (2010).

Capacity building is a slow process that can be significantly aided by reference to the experiences of others. An attempt has been made to establish a tool for systematic accessibility to experiences from capacity building on chemical safety. This tool has been transferred to the Secretariat of the SAICM (SAICM 2010).

12.3 The Future of Global Chemical Safety

Over the next decades developed countries are likely to continue to accomplish what they have begun in global chemical safety, and developing countries will likely continue trying to catch up. What new developments are waiting beyond the horizon?

There will always be new types of chemicals and new exposure situations, but these are not really a major challenge. They can be exemplified by the 'emerging issues' before the ICCM-2: nanotechnology, chemicals in articles, lead in paint and electronic waste. These are not really new issues but have been discussed for a decade or more. It is likely that the global system will be able to deal with similar issues through minor modifications to the established system. As a precedent, when the endocrine disruptors emerged in the beginning of the 1990s, they were dealt with mainly by adding some tests and changing priorities (see for instance USEPA 2009a).

But there are other challenges that will be more difficult to meet. These include:

- Securing commitment over a broad range of stakeholders in ever more complex work with chemical safety
- An increasing number of international instruments for chemical safety
- An increasing number of countries participating in international cooperation
- An increasing risk due to increasing volumes and increasing numbers of substances of concern

12.3.1 Developed Countries Accomplish What They Started

Over the next decades developed countries will continue to implement what has been accomplished since Rio 1992. For instance:

- The timeline for the European Union extensive program on assessing industrial chemicals is to end by 2019.
- The OECD countries have adopted a vision that by the end of 2014 governments will routinely accept pesticide 'dossiers' prepared by stakeholders in the OECD format.
- The Globally Harmonised System for classification and labelling will be implemented in many countries over the next few years.
- In cooperation within OECD, emerging issues, such as nanomaterials, will continue to be addressed.

With the obvious effects being regulated and assessments and approaches harmonised, it is natural that attention be turned even further to ubiquitous environmental contamination with chemicals. The most obvious substances with persistent, toxic and bioaccumulating properties have been addressed in the Stockholm Convention, and there is a working process to review additional substances for possible inclusion. Such additions may not come easily, as shown by the example from the Rotterdam Convention with contention around inclusion of additional 'live' chemicals, such as chrysotile asbestos and endosulfane. The great attention paid to these types of substances in, for instance, the United States and European Union points towards a continued struggle to control a large number of substances that each may contribute only a small share to the overall risks to health and the environment.

Another set of controversies has been raging around the concept of substitution, that is replacing hazardous chemicals with less hazardous ones. This is an important constituent in chemical safety work, for instance under the name of 'Green chemistry'. The controversies concern the formalisation of this process. Rather strong requirements for substitution concerning plant protection products and biocides have been included in European Union legislation. Substitution was given a prominent place in the SAICM outcome (Section 12.2.4.1) in 2006. In some countries this has been aided by lists of chemicals of particular concern. A renewed interest and controversy has appeared around the SIN (Substitute It Now) list of 267 substances presented by a Swedish non-governmental organisation (ChemSec 2009).

12.3.2 Lagging Implementation, Few New Agreements, Calls for Coordination

It is easy to predict that this increasing complexity will not make international cooperation easier. In fact, with upwards of 100 international agreements, programs and initiatives on chemical safety, a lower degree of implementation is likely. For instance, part of the work will have to be done via the foreign ministries. Considering that overall the United States has around 12,000 and Canada around 4,000 international agreements to deal with, the manpower that can be devoted to each of them will not be very great. Participation in international negotiations will be a burden for many smaller countries with limited resources for chemical safety. Also, non-governmental organisations will find it necessary to set tough priorities for where they interact in the international processes. In particular, this will hold for academia, labour and public interest organisations, with their very limited resources.

There is also bound to come a time when countries will be reluctant to sign further agreements. In Sweden in the 1990s, representatives of the Environment Ministry had instructions to be very restrictive with new international commitments and to make sure that they could be implemented and periodically reported on within the limits of existing resources before anything was agreed upon. Arrangements for coordination of international work, such as those described in Section 12.2.2, are also likely to gain more weight.

12.3.3 Developing Countries Will Not Keep Up To Speed

With developing countries playing an increasing role in international negotiations on chemical safety, funding issues have become among the most critical elements in the negotiations. As an indication of funding needs, China has made a first assessment that it may need 1.3 billion USD for implementation of the Stockholm Convention over the years 2007–2010, or some 300 million USD annually. With China having one-fifth of the world's population, a first estimate of global needs would be the fivefold amount, or some 1,500 million USD annually. This can be compared with the expected total allocation of funds for POPs from the Global Environment Facility of about 100 million USD per year for the same period, with approximately the same amount leveraged in co-funding. A similar level can be inferred for all chemical safety assistance over the first years of the new millennium, according to an OECD survey (OECD 2003). The SAICM quick start programme trust fund allocations are expected to be much smaller, around 4 million USD per year for 2006–2011. Contributions for assistance from the other two conventions, Basel and Rotterdam, are even smaller, each being less than 1 million USD per year.

Even though all of these numbers are uncertain estimates, it is obvious that development aid will be on an order of magnitude less than what is needed. The lack of resources was a major issue at ICCM-2, but resolutions were vague, encouraging research and further funding, including requesting GEF to consider further chemicals management. Consequently, it is highly unlikely that the commitments of developing countries under the chemicals agreements will be met at the agreed pace, and it may take ten times longer than anticipated, thus running into centuries rather than decades. It will be a great challenge for international chemical safety to falsify these gloomy predictions. As one option, a tax has been proposed on the global chemicals producing industry (IPEN 2005). A tax of 0.1% was determined to yield 1.5 billion USD annually, a number that would be of the right order of magnitude to meet developing country needs. In the SAICM Overarching Policy Strategy, the text on Financial considerations contained a follow-up of the proposal in the form 'Where appropriate, assessing and adopting at the national and sub-national levels economic instruments intended to internalise the external costs of chemicals'. It remains to be seen whether this instrument will be used; there are a few previous examples at the national level (OECD 2009e).

Fortunately, general economic growth provides greater resources for some countries. Several former developing countries and countries with economies in transition are now members of the OECD (Korea, Mexico, Turkey, Czech Republic, Slovak Republic, Poland, Hungary and Chile) and the OECD has invited Estonia, Israel and Slovenia to become members and offered enhanced engagement, with a view towards possible membership, to Brazil, China, India, Indonesia and South Africa. Similarly, the European Union now comprises more than half a dozen countries that used to be among countries with economies in transition. In addition, there is a slow migration of countries from recipients to donors of assistance, with the Czech Republic, Hungary and Thailand being examples in relation to chemical safety.

12.3.4 New Approaches Needed To Meet Increasing Risks

The OECD report (2001) predicted almost a doubling of chemicals production in the 25 years between 1995 and 2020, and a similar rate of increase was confirmed by CEFIC (2008). With lagging implementation and convention fatigue, what can be done to keep pace with this potential increase in risks?

12.3.4.1 A More Complex Chemical Safety Landscape Takes Time To Master

The obvious and immediate effects of chemical substances were regulated long ago in developed countries, and international cooperation began already in the nineteenth century. As more subtle effects, such as cancer, hereditary effects and effects on the developing organism, became known, the hazards and risks became more difficult to identify and quantify. Environmental pathways were also added to the exposure routes of interest. There was also an increased understanding that many chemicals in manufactured goods will eventually lead to exposures from the use or disposal of the goods. These developments led to a complexity in international cooperation, with developed countries worrying about subtle and long-term effects, while developing countries struggled with managing obvious and immediate effects.

The measures taken in developed countries are already quite complex. For instance:

- The European chemicals legislation called REACH encompasses some 850 pages of legal text
- For one single substance, trichloroethylene, 29 different working groups have addressed the issue of its carcinogenicity without reaching any unambiguous answer (Rudén 2001)
- There are several thousands of high production volume substances, and tens of thousands more, with lower production volumes

The complexity and the difficulties in obtaining proof entail that international regulation of chemical safety takes time and develops over decades, as shown by the two half-century examples of GHS (Section 12.3.3.2) and PCBs (Section 12.2.3.4). The time scale for risk abatement is, thus, of the same order of magnitude as the time scale for significant risk increase, a few decades.

12.3.4.2 Controlling Total World Emissions To Be Below Natural Ones?

Efforts to control a substance are often hampered by a lack of knowledge about its harmful effects. One simple policy instrument that obviates the need for knowledge has been adopted as a major principle by some countries and by the organisation 'The natural step', but has not been applied in international negotiations. It concerns the warning signal that may come from a significant accumulation of substances above the natural levels. In the long run, the average global accumulation will be governed by the anthropogenic flow in relation to the natural one. At present this ratio is, for instance, between 100 and 2,000 for copper, lead, nickel and zinc (based on OECD 2008 and Ayres and Simonis 1994). This may be of concern in the long run, if production continues at the current level over a period of time longer than the time of residence of these substances in the technosphere or biosphere; theoretically one could expect average levels of up to 2,000 times the natural ones. However, this is not likely to occur, since the existing reserves generally have a life expectancy of less than 100 years (OECD 2008), and even if new reserves are found, production rates are likely to decrease due to price increases. Still, a high ratio may be an indication that further studies on health and environmental consequences of long-term use of the substance are justified.

For strongly toxic organic substances, there is little information on natural flows even though many organic substances occur in nature. For instance, a wide range of chlorinated substances is produced in ordinary wood, and forest fires produce similar combustion products as are produced due to the combustion of petroleum products. The complexity of the thousands of substances involved makes it likely that references to natural flows of organic substances will rarely be able to provide perspectives on the degree of concern that would be reasonable in relation to anthropogenic emissions.

12.3.5 Control at the Source Instead of Cleaning Up Later

There are tens of thousands of substances that are harmful, as mentioned in Section 12.1.5. With widespread environmental exposures comes the risk that many of these will influence the same mechanism for injury to humans or other organisms as may have been the case for the previously mentioned tadpole example. Even though effects of exposures to multiple substances are systematically used to advance studies on the effects of pharmaceuticals on humans (Lehár et al. 2008), the complexities have prevented broader generalisations about deleterious effects on health and the environment. There are, however, some examples of potential rules of thumb. For instance, it has been shown (Silva et al. 2002) that the response of several substances may be best described by adding their amounts weighted by their toxicity equivalency factors.

In contrast to the existence of tens of thousands of harmful substances, major international agreements on chemical safety target only some 40 substances or substance groups, and the rate of addition of substances is of the order of one per year. New approaches will be needed to cope with the total effects of all hazardous substances. There is, however, still little policy available for chemicals concerning the combined effects of all substances from all sources via all pathways, although some efforts have been made in this direction. For instance, the United States Environmental Protection Agency is developing tools to address aggregate exposures (exposure to a single pollutant via multiple pathways) and cumulative exposures (aggregate exposure from multiple pollutants) (USEPA 2009b) and applying a life stage perspective (USEPA 2006). A case in point concerns phthalates (NRC 2008).

A global policy for all substances from all sources is, in contrast, being implemented for radioactive substances (ICRP 2008). Two characteristics of this policy are source control and the use of economic instruments. To implement a similar policy for chemical substances will require simplifications with respect to the relationships between releases from sources and exposure and between exposure and response. In the following, two potential simplifications are suggested: the intake fraction for exposures and assigned linearity for responses. Thereafter, some potential policy instruments for global use are discussed.

12.3.5.1 The Intake Fraction Links Release and Exposure

It is, of course, extremely difficult to get an idea of the exposures and effects of tens of thousands of substances arising from hundreds of thousands of different preparations and millions of different articles manufactured using these preparations. The concept of intake fraction, however, promises to yield approximations to exposures. The intake fraction is the fraction of a released substance passing through any human being at any time. This concept has attracted increasing attention as a potential tool for risk assessment for hazardous chemicals. One very interesting feature is the relatively low variability of the intake fraction for substances that are relatively persistent in nature (Jantunen et al. 2008). According to a case study (Bennett et al. 2002), the intake fraction distributions for 308 substances are very similar, whether the release is to the air or to water. The log-normal distributions are relatively narrow, with a standard deviation of about 11 times, meaning that most intake fractions are within a factor of 10 from the geometric mean of about 5 parts per million (ppm). Similar values were found using monitoring data for four radioactive substances with global dispersion (Bengtsson 1985).

12.3.5.2 Equitable Responsibility for Releases Through Assigned Linearity

In toxicology, there is a long standing tradition that more often than not dose response relationships have thresholds. A linear dose-response relationship would, accordingly, have little value in describing harm; for most exposed individuals, there would be no harm, and protection should aim at keeping exposures below certain thresholds.

In contrast, radiation protection policy (ICRP 2008) benefits from an assumption of linearity as a tool to attribute causation in an equitable way. Linear dose-response relationships could also serve a useful purpose for chemical safety, with the following qualifications:

- They should be used as one facet, probably only to protect against human health effects; other paradigms will be necessary to address other areas of protection, say, of the environment
- They would only be useful in protection against widespread exposures at low levels (obviously higher exposures exceeding certain thresholds, say, at the workplace or in conjunction with accidents would lead to acute symptoms)

- It should be recognised that they are assumptions for attributing potential harm and protective needs among sources – not descriptions of biological consequences

Arguments, within these qualifications, to assume linear dose-response relationships or to assume that increments of exposure cause a proportionate increment of response, include the following:

- Some biological dose-response relationships are nearly linear, e.g. those involving mutations or cancer initiation. The quasi-linear dose-response relationship for proteinuremia following environmental cadmium exposure to human populations may be an example. Reproductive and developmental toxicity, including neurotoxic effects on children, may also provide examples; incremental impairment of IQ in children may be linearly related to increments of low lead exposures (WHO/IPCS 2000). There is, however, a continuing debate concerning the applicability of linear or threshold relationships (Swenberg et al 2009).
- Some exposures may involve additions to considerably high levels previously existing.
- Exposures from other substances may affect the same target mechanism.
- Threshold relationships will tend to be smoothed when applied to a strongly heterogeneous population; the response may depend strongly on e.g. age, gender, genetic constitution and health status.
- If regulation is based on a designated linear dose-response relationship, responsibility will automatically be assigned in relation to the magnitude of exposures. There are no other practically useful options for attribution. It is, for instance, extremely difficult, and in practice impossible, to make a detailed assessment of potential synergistic or antagonistic effects of simultaneous exposures. Fair regulation would require that responsibility be assigned in relation to the exposures, independent of the order in which the exposures occur.

12.3.5.3 Screening Tools Can Elucidate the Need for Source Control

One simple policy application using intake fraction and presumed linearity would be to screen for potentially troublesome amounts of substances produced globally. This would require assessing the chain emission-intake-risk.

For the emission term, it can be assumed that in a longer time perspective (decades or more) all substances produced will be emitted; this will overestimate the risks, since there will be losses e.g. due to chemical transformations.

For the intake term, the intake fraction can be used. As a default value, an intake fraction of approximately 100 ppm (arithmetic mean) can be assumed; it should be remembered that the variation is large and, for instance, the geometric mean is likely to be around 5 ppm.

For the risk, the Threshold of Toxological Concern of 0.15 μg per day (Kroes et al 2005) can be applied. For comparison, it is about 60 times stricter than the guideline values for known poisons such as benzene and arsenic and 5,000 times stricter than the geometric mean of Acceptable Daily Intakes in a database of 588 substances (Australian Government 2009).

Alternatively, a threshold of concern could be assumed for the risk of serious effects, such as cancer and hereditary disease or serious impairment of the intellectual ability (loss of more than 10 IQ points). One option might be that lifetime exposure entails a lifetime risk of 10 in one million of suffering such a serious effect. This is the level of cancer risk associated with the WHO guidelines for drinking water; for the sake of comparison, it is about 400 times stricter than the risk for cancer and hereditary disease associated with radiation exposures at the dose limit for the public (ICRP 2008).

The total global emission of a substance that is tolerable according to the Threshold of Toxicological Concern is, then, about 400 t per year (assuming 7 billion people each ingesting 0.15 μg per day and an intake fraction of 100 ppm). The corresponding number according to the 10 in a million cancer risk criterion is 200 t per year (using the arithmetic mean cancer risk factor for oral intake in the database IRIS (2009), which is 54 times the value for arsenic, assumed to be 1 case per ingested kilogram by Spadaro and Rabl 2004). Considering the uncertainties in the assumptions of several orders of magnitude, the similarity of the tolerable emissions is fortuitous.

Finally, an emission of 2 million tonnes of lead per year might in equilibrium cause an average loss of 10 IQ points to the world's population (an emission of 17 kilograms of lead would cause a collective loss of 10 IQ points, according to Spadaro and Rabl 2004). If the probability of the effect were to be 1/10,000, then 20 t would be the tolerable emission.

All of these numbers are the result of a long series of assumptions and, for instance, do not account for the distribution of exposures or individual sensitivities among the population, intake fractions among substances or emission factors per produced volume among substances.

These examples indicate that there should be no concern for emissions on a global scale in relation to chemicals that are produced in production volumes below 10 t per year. Volumes below 10 t per year are subject to the lowest degree of requirements in the new REACH regulation in the European Union. This is certainly necessary for exposure situations in occupational or other use but not when it comes to global dispersion.

Above 10 t per year, relevant to some 10,000 substances in the European Union, a more detailed analysis should be considered to see if restrictions on total global production might be warranted for the case of very sensitive endpoints, such as cancer or influence on the central nervous system.

12.3.5.4 Paying for Unnecessary Emissions

In considering economic instruments, what are reasonable prices to pay for emissions? The economic detriment due to risks can be calculated using a series of assumptions (Spadaro and Rabl 2004). In two examples, the assumed numbers are 10,000 Euros per lost IQ point and 2 million Euros per case of cancer, corresponding to 6,000 Euros and 2 Euros per emitted kilogram of lead and carcinogenic substances, respectively. Estimates of detriments such as these can be used in two different ways.

Directly, the cost calculations can be a basis for restricting emissions using economic instruments such as emission fees or trading of emission permits. This might, for instance, have important applications for emissions of dioxins and other substances from combustions. The examples above indicate that the costs of a detriment can be substantial in relation to the value of the product, which is usually in the range of 1–10 Euros per kg. Consequently, such economic instruments might be expected to lead to substantial reductions in emissions.

The cost calculations can also be used in a more indirect way, following the example from global radiation protection recommendations (ICRP 2008). According to this philosophy, even small exposures should be kept as low as reasonably achievable. Legislation, then, requires that enterprises wishing to use hazardous emissions should calculate the worldwide detriment caused by them translated to economic terms. Emission reduction should be sought as long as the costs of the reduction activities are less than the costs of the prevented detriment. For instance, a scrubber preventing the emission of 10 kg of lead annually should be installed as long as the annual operating costs are below 10*6,000 = 60,000 Euros.

12.3.5.5 Policy Developments Overdue

There are already documented effects from environmental exposures on, for instance, children's health and reproduction in aquatic organisms. With chemicals production and thereby releases doubling over the next generation, the time to take action to prevent the future occurrence of even worse effects from a multitude of substances and sources is well overdue. Tools, such as intake fraction and assumed linearity, and economic instruments need to be further developed and applied. The weak trends in this direction that have been observed during the past decade need to be strengthened in order to achieve the globally agreed goal that by 2020 chemicals be used and produced in ways that lead to the minimisation of significant adverse effects on human health and the environment.

References

Australian Government (2009) Acceptable daily intakes for agricultural and veterinary chemicals. Department of Health and Ageing. http://www.health.gov.au/internet/main/publishing.nsf/content/ocs-adi-list.htm. Cited 22 February 2009.

Ayres, R.U., and Simonis, U.E. (1994) *Industrial Metabolism: Restructuring for Sustainable Development*. United Nations University Press, New York.

Bengtsson, G. (1985) Judging the long-term acceptability of radioactive waste disposal practices from the radiation protection point of view and other perspectives. Proceedings of Seminar on Interface Questions in Nuclear Health and Safety, OECD, Paris.

Bennett, D. H., et al. (2002) Intake fraction for multimedia pollutants: a tool for life cycle analysis and comparative risk assessment. *Risk Analysis* 22: 905–918.

Buccini, J. (2004) Global pursuit of the sound management of chemicals. The International Bank for Reconstruction and Development/The World Bank, Washington DC. http://siteresources.worldbank.org/INTPOPS/Publications/20486416/GlobalPursuitOfSoundManagementOfChemicals2004Pages1To67.pdf. Cited 12 January 2009.

Carballo, M., et al. (2008) Organochlorine residues in the blubber and liver of bottlenose dolphins (Tursiops truncatus) stranded in the Canary Islands, North Atlantic Ocean. *Environmental Toxicology* 2008: 200–10. http://www.ncbi.nlm.nih.gov/pubmed/18214918. Cited 28 January 2009.

CDC (2005) Third national report on human exposure to environmental chemicals executive summary. Centers for Disease Control and Prevention, Atlanta. http://www.cdc.gov/exposurereport/pdf/thirdreport_summary.pdf. Cited 29 January 2009.

CEFIC (2008) The European chemical industry: competing in global markets. Presentation at Ad hoc Group Meeting on Trade and Competitiveness with Other Regions, 2 June, 2008. http://ec.europa.eu/enterprise/chemicals/hlg/hlg2/pdf_docs/wg_02june08/magdalenic_meincke_hadhri_gronych_competing_global_market.pdf. Cited 12 January 2008.

CIEN (2009) Chemical information exchange network. http://jp1.estis.net/communities/cien/default.asp?community=cien&page_id=A03A2E9E-9330-44E9-886B-CDD8537A9DAA. Cited 14 January 2009.

Chemsec (2009) The international chemical secretariat. http://www.chemsec.org/index.php. Cited 21 January 2009.

Colborn, T., Dumanoski D., and Myers, J. P. (1996) *Our Stolen Future*. Penguin Books, New York.

CSIRO (2009) Trace elements in Australian export thermal coals. Commonwealth Scientific and Industrial Research Organisation, Australia. http://www.csiro.au/partnerships/TraceElements.html. Cited 7 March 2009.

DiGangi, J. (2003) US intervention in EU chemical policy. Environmental Health Fund, Chicago. http://www.cleanproduction.org/library/USIntervention.pdf. Cited 17 June 2010.

Eckley, Selin, N., and Selin, H. (2006) Global politics of mercury pollution: the need for multi-Scale governance. *RECIEL* 15: 258–269. http://www.ingentaconnect.com/content/bpl/reel/2006/00000015/00000003/art00003. Cited 10 February 2009.

Fei, C., et al. (2008) Fetal growth indicators and perfluorinated chemicals: a study in the Danish national birth cohort. *American Journal of Epidemiology* 168(1): 66–72.

GEF (2007a) Sound chemicals management framework strategy and strategic programming for GEF-4. Global Environment Facililty. http://www.gefweb.org/uploadedFiles/Policies/Focal_Area_Strategies/GEF_4_strategy_SCM_Oct_2007.pdf. Cited 19 January 2009.

GEF (2007b) China - strengthening institutions, regulations and enforcement capacities for effective and efficient implementation of the National Implementation Plan (NIP). China Global Environment Facility. http://gefweb.org/uploadedFiles/Documents/Council_Documents__(PDF_DOC)/GEF_31/UNIDO-China%20SIRE%20PES%20final%2027Apr07.pdf#. Cited 11 February 2009.

Hayes, T., et al. (2006) Pesticide mixtures, endocrine disruption, and amphibian declines: are we underestimating the impact? *Environmental Health Perspectives* 114(9).

Hovander, L., et al. (2002) Identification of hydroxylated PCB metabolites and other phenolic halogenated pollutants in human blood plasma. *Archives of Environmental Contamination and Toxicology* 42: 105–117.

Houlihan J, et al. (2005) Body burden – the pollution in newborns: a benchmark investigation of industrial chemicals, pollutants and pesticides in umbilical cord blood. Environmental Working Group. http://archive.ewg.org/reports/bodyburden2/execsumm.php. Cited 28 January 2009.

ICCA (2009) The International Council of Chemical Associations (ICCA) initiatives. http://www.icca-chem.org/section02.html. Cited 13 January 2009.

ICCM (2009) Recommendation of the ad hoc joint working group on enhancing cooperation and coordination among the Basel, Rotterdam and Stockholm conventions. Meeting document SAICM/ICCM.2/INF/2. http://www.saicm.org/documents/iccm/ICCM2/meeting%20documents/ICCM2%20INF2%20AHJWG.doc. Cited 13 January 2009.

ICRP (2008) ICRP Publication 103: Recommendations of the ICRP. Annals of the ICRP Volume 37/2-4, International Commission on Radiological Protection Elsevier, Oxford.

IFCS (2003a) Protecting children from harmful chemical exposures. Information document at Fourth Session of the Intergovernmental Forum on Chemical Safety, Bangkok, Thailand, 1–7 November 2003. http://www.who.int/ifcs/documents/forums/forum4/en/11inf_en.pdf. Cited 19 January 2009.

IFCS (2003b) Current practices and proposals for generation, assessment and accessibility of hazard data. Fourth Session of the Intergovernmental Forum on Chemical Safety Bangkok, Thailand, 1-7 November 2003. http://www.who.int/ifcs/documents/forums/forum4/en/09w_an_en.doc. Cited 14 January 2009.

IFCS (2003c) Assistance for capacity building. Report for Forum IV Fourth Session of the Intergovernmental Forum on Chemical Safety Bangkok, Thailand, 1–7 November 2003. http://www.who.int/ifcs/documents/forums/forum4/en/11w_f4_en.doc. Cited 20 January 2009.

IFCS (2009) The Intergovernmental Forum on Chemical Safety. http://www.who.int/ifcs/en/. Cited 13 January 2009.

IOMC (2009) National implementation of SAICM: a guide to resource, guidance, and training materials of IOMC participating organisations. IOMC and UNITAR, Geneva. http://www.who.int/iomc/saicm/resource_guide.pdf. Cited 2 March 2009.

IOMC (2010) IOMC Strategy for Strenthening National Chemicals Management Capacities. December 2008. http://www.who.int/iomc/publications/strategy_english.pdf. Cited 17 June 2010.

IPCS (2009) International programme on chemical safety. http://www.who.int/ipcs/en/. Cited 13 January 2009.

IPEN (2005) Paper submitted by the international POPs elimination network on internalization of SAICM costs within relevant producer industries. Document SAICM/PREPCOM.3/INF/12 at the Third Session of the Preparatory Committee for the Development of a Strategic Approach to International Chemicals Management, Vienna, 19–24 September. http://www.saicm.org/documents/meeting/prepcom3/en/INF12%20IPEN.doc. Cited 10 February 2009.

IPEN (2009) The international POPs elimination network (IPEN). http://www.ipen.org/ipenweb/firstlevel/about.html. Cited 9 January 2009.

IRIS (2009) Integrated Risk Information System (IRIS). US Environmental Protection Agency. http://cfpub.epa.gov/ncea/iris/index.cfm. Cited 18 February 2009.

Jantunen M.J., et al. (2008) Final report for iF-project: the intake fraction (iF) concept; expansion and application tests. Report of Cefic-LRI contract B3.3-KTLE-0511 KTL – Department of Environmental Health, Kuopio. http://www.cefic-lri.org/uploads/ModuleXtender/MCLibrary/1224664022/documents/iF_final_report.pdf. Cited 10 February 2009.

KemI (2009) Overview of chemicals. Danger class distribution – products. http://www.kemi.se/templates/Page____3921.aspx#Danger%20class%20distribution%20-%20products. Cited 10 February 2009.

Kroes, R., et al. (2005) The threshold of toxicological concern concept in risk assessment. *Toxicological Sciences*. DOI: 10.1093/toxsci/kfi169.

Latini G., et al (2003) In-Utero Exposure to Di-(2-ethylhexyl)-phthalate and Human Pregnancy Duration. *Environmental Health Perspectives* 111: 1783–1785. http://www.pubmedcentral.nih.gov/articlerender.fcgi?artid=1241724. Cited 29 January 2009.

Lehár J., et al. (2008) Combination chemical genetics. *Nature Chemical Biology*. DOI: 10.1038/nchembio.120

Lönngren, R. (1992) International Approaches to Chemicals Control. An Historical Overview. Kemikalieinspektionen, Stockholm.

Massart, F., et al. (2008) Chemical biomarkers of human breast milk pollution. *Biomarker Insights* 3: 159–169. http://www.la-press.com/chemical-biomarkers-of-human-breast-milk-pollution-a671. Cited 25 January 2009.

Mitchell, R.B. (2009) Data from Ronald B. Mitchell, 2002–2007. International Environmental Agreements Database Project (Version 2007.1). http://iea.uoregon.edu/. Cited 13 January 2009.

MSDSonline (2009) Where to find material safety data sheets online. http://www.ilpi.com/msds/index.html. Cited 14 January 2009.

NRC (2008) *Phthalates and Cumulative Risk Assessment: The Task Ahead*. Committee on the Health Risks of Phthalates, National Research Council. The National Academies Press, Washington DC. http://books.nap.edu/catalog.php?record_id=12528#orgs. Cited 29 January 2009.

NTP-CERHR (2008) NTP-CERHR Monograph on the potential human reproductive and developmental effects of bisphenol A. September 2008 NIH Publication No. 08 – 5994. http://cerhr.niehs.nih.gov/chemicals/bisphenol/bisphenol.pdf. Cited 25 January 2009.

OECD (2001) Environmental outlook for the chemicals industry. Organisation for Economic Co-operation and Development, Paris. http://www.oecd.org/dataoecd/7/45/2375538.pdf. Cited 12 January 2009.

OECD (2003) Fourth Survey of OECD Member Country Development Assistance Activities for Capacity Building. Document IFCS/FORUM-IV/4 INF Forum IV Fourth Session of the Intergovernmental Forum on Chemical Safety Bangkok, Thailand, 1–7 November 2003. http://www.who.int/ifcs/documents/forums/forum4/en/4inf_en.doc Cited 17 June 2010.

OECD (2008) Measuring material flows and resource productivity. Synthesis report. OECD, Paris. http://www.oecd.org/dataoecd/55/12/40464014.pdf. Cited 6 March 2009.

OECD (2009a) Chemical safety. http://www.oecd.org/topic/0,3373,en_2649_34365_1_1_1_1_37465,00.html. Cited 13 January 2009.

OECD (2009b) eChemPortal: the global portal to information on chemical substances. http://webnet3.oecd.org/echemportal/. Cited 14 January 2009.

OECD (2009c) Chemicals testing – guidelines. http://www.oecd.org/department/0,3355,en_2649_34377_1_1_1_1_1,00.html. Cited 14 January 2009.

OECD (2009d) Agricultural chemical pesticide registration. http://www.oecd.org/document/15/0,3343,en_2649_34383_32167055_1_1_1_1,00.html. Cited 14 January 2009.

OECD (2009e) OECD/EEA database on instruments used for environmental policy and natural resources management. http://www2.oecd.org/ecoinst/queries/index.htm. Cited 5 March 2009.

Prevent (2009) Chemical substances database. http://kemi.prevent.se/default_eng.asp. Cited 10 February 2009.

Rudén, C., (2001) Interpretations of primary data in 29 trichloroethylene carcinogen risk assessments. *Toxicology* 169(3): 209–225.

SAICM (2006) Strategic approach to international chemicals management. http://www.saicm.org/documents/saicm%20texts/SAICM_publication_ENG.pdf. Cited 10 February 2009.

SAICM (2010) SAICM Information clearing house. http://www.saicm.org/ich/Cited 17 June 2010.

Selin, H. (2009) *Managing Hazardous Chemicals: Longer-range Challenges*. The Pardee Papers No. 5. The Frederick S. Pardee Center for the Study of the Longer-Range Future, Boston.

Silva, E., et al. (2002) Something from 'nothing'-eight weak estrogenic chemicals combined at concentrations below NOECs produce significant mixture effects. *Environmental Science Technology* 36: 1751–1756.

Spadaro, J.V., Rabl, A. (2004) Pathway analysis for population-total health impacts of toxic metal emissions. *Risk Analysis* 24: 1121–1141.

Sterner, T. (2002) *Policy Instruments for Environmental and Natural Resource Management*. Resources for the Future Press, co-published with World Bank and Swedish International Development Cooperation Agency, Washington.

Swenberg J.A., et al. (2009) Response to Keshava et al. (2008). *Chemical Research in Toxicology* 22(1).

Trasande, L., Landrigan, P.J., and Schechter, C. (2005) Public health and economic consequences of methyl mercury toxicity to the developing brain. *Environmental Health Perspectives* 113: 590–596. http://www.ehponline.org/members/2005/7743/7743.pdf. Cited 19 January 2009.

UNDP (2009) UNDP Technical Guide for Integrating the Sound Management of Chemicals in MDG-Based Policies & Plans UNDP New York May 2009. http://www.undp.org/chemicals/documents/UNDP%20Technical%20Guide%20on%20Mainstreaming.pdf. Cited 17 June 2010.

UNECE (2003) Convention on access to information, public participation in decision-making and access to justice in environmental matters, Kiev protocol on pollutant release and transfer registers. http://www.unece.org/env/pp/prtr/docs/prtrtext.htm. Cited 10 February 2009.

UNECE (2007) Globally harmonized system of classification and labelling of chemicals (GHS). Second revised edition. UNECE, Geneva. http://www.unece.org/trans/danger/publi/ghs/ghs_rev02/02files_e.html. Cited 14 January 2009.

UNEP (2009a) Stockholm Convention on persistent organic pollutants. http://chm.pops.int/Convention/tabid/54/language/en-US/Default.aspx. Cited 14 January.

UNEP (2009b) Notice concerning a new mercury treaty. http://www.chem.unep.ch/mercury/148-000000003F53D720F8C9E34FBC565730B3736318070016BCB.pdf. Cited 21 February 2009.

UNITAR and ECB (2009) National profile homepage. http://www2.unitar.org/cwm/nphomepage/index.html. Cited 17 January 2009.

United Nations (2008) The Millennium Development Goals Report 2008. Published by the United Nations Department of Economic and Social Affairs (DESA). http://www.un.org/millenniumgoals/pdf/The%20Millennium%20Development%20Goals%20Report%202008.pdf. Cited 10 January, 2009.

USEPA (2006) A framework for assessing health risk of environmental exposures to children (Final). U.S. Environmental Protection Agency, Washington. http://cfpub.epa.gov/ncea/cfm/recordisplay.cfm?deid=158363. Cited 29 January 2009.

USEPA (2009a) Endocrine disruptor screening program (EDSP). United States Environmental Protection Agency, Washington. http://www.epa.gov/scipoly/oscpendo/index.htm. Cited 28 January 2009.

USEPA (2009b) Aggregate and cumulative Risk. United States Environmental Protection Agency, Washington. http://www.epa.gov/heasd/risk/. Cited 29 January 2009.

WHO/IPCS (2000) Safety evaluation of certain food additives and contaminants. WHO food additive series 44, WHO/IPCS, Geneva.

WHO (2002) The World Health Report 2002. World Health Organization, Geneva. http://www.who.int/whr/2002/en/. Cited 29 January 2009.

WHO (2005) Toxic hazards. http://www.who.int/heli/risks/toxics/chemicals/en/index.html. Cited 18 January 2009.

WHO (2008) World Malaria Report 2008. World Health Organization, Geneva. http://malaria.who.int/wmr2008/. Cited 19 January 2009.

World Bank (2002) Toxics and poverty: the impact of toxic substances on the poor in developing countries. http://web.worldbank.org/WBSITE/EXTERNAL/TOPICS/ENVIRONMENT/EXTPOPS/0,,contentMDK:20486400~menuPK:1165908~pagePK:148956~piPK:216618~theSitePK:408121~isCURL:Y~DIR_PATH:WBSITE/EXTERNAL/TOPICS/ENVIRONMENT/EXTPOPS/,00.html. Cited 18 January 2009.

WSSD (2002) World summit on sustainable development (WSSD) Johannesburg Plan of Implementation, Johannesburg. http://www.un.org/esa/sustdev/documents/WSSD_POI_PD/English/POIToc.htm. Cited 10 March 2009.

Chapter 13
Regulating Chemical Risk: REACH in a Global Governance Perspective

Veerle Heyvaert

Abstract This chapter analyses the EU REACH Regulation as a blueprint for an international model of risk governance. It reviews the institutional set-up of REACH, documenting a shift of decision-making authority away from the State level towards the private and European level, and explains why the Member States of the EU agreed to limit their decision-making power. It then considers the potential for REACH to be exported beyond EU boundaries, contemplates two globalisation models, and discusses one of the challenges of the proliferation of REACH as a global standard. The chapter argues that, while the adoption of REACH abroad may bring improvements in trade relations and health and environmental protection, these benefits risk to be substantially reduced and even reversed if REACH is incorporated in an institutional setting that is not equipped to deal with its managerial and administrative demands.

Keywords Centralisation • Global governance • Privatisation • REACH • Risk regulation

13.1 The Evolution of European Risk Regulation: The Road to REACH

Past experience informs today's choices. As one of the first areas for market harmonisation, the regulation of chemicals has a rich history in EU law (Heyvaert 1999). It was also one of the first fields where regulation targeted a dual objective of market liberalisation and health and environmental risk control, casting the mould for future EU risk regulation across product sectors. The following sections sketch the main building blocks of the EU chemicals control regime before REACH,

V. Heyvaert (✉)
Department of Law, London School of Economics (LSE), Houghton Street, London, WC2A 2AE, United Kingdom
e-mail: v.heyvaert@lse.ac.uk

J. Eriksson et al. (eds.), *Regulating Chemical Risks: European and Global Challenges*, DOI 10.1007/978-90-481-9428-5_13,

analyse the reasons for reform, and shed some light onto the reform process resulting in the adoption of REACH, thus enabling a deeper, but also more critical, understanding of the institutional and normative choices made in REACH.

13.1.1 Chemicals Control in the EU Before REACH

Prior to June 2007, chemicals control in the EU was governed by a network of Directives and Regulations. Community law identified categories of 'dangerous' substances and preparations, for which harmonised classifications (substances) and packaging and labelling rules (substances and preparations) applied.[1] EU law furthermore distinguished between 'existing substances', which were those circulating on the Community market prior to September 1981, and 'new' substances. A 1979 EU Directive conditioned market access for new substances upon the notification of a comprehensive technical dossier to the national regulatory authority (NRA) of the Member State where access was sought.[2] In exchange for the information supplied, manufacturers and importers gained a one-stop shop facility, whereby a single notification was passed around to the Commission and other EU NRAs, and was in principle recognised throughout the EU. For the aforementioned 'existing' chemicals however, the imposition of post-marketing information supply and testing requirements was considered too onerous and potentially disruptive to the economy. Hence, during the first decade after notification duties were introduced, EU law did not foster information supply concerning existing chemicals in a systematic way.

As of the early 1990s, EU law instructed NRAs to perform a risk assessment on the basis of the information in the notification dossier, in accordance with newly enacted Community risk assessment standards.[3] The risk recommendations flowing from the risk assessment could constitute a basis for regulatory action at the EU or, residually, the national level (Heyvaert 2001).[4] At the EU level, such regulatory intervention would typically take the form of a marketing and use restriction, adopted via legislative amendment of the Marketing and Use Restrictions Directive.[5]

Notification ensured information about the risks posed by new chemicals, but nearly nothing was known about the 100,000 existing chemicals, of which 70,000 are still traded, and 30,000 traded in significant volumes. In 1993, the Council

---

[1]Directive 67/548/EEC on the approximation of laws, regulations and administrative provisions relating to the Classification, Packaging and Labelling of Dangerous Substances [1967] OJ 196/1.

[2]Sixth Amendment to Dir. 67/548/EEC [1979] OJ L259/10.

[3]Directive 93/67/EEC laying down the Principles for Assessment of Risks to Man and the Environment of Substances Notified in accordance with Council Directive 67/548/EEC [1993] OJ L227/9.

[4]Cx [2000] ECR I-9741.

[5]Directive 76/769/EEC on the approximation of the laws, regulations and administrative provisions of the Member States relating to Restrictions on the Marketing and Use of Certain Dangerous Substances and Preparations [1976] OJ L 262/201.

adopted a Regulation (Existing Substances Regulation) to fill this data gap.[6] Following the Existing Substances Regulation, manufacturers and importers were to report available chemical data directly to the Commission. The European Chemicals Bureau (ECB) collected the various submissions[7] and processed everything into an EU-wide database (EUCLID). Based on EUCLID, the Commission drew up priority lists enumerating those substances with the highest risk potential, and assigned them to different Member States for further data gathering and assessment. The Member States authorities reported their findings back to the Commission in the form of a risk recommendation. If the risk recommendation indicated that regulatory interventions were necessary to control identified health and/or environmental risks, the Commission would draft a legislative proposal for risk reduction, either under the Marketing and Use Restrictions Directive or under an alternative EU measure (for instance, the Control of Chemical Agents at Work Directive).[8]

To recap, before REACH national authorities took charge of data gathering for new substances, and Community authorities received data on existing substances. NRAs performed risk assessments and risk evaluations. Marketing and use restrictions were adopted in a 1976 Community Directive and ensuing legislative amendments, or residually at the national level. REACH significantly alters many of these operational and institutional arrangements.

13.1.2 Why Reform?

By the mid-1990s, EU trade in chemicals was covered by an expansive network of Community legislation. Harmonised product standards facilitated free trade, and the regime was one of the most proactive in terms of detecting and controlling chemical risks, certainly when compared to its US counterpart, the 1976 Toxic Substances Control Act. Why, then, did the Commission decide in 1998, a mere 5 years after the enactment of the Existing Substances Regulation, to launch a 360° review of the keystones of EU chemicals regulation?[9]

A first reason is that some parts of the regulatory framework were not performing adequately. The notification procedure did foster trade across borders and generate information on new substances, but the Existing Substances Regulation spectacularly failed to deliver, mostly because industry had no incentive to cooperate, Member States proved reluctant to enforce the provisions or back them up with penalties, and risk evaluations progressed too slowly (Heyvaert 1999b). Since existing substances

---

[6]Regulation 793/93/EEC on the Evaluation and Control of the Risks of Existing Substances [1993] OJ L84/1.

[7]See Commission Communication to the Council and the EP – The European Chemicals Bureau [1993] OJ C1/3.

[8][1998] OJ L131/11.

[9]Commission Working Document – Report on the operation of Directive 67/548/EEC, Directive 88/379/EEC, Regulation (EEC) 793/93, Directive 76/769/EEC SEC(1998)1986, 18 Nov. 1998.

were less, and less effectively, regulated than new substances, the EU regime created incentives for manufacturers and importers to stick to the old rather than invest in the new (Stewart 1981). In addition to stifling innovation, the approach was undesirable from a health and environmental protection perspective, since old chemicals usually pose greater risks than newer generations.

Second, chemical risk awareness had risen considerably. This chapter will not detail the many factors influencing chemical risk tolerance in the 1990s, but various developments, such as heightened public risk sensitivity in the wake of the BSE crisis and the adoption of the precautionary principle in the 1992 Maastricht Treaty, certainly played a part. On the latter point, chemicals reform constituted a major opportunity for the EU to express its vision on the meaning and application of the precautionary principle. Throughout its gestation, REACH would become a calling card for the EU interpretation of precaution, thus helping to lend a stronger identity and cohesion to the European style of risk regulation (Heyvaert 2006).

The effectiveness of harmonised product regulation as a trade liberalisation mechanism and a public protection instrument crucially hinges on its decision-making speed. Here, too, the EU approach left room for improvement. Any results generated from the Existing Substances Regulation came at a snail's pace. Similarly, marketing and use restrictions stayed in the pipeline for too long, largely because any new restriction required an amendment of the 1976 Directive, which in turn needed Council and European Parliament approval. Increasing decision-making speed was at the forefront of considerations for reform.[10]

Reform also created an opportunity to streamline the near bewildering tangle of Directives and Regulations that governed chemicals control, in line with the ideas inspiring the EU's Better Regulation Agenda.[11] Greater simplicity could be achieved, first, by replacing the multitude of existing measures with a single instrument and,[12] second, by switching from a devolved, indirect approach to chemicals management, where EU Directives stipulate the basic requirements and the Member States transpose and implement, to a more direct form of control that increases the uniformity of manufacturing, marketing, and trading conditions for chemicals across the EU, and that is centrally administered. The Commission asserted that the call for streamlining and greater uniformity was supported by 'many stakeholders', who were concerned that the new regime would be too bureaucratic and inefficient if the distribution of tasks between Member States authorities, the Commission, and ECHA were too complex.[13]

---

[10]Commission Working Document – Report on the operation of Directive 67/548/EEC, Directive 88/379/EEC, Regulation (EEC) 793/93, Directive 76/769/EEC SEC(1998)1986, 18 Nov. 1998.

[11]See Commission Communication on Implementing the Lisbon Programme: A Strategy for the Simplification of the Regulatory Environment COM (2005)535, 25 Oct. 2005.

[12]Commission Communication on Updating and Simplifying the Community Acquis COM (2003)71, 11 Feb. 2003.

[13]Proposal for a Regulation of the European Parliament and of the Council concerning the Registration, Evaluation, Authorisation and Restriction of Chemicals (Reach), establishing a European Chemicals Agency and amending Directive 1999/45/EC and Regulation (EC) {on Persistent Organic Pollutants} {SEC(2003 1171} COM (2003) 644 , 29 Oct. 2003.

EU enlargement from 15 to 25 and then 27 States was also a factor. The Commission observed that homogenous and directly applicable rules, as opposed to a legal framework leaving extensive scope for national implementation and diversity, would be of particular benefit in the enlarged Community. The acceding Member States, still recovering from the Herculean effort of adopting the *acquis communautaire* in preparation of EU Membership, might particularly value a regulatory framework that was simpler, more accessible, and that left a greater proportion of regulatory responsibilities to the EU rather than to domestic, already seriously overtaxed administrations.

A final factor to consider in the EU's decision to overhaul its chemicals regulation relates to the challenges posed by globalisation. The increased circulation of goods, services, labour, and capital across the globe puts pressure on domestic and regional regulation, which can be recast, scrutinised, and potentially condemned as an illegitimate obstacle to free trade. Authors such as Meunier and Jacoby argue that, to withstand the eroding force of market globalisation, the EU engages in 'defensive globalisation management,' meaning that it seeks to protect its authority and autonomy as a risk regulator by developing robust, cohesive, and efficiently functioning regulatory frameworks that can stand up to external scrutiny and be promoted as the gold standard for risk regulation (Meunier and Jacoby 2007). Globalisation concerns have most likely affected the drive towards chemicals reform,[14] and have most certainly influenced the shape of the reformed framework, as will be discussed further below.

13.1.3 Negotiating and Adopting REACH

The reform process, kick-started in 1997 with broad Commission consultation on the strengths and weaknesses of the key EU legal instruments for chemical control,[15] culminated in December 2006 in the adoption of the REACH Regulation.[16] The Regulation demands registration of all chemical substances, on their own, in preparations, or in articles, that are produced, traded, or imported onto the EU market. The Regulation sets a standard of 'no data, no market'. Chemicals for which technical data have not been submitted, should be taken out of circulation. The technical data submitted in the course of registration are used to identify chemicals that may pose

[14]In a Q&A on REACH, the Commission explicitly confirmed that 'the EU has taken a constructive international leadership role on chemicals safety and REACH has the potential to inspire new standards worldwide. European Commission (2006) Q&A on the New Chemicals Policy, REACH, MEMO/06/488. http://europa.eu/rapid/pressReleasesAction.do?reference=MEMO/06/488.

[15]Commission Working Document – Report on the operation of Directive 67/548/EEC, Directive 88/379/EEC, Regulation (EEC) 793/93, Directive 76/769/EEC SEC(1998)1986, 18 Nov. 1998.

[16]Regulation (EC) No 1907/2006 of the European Parliament and of the Council of 18 December 2006 concerning the Registration, Evaluation, Authorisation and Registration of Chemicals (REACH), establishing a European Chemicals Agency, amending Directive 1999/45/EC and repealing Council Regulation (EEC) No 793/93 and Commission Regulation (EC) No 1488/94 as well as Council Directive 76/769/EEC and Commission Directives 91/155/EEC, 93/67/EEC, 93/105/EC and 2000/21/EC [2006] OJ L396/1 ('REACH').

a risk and should be investigated, which happens in a process called 'substance evaluation'. REACH furthermore makes the use of certain categories of highly dangerous chemicals, such as CMRs,[17] subject to authorisation, which will only be granted if the applicant shows that the risks of the chemical are adequately controlled, or that they are managed and no less dangerous substitutes are available. The final pillar of REACH incorporates all the marketing and use restrictions enacted to date, and sets out a procedure for Commission adoption of further restrictions.

The rich and dynamic history of REACH's genesis, of the vociferous responses to the 2001 Commission White Paper that outlined its vision on the future of chemicals management, of the intense debates between the legion interest groups affected by the impending rules, and of the ultimate compromises struck to enable political agreement on REACH, has been consummately described and analysed elsewhere, and does not need repeating (Molyneux 2005; Fischer 2008; Pesendorfer 2006; Selin 2007). It suffices to observe a few pertinent points. First, REACH does live up to its objective of combining the subject matter of a variety of Regulations and Directives, and could therefore be seen as a simplified, more direct form of regulatory intervention.[18] Second, starting with the release of the 2001 White Paper on the Strategy for a Future Chemicals Policy, the Commission strongly profiled the new chemicals strategy, which would blossom into REACH, as a precautionary strategy, overtly in compliance with the precautionary principle as stated in Article 191 of the Treaty on the Functioning of the European Union (TFEU). Finally, while, particularly at the start of the negotiation process, the reaction of the chemicals industry to the REACH proposal, and the new regulatory burdens it entailed, was vehement, perhaps the strongest opposition to REACH came from third countries, who saw its regulatory prescriptions as unnecessary obstacles to international trade in chemicals, and called into question the legitimacy of the precautionary principle on the basis of which the most stringent of those restrictions were justified.[19]

13.2 REACH and the Transformation of Regulatory Decision-making

The REACH Regulation is characterised by, first, a strong emphasis on the privatisation of risk management functions and, second, centralisation of public decision-making (Fischer 2008). Both features shift risk responsibilities away from the national level. The following sections describe how privatisation and centralisation of decision-making occur within REACH, and discuss the reasons behind the transformation.

---

[17]Carcinogens, mutagens, and reprotoxins.

[18]Although some aspects of chemicals control, such as classification, packaging and labelling, remain outside its remit.

[19]EUOBserver (2006) EU Chemicals Bill Under Fire From US-Led Coalition. http://www.euobserver.com.

13.2.1 Privatisation of Chemical Control Responsibilities Under REACH

The privatisation of risk control obligations under REACH is most tellingly reflected in the chemical sector's responsibilities for risk identification and assessment. Chemicals manufacturers, importers, and downstream users must generate the body of information on which later private or public risk decisions rest. Industrial data production and supply commitments already existed under the previous framework, however they have been considerably extended, first, by the inclusion of broader categories of chemicals (chemical preparations and chemicals in articles) for which data must be generated, and, second, by the introduction of residual information supply duties for downstream users. Data supply duties are further strengthened by the introduction of more exacting information duties for old (existing) chemicals and the establishment, through the 'no data, no market' provision, of clearer incentives for producers, importers, and users of old chemicals to comply with their reporting duties.

Even more significantly, risk assessment responsibilities have been devolved to the private sector. Registration of chemicals produced or imported in quantities of at least 10 t per manufacturer and per year (pm/py) must be accompanied by a Chemical Safety Report (CSR), which is a renamed but other otherwise largely unchanged risk assessment as formerly required from NRAs under Directives 67/548/EEC[20] and 93/67/EEC.[21]

Beyond identification and assessment duties, private parties are more prominently involved in risk management than before. Preliminarily, we recall that the boundaries between risk assessment and management are somewhat artificial, and that the development and implementation of risk assessment protocols is preconditioned on risk management choices (for instance, determining safety factors to extrapolate safe use levels from laboratory testing results) (Finkel 1994). Moreover, the requirement for private registrants to draw up guidance notes for safe use and, for chemicals above a 10 t threshold, a CRS and safety data sheets, implies that contextual factors pertaining to the use and anticipated or known exposure of the substance emerge in and can influence risk identification processes. The need to contemplate use and exposure, and to formulate safe use protocols, may affect decisions on whether to pursue commercialisation of a new substance at an early stage in the chemical engineering process, before the costs are sunk, and decisions on whether to apply for registration of older chemicals, thus hopefully fostering a higher level of self-selection within the industry (Koch and Ashford 2006). Finally, use information is not only submitted to public authorities, but is passed through the supply chain, with producers and importers furnishing guidance notes and other

[20]Directive 67/548/EEC on the approximation of laws, regulations and administrative provisions relating to the Classification, Packaging and Labelling of Dangerous Substances [1967] OJ 196/1.

[21]Directive 93/67/EEC laying down the Principles for Assessment of Risks to Man and the Environment of Substances Notified in accordance with Council Directive 67/548/EEC [1993] OJ L227/9.

safety information to downstream users, and the latter giving feedback to their suppliers on the basis of working experience. Each of these dynamics strengthens the risk managerial impact of private action.

13.2.2 Centralising Regulatory Decision-making

The private risk management incentives in REACH should ideally reduce the need for public risk control, but chemicals control under REACH remains an essentially and predominantly public mandate. It is also a strongly centralised task, as the two pivotal institutions in charge of implementing, managing, and administering REACH are EU bodies: the European Commission, and the newly established European Chemicals Agency (ECHA). In fact, compared to the preceding framework, REACH significantly disempowers the Member State as a risk regulator. This disempowerment is effectuated in four ways, conveniently illustrated in the four main segments of REACH: registration, evaluation, authorisation, and restriction.

Centralisation occurs within the registration process, which reallocates some of the tasks formerly carried out by NRAs to ECHA. Following Article 5 of the REACH Regulation, all chemical substances, chemicals in preparations and in articles must be registered (provided they are not subject to listed exemptions). After a transition period, any chemical that does not bear a registration number should be taken out of circulation.

To complete a registration, applicants must submit a voluminous dossier with technical data, ranging from the name of the substance to its physico-chemical properties and a base set of toxicity and eco-toxicity test results. While notifiers of new substances under the Sixth Amendment to Directive 67/548/EEC submitted this dossier to NRAs, new and existing substance producers and importers alike are now expected to report directly to ECHA. This shift deprives Member States of an early opportunity to communicate with the registrant and, where indicated, stage an early formal or informal intervention.[22]

Second, the more centralised organisational features of the old regime are precisely those that survived the REACH reform. The Commission and the ECB used to receive and manage existing chemicals data. REACH undoes the distinction between new and existing chemicals, meaning that existing chemicals now also need to be registered with ECHA. This does mark an institutional change from the previous setup, however it is a change carried out within the same (centralised) governance level.

The Substance Evaluation process, in turn, rests on a near identical institutional configuration as the one formerly supporting existing substances prioritisation and

[22]Even though, at the stage of notification and now registration, the competent authority is only expected to perform a completeness check, ample documentary evidence reveals that the responsibilities of NRAs in the course of notification went far beyond box-ticking. Notification of problematic new substances would often lead to intense exchanges between the notifier and the NRA, at times causing the former to rethink its marketing plans for the new substance. See Commission Working Document, n. 15 above, p. II-1 and II-18.

evaluation.[23] The identification of chemicals that should be prioritised for further investigation, assessment and, possibly, intervention, happens at Community level within ECHA. Member States evaluate assigned chemicals, draft recommendations, and report back to ECHA. It would be wrong to portray the Member States as hapless subcontractors in this process, since they have a vital influence on the development of criteria for prioritisation, on the composition of the Community rolling action plan for evaluation of priority substances, on the selection of the Member State rapporteur and, obviously, on the evaluation and ensuing risk recommendations,[24] but it is ultimately ECHA, not the Member States, that adopts the action plan.

The third form of centralisation under REACH occurs within the newly launched authorisation process. The Commission, rather than the NRAs, decides on authorisation requests, and determines EU-wide authorisation conditions. Moreover, ECHA's influence is clearly felt at every stage. First ECHA is the lead institution preparing the technical dossier for the identification of substances for inclusion in Annex XIV of the REACH Regulation, which contains the list of substances subject to authorisation.[25] Member State interests in this process are represented via ECHA's Member State Committee. The Member State Committee gives its opinion on the recommendation for inclusion, which is passed on to the Commission together with the recommendation itself, as well as any public comments on the recommendation. The Commission decides on inclusion following a comitology procedure (the regulatory committee procedure with scrutiny).[26]

Once included in Annex XIV, the manufacturers, importers, and/or downstream users of the affected substance must request authorisation. The application is directly submitted to ECHA, which forwards the dossier to its Committee for Risk Assessment (CRA) and its Committee for Socio-Economic Analysis (CSEA) to produce a draft opinion within 10 months of submission. Applicants and interested third parties receive an opportunity to comment on the draft, whereafter the now finalised opinion is forwarded to the Commission, the Member States, and the applicant. The Commission then decides on the authorisation request following the regulatory committee procedure,[27] which gives the Council veto power, but only in the rare instances where it can muster a qualified majority against the Commission measure.[28]

While authorisation is a new instrument, the REACH provisions on marketing and use restrictions rearrange existing arrangements by moving the primary locus of decision-making from the Council of Ministers to the Commission, and by whittling away Member States' residual powers to deviate from Community norms.

---

[23]Regulation 793/93/EEC on the Evaluation and Control of the Risks of Existing Substances [1993] OJ L84/1.

[24]See, e.g., REACH, Art. 44 *in fine*.

[25]REACH, Art. 58. Alternatively, member states may (but are not obliged to) conduct own initiative identifications (REACH, Art. 59).

[26]Council Decision 1999/468/EC laying down the procedures for the exercise of implementing powers conferred on the Commission [1999] OJ L184/23, Art. 5(a).

[27]Ibid., Art. 3.

[28]Council Decision 1999/468/EC laying down the procedures for the exercise of implementing powers conferred on the Commission [1999] OJ L184/23, Art. 5.

This constitutes the fourth centralisation mechanism under REACH. Up to now, new restrictions were introduced via Council amendments to the Marketing and Use Restrictions Directive,[29] whereas under REACH they will be adopted by the Commission following a regulatory committee procedure with scrutiny.[30] As for authorisation, ECHA will prepare the dossier and suggest restrictions. Moreover, whereas formerly the restrictions regime was non-exhaustive and left residual powers to the Member States, first, to adopt restrictions concerning substances not (yet) covered by Directive 76/769/EEC, and, second, to derogate from the Community framework by maintaining or introducing stricter standards, it now appears replaced by an exhaustive mechanism of EU decision-making. Article 67(3) of the REACH Regulation sets 1 June 2013 as a sunset date until which Member States may 'maintain any existing and more stringent marketing and use restrictions than the harmonised ones.' Although it is not explicitly confirmed in the Regulation, this provision certainly implies that, thereafter, the marketing and use restrictions contained in the Regulation will be deemed exhaustive, and additional or more stringent national restrictions will be considered in violation of Community law. The compatibility of this approach with the guarantees of Articles 114(4) to (6) TFEU, which allow Member States to opt up from harmonised standards provided certain procedural and substantive conditions are complied with is very questionable (De Sadeleer 2003). However, there can be no doubt about the Regulation's intention to reduce the scope for national deviations from Community criteria for the marketing and use of dangerous chemicals.

13.2.3 Understanding the Transformation of Regulatory Decision-making

The institutional choices made in REACH have obviously been vetted by at least the qualified majority of Member States required to pass Community law. Why, then, did the EU Member States agree to an institutional set up and decision-making rules that narrow their competencies in favour of either private or supranational decision-making? The privatisation move is relatively easier to explain, since it encourages a 'soft' form of self-regulation based on information creation and sharing, and, importantly, may pre-empt but does not preclude further regulatory action. Also, placing the onus of data production and assessment on the chemical industry is in line with both the polluter pays principle, established in Article 191 TFEU, as well as the ideas underscoring the subsidiarity principle in Article 5(3) of the Treaty on European Union (TEU), articulating that decisions should be taken as closely to the citizen as possible. However, this same subsidiarity principle warrants against moving decision-making from the national to the EU level, unless 'the objectives of the proposed action cannot be sufficiently achieved by the Member States,'[31] thus inviting closer scrutiny of the centralising aspects of REACH.

---

[29]And, post-Maastricht, via Council and European Parliament amendments pursuant to Article 251 EC.

[30]Council Decision 1999/468/EC laying down the procedures for the exercise of implementing powers conferred on the Commission [1999] OJ L184/23, Art. 5.

[31]Art. 5(3) TEU.

In response, we first recall that simplification was an important motivation for reform. A single, comprehensive regulation that directly imposes uniform obligations on all chemicals manufacturers, traders, and downstream users active on the EU market is more accessible and manageable than the patchwork of EU and national measures that previously governed chemicals production and trade. Moreover, Member States may have considered that the opportunities for expertise-pooling under the substance evaluation and authorisation frameworks, and the alleviation of the domestic administrative burden by the establishment of a centrally managed and administered registration regime, are worth the sacrifice of domestic sovereignty over chemicals risk control, freeing up more resources to devote to helpdesking, inspection, and enforcement at the national level,[32] and facilitating the overall speed and quality of decision-making. This argument has been made forcefully with respect to recently acceded Member States, however the older EU Members, too, could easily find themselves technically and financially overtaxed by the demands of chemical risk assessment and evaluation. Through its rapporteur system for the evaluation of registered substances, REACH enables Member State authorities to specialise in the study of particular groups of chemicals, often coinciding with the chemicals that are produced domestically and concerning which they therefore have the most extensive experience. The rapporteur system gives Member States a reasonable expectation of retaining a significant input for precisely those chemicals that are most relevant for the local economy, or that pose the most serious health and environmental risks domestically, thus diluting the perceived loss of risk regulatory autonomy.[33]

Equally relevant in understanding the choice for centralised decision-making is past EU experience with product authorisation regimes, where authorisations were granted domestically and then mutually recognised across the EU. To function effectively, trade liberalisation through mutual recognition requires trust between the Member States in each other's assessments. Giandomenico Majone's seminal study of mutual recognition in the pharmaceuticals sector indicates that the assumption of trust is often misplaced, and vividly illustrates the disruptive consequences of conflicts between Member States over the reliability and universalisability of national safety assessments (Majone 1995). This explains the present, widely supported view that mutual recognition as a trade liberalisation technique works best for simple products and technologies, and that complex products, processes, and services often require a higher level of institutional orchestration (Majone 2008). Let us consider, for instance, the authorisation regime under REACH. Hopefully, the development of authorisation proposals by ECHA, under the auspices of which national disagreements can be revealed, negotiated and ironed out within the Member State Committee *prior* to the adoption of a final opinion, rather than come to light *after* an authorisation has been granted or refused, will pre-empt or resolve potential inter-state conflicts. This should avoid the fallacies and stalemate results to which the pharmaceuticals framework was so vulnerable.

---

[32]See, e.g., information on the UK Environmental Ministry website. Available at http://www.defra.gov.uk/environment/chemicals/reach/qanda/implementation.htm.

[33]I am grateful to Elizabeth Fisher for this observation.

The potential for speedier and more robust decision-making under institutional centralisation is moreover highly relevant when we factor in the EU's globalisation-related concerns. As argued earlier, globalisation has not only affected the choice *to* reform, but even more so influenced the choice *how to* reform. The centralised structure of REACH allows the EU to present a united front vis-à-vis the rest of the world. Registration is required and recognised throughout the EU; authorisations are granted on an EU-wide scale; and national variations in marketing and use restrictions are destined to be phased out. For a regime that explicitly portrays itself as environmentally ambitious and precautionary, unity is an attractive feature (Heyvaert 2006). Non-EU countries wishing to challenge any aspect of the REACH Regulation as incompatible with international trade law[34] know that they will face the full force of the European Commission and the Member States, who are all heavily invested in the success of REACH. In this context, it is important to remember the symbolic importance of REACH as the seminal exponent of the contemporary European approach to the management of complex risks, an approach that defines itself increasingly by reference to principles such as precaution and substitution, and profiles itself against more conservative approaches to risk assessment and management as followed in, for instance, the United States (Vogel 2003). An international challenge of the legality of REACH would (perhaps will) be a challenge to the legitimacy of European risk governance in a much more pronounced and generalised way than the *Beef Hormones* and *Biotech* cases ever were.[35] The unity of decision-making, endorsed and accepted by no less than 27 Member States, may help the Commission to justify REACH, and by extension European risk regulation, as a genuinely equivalent alternative to conservative risk-based decision making, rather than an aberration of 'normal' risk decision-making in accordance with internationally recognised standards of sound science and good governance (Trachtman 2006). On a less defensive note, EU trading partners may be much more willing to put up with exacting health and environmental requirements if the regulatory machinery delivers predictable outcomes. Centralised decision-making processes, with built-in mechanisms to moderate, resolve and, if necessary, override inter-state conflicts, promise a greater level of predictability than decentralised ones.

13.3 REACH as a Model for Global Risk Governance

The relevance of REACH goes beyond the transformation of EU chemicals safety management. Internally, REACH's institutional design and procedural sequences may well become a model for EU risk regulation generally (Pesendorfer 2006).

[34] EUOBserver (2006) EU Chemicals Bill Under Fire From US-Led Coalition. http://www.euobserver.com.

[35] Case (WT/DS26/AB/R and WT/DS48/AB/R) European Communities – Measures Affecting Meat/Livestock and Meat Products (Hormones), and Case (WT/DS291, WT/DS292 and WT/DS293) EC – Measures Affecting the Approval and Marketing of Biotech Products.

After all, the justifications for centralisation discussed in the previous section are not unique to chemicals regulation, but characterise most areas of risk regulation. The possibility of using REACH as a blueprint is already being discussed in the context of nanotechnology regulation (Van Calster 2006). Thus, we may be looking at a future for European risk regulation that is characterised by increased supranational regulatory decision-making following comitology procedures; by the presence of more specialised and highly influential independent Agencies, which do not only function as a source of expertise but also as the primary interlocutor through which Member States defend their interests and negotiate compromises; and by decision-making structures that aspire to mediate expert input and interest representation (Heyvaert 2008).

At least as important is the role of REACH in international risk regulation. The Commission makes no secret of its aspirations to promote the REACH approach to chemical safety beyond EU borders. Countries such as Switzerland, Norway, Japan, Canada, Korea, New Zealand and China have apparently expressed a keen interest to 'learn from REACH.'[36] And even in countries where governments have declared no such intention, such as the USA,[37] REACH is being used as a yardstick against which to assess domestic chemical risk regulation (Hogue 2007).

13.3.1 Why Would Non-EU Countries Adopt REACH?

A range of policy and economic considerations might sway non-EU countries to adopt REACH as a standard for chemicals management. First, a desire to improve domestic standards for health and environmental protection may persuade countries to strengthen their data reporting, assessment, and approval processes in step with REACH. The fact that the EU is actively lobbying foreign governments to contemplate the adoption of REACH might even be used by the latter to 'sell' a beneficial but expensive and controversial measure to its domestic constituencies (Moravscik 1994). Furthermore, Lazer points out that the export of regulatory regimes can have a self-perpetuating effect. As proliferation is equated with success, adoption by some countries triggers further expansion (Lazer 2006).

Economic considerations are equally relevant. Notwithstanding the cost, large chemical industries located outside the EU may be moved to put pressure on their governments to lift local standards to the EU level. The 'race to the top' or 'California effect' has become a familiar term in political studies, and refers precisely to the phenomenon of countries tightening up health and environmental standards to match stricter foreign standards (Vogel 1997). For a race to the top to occur, first, the country upholding the more stringent standards must be able to close its borders to

[36]European Comission (2006) Q&A on the New Chemicals Policy, REACH, MEMO/06/488. http://europa.eu/rapid/pressReleasesAction.do?reference=MEMO/06/488.

[37]Although, see Euractiv (2008) US Eyes REACH-Style Law for Chemicals. http://www.euractiv.com/en/environment/us-eyes-reach-style-law-chemicals/article-172968.

products that do not meet its regulatory prescriptions (Golub 2000). Second, the country with the toughest regulation must constitute a desirable export market. In the case of REACH, both conditions are fulfilled: the EU is a highly desirable export market for chemicals, and the REACH provisions apply both to domestically manufactured and imported goods. As industries located in third countries that plan to export, or continue exporting, to the EU cannot escape REACH's grasp, it is in their interest to lobby for the adoption of identical, or at least compatible, standards domestically. One reason is to avoid the dual or multiple burdens of dealing with different regulatory regimes internally and externally, which causes great inefficiencies in industrial production and management. A second reason is remove the indirect advantage that non-EU exporting companies might gain over their competitors on the domestic market, or other export markets located outside the EU, as a consequence of their lighter regulatory burden. Finally, adoption of a REACH approach may bring indirect economic advantages by tightening relations with the EU which, in turn, could improve trade flows, create opportunities for exchanges of expertise and technology, and help countries secure international funding for chemical safety capacity building projects.

13.3.2 Globalisation Models

How would processes of regulatory globalisation unfold, and what could we expect of, say, a Korean or Canadian version of REACH? The question explored in this section is not just theoretical, since REACH has already been transposed in Norway, with other EFTA countries planning to follow suit in the near future.[38] The case of Norway and other EFTA countries is, admittedly, idiosyncratic since the Treaty of Oporto requires EFTA members to adopt all EU legislation in the fields of the internal market, research and development, social policy, consumer protection and environmental protection (Chalmers et al. 2006). The Norwegian model of rules importation, discussed in greater detail in the paragraphs immediately below, amounts to a full normative and institutional assimilation. Outside of the EEA, the diffusion of REACH as a global norm is much more likely to assume the form of an approximation of rules. The final paragraphs of this section consider the likely features and outcomes of approximation processes.

13.3.2.1 First Model: Full Assimilation

The Norwegian approach to REACH constitutes the fullest possible assimilation to the EU regulatory framework for chemicals management (Norwegian Pollution Control Authority [SFT] 2007). The REACH Regulation has been translated into

[38]Also, see ENDS Europe Daily (2007) Switzerland Mulls Adoption of REACH-Lite. http://www.endseuropedaily.com/articles/index.cfm.

Norwegian and the Norwegian Parliament has adopted this version in May 2008. The Act incorporates not only the normative and procedural content of the REACH Regulation, but importantly recognises the authority of EU institutions to make decisions on chemicals produced in or imported into Norway.

Norwegian chemical manufacturers, importers, and downstream users need to register any chemical produced, imported, or used in quantities of at least 1 t pm/py with ECHA. It is ECHA, rather than the Norwegian Pollution Control Authority, that will administer Norwegian registration files, perform completeness checks, levy and receive registration fees, and assign registration numbers. Norway will participate in the ECHA Management Board and all ECHA Committees, and has committed to cooperate in REACH's substance evaluation scheme. This means that it may be called to perform risk evaluations for chemicals that have been identified as 'of concern' by the Commission, and then submit these evaluations to the Commission, where they might constitute the basis for risk reduction measures. Thus, the fate of a chemical manufactured in Spain and sold overwhelmingly on the domestic market could potentially be determined by an expert evaluation conducted in Norway. Conversely, Norway's chemical industry and importers will be affected by evaluations performed in the UK, Germany, Sweden, or any of the other Member States that are set to play an active role in the substance evaluation scheme. The Commission will decide on authorisations as well as restrictions pertaining to Norwegian dangerous substances. Commission authorisations and refusals must be adopted in Norway within 30 days.

13.3.2.2 Second Model: Approximation of Rules

For countries located outside the EEA, a full assimilation to REACH is not a plausible scenario; they are much more likely to approximate the REACH format in domestic legislation. Recalling the public policy and economic motivations that drive the globalisation of REACH, it is reasonable to assume that the most likely feature of REACH to be exported would be its registration regime. In light of the broad range of chemicals (and, hence, enterprises) affected, registration will have the strongest impact on trade of all chemicals management mechanisms within REACH. Moreover, the data generated pursuant to registration can constitute a key building block for improved health and environmental decision-making.

Similar considerations affect the marketing and use restrictions regime. Having to contend with, for instance, different sets of maximum concentration limits for identical substances traded on different markets is inefficient and disruptive to the free flow of trade in chemicals. Moreover, it sharpens concerns that industries located in the most desirable markets gain a competitive advantage over others, by virtue of being less affected by dual regulatory burdens. Rules importing states might therefore contemplate the adoption of marketing and use restrictions provisions, or, to the extent that the rules importing state already has chemicals restrictions in place, work towards their gradual convergence with EU standards.

Authorisation and substitution, too, require a significant commitment from the affected industries, since the burden of proving, first, that a chemical of high concern poses no unacceptable risk and, second, that no adequate alternatives are available, rests firmly on the applicant's shoulders. Hence, to the extent that exporting industries feel disadvantaged vis-à-vis traders who predominantly sell on the less regulated domestic market, they may support the introduction of similar authorisation and substitution requirements at home. Yet the arguments for approximation are less forceful here than in the case of registration. For one, a much smaller category of chemicals is involved, which reduces the frequency with which traders would find themselves in a situation of competitive disadvantage caused by different regulatory standards. Additionally, even though the burden of proof is shifted towards the private applicant, the establishment of an authorisation regime still requires a massive investment in public administration. Regulatory authorities need to organise the identification of substances for which authorisation is required, evaluate applications, review substitution plans, and ultimately decide whether to authorise or reject the use of the chemical concerned, and determine the conditions for use. All of these tasks are enormously time- and resource-intensive, and hinge on the input of an abundance of state-of-the-art expertise covering fields as diverse as toxicology, ecotoxicology, engineering, economics, statistics, biology, geography, climatology, etc.[39] To orchestrate authorisation processes may be beyond the capacity or inclination of states that are contemplating the adoption of REACH.

In addition to the standards for registration, authorisation and restriction, REACH contains a wealth of provisions stipulating the powers and responsibilities of the institutions that govern REACH. The institutional and operational clauses are closely tailored to suit EU conditions, which implies that, beyond the objective of establishing an institutional framework to implement, manage, administer, and enforce the REACH policy, many of the Regulation's institutional provisions address the ever delicate balance between supranational and Member State involvement in risk decision-making. Consequently, the Regulation's institutional design does not travel as well as its normative principles and risk management strategies. Third countries are hardly likely to copy (or request accession to) the REACH Regulation's institutional framework, but will sooner seek to anchor the REACH principles and standards within the domestic institutional regulatory landscape (Fisher 2008). The reception of REACH norms in different institutional environments may cause greater divergence in their application than is discernible on paper, as each administration will refer to its own institutional and operational principles and practices to interpret and implement the REACH regime (Fisher 2008; Selin 2007). Moreover, the differences are likely to become more pronounced with the passing of time, as regulatory authorities gradually streamline the procedural prescriptions of REACH with domestic standards and protocols for risk decision-making, fostering a cross-fertilisation between domestic and imported risk governance.

[39]To an extent, similar considerations apply to the adoption of marketing and use restrictions. However, regulatory authorities retain greater control over the development, timing and pacing of proposals for new restrictions than they have in application processes instigated by private parties.

13.3.2.3 Global Risk Governance Considered

The development of REACH into a global norm for chemical risk regulation opens new and exciting vistas. Its extensive requirements for chemical data production and supply, as well as the increased opportunities it creates for regulatory intervention through the authorisation regime for categories of highly dangerous chemicals, could assist countries across the world to control the health and environmental risks caused by chemicals more effectively than has been the case so far. Its inclusion of old substances in the registration scheme could spur global innovation in the chemicals sector by eliminating the advantage of incumbents over newer, and therefore often more fine-tuned, alternatives. Last but not least, the standardisation of production, marketing and use conditions could facilitate international trade flows and reduce the potential for trade conflicts by equalising conditions for market access across borders.

The blessings are, however, not to be taken for granted, since the globalisation of the EU's chemical risk regulation regime will undoubtedly unleash an abundance of new challenges to both the effective management of chemical safety, and to global trade in chemical products. For instance, the adoption of similar regulatory formats across the globe will reduce discrepancy but also diversity in regulation, which is an important source of learning, and will tend to amplify the particular strengths and weaknesses of the favoured approach. Also, while the adoption of similar trading standards may facilitate international trade, it could on the other hand stir up a host of new conflicts about the degree of similarity required before two sets of regulatory standards can legitimately be recognised as equivalent. This chapter does not offer a full overview of expected globalisation challenges, but focuses on the tensions between REACH's normative content and its institutional design in a context of globalisation (Heyvaert 2009).

As discussed above, in the case of EEA countries the adoption of REACH results in a full assimilation, including an accession to REACH's governance structure. This approach raises a serious concern of loss of national sovereignty over risk decision-making. We recall that the chemicals management regime of Norway will be administered by ECHA, an institution that was neither established by nor accountable to Norway. Decisions on, inter alia, authorisations and restrictions are handed over to the European Commission, an institution in which Norway does not wield any decision-making power, and which has as its primary mandate safeguarding the interests of the EU, a entity to which Norway, by popular choice, does not belong. Admittedly, the erosion of sovereignty is balanced by a number of political and economic accommodations. Generally, although they do not hold voting rights on EU matters, EFTA countries have 'decision-shaping' opportunities: they have guaranteed access to Commission committees and working groups during the preparation of new pieces of legislation, and can and frequently do submit comments to EU legislative bodies (Tovias 2006). Within REACH, Norway's accountability is strengthened through its participation in the ECHA Management Board and in all ECHA Committees. The latter arrangements improve Norway's opportunities to exert influence in the expert opinion formation process on the basis of which REACH decision-making is premised, however, they still fall short of full participation in decision-making.

The alternative model of approximation puts less stress on domestic sovereignty, but presents other, equally pressing challenges. This contribution has already drawn attention to the heavy demands of managing and administering REACH. As recent reports indicate that the organisation of REACH is already stretching ECHA to its limits,[40] a mere year after its launch, the question arises how manageable this form of risk management really is for other and particularly poorer countries with fewer administrative resources and less regulatory experience. It is moreover important to recall that the administrative burdens of REACH on individual EU countries are extenuated by the centralisation of registration, substance evaluation, and decision-making, which generates some economies of scale and gives national competent authorities more space to focus on the crucial task of enforcement. The organisation of substance evaluation in the EU enables functional differentiation between Member States, where expertise on distinct categories of chemicals is fostered in different Member States and then pooled as a collective decision-making basis (Kjaer 2007). But it is precisely those features of REACH that will not be transported in the case of regulatory approximation, making REACH a more onerous, less manageable format for States that are not part of either the EU or a similar regional governance structure.

Possible consequences of a mismatch between the normative and institutional features of chemical risk management are that decision-making slows down or even comes to a halt, that certain aspects of the framework remain unimplemented or are not enforced, or that drastic short-cuts are developed in an effort to keep the regulation afloat, which might preserve the regulation's productivity but taint its quality. Each of these consequences would undo the anticipated benefits of improved health and environmental protection, and could moreover cripple the credibility and legitimacy of REACH on the global trade market since, as discussed earlier, REACH's potential to deliver speedy, efficient and procedurally sound outcomes may be a crucial factor in determining the regime's compatibility with WTO law.[41]

Perhaps these predictions are too bleak. Non-EU or -EFTA countries interested in adopting REACH might establish transnational regulatory frameworks and thus gain the significant advantages from centralisation that the EU version displays. Or, domestic environments may be successful in adapting and transforming REACH into a variety of chemical risk regulation more suited to local needs and circumstances. It is however quickly apparent that either development entails its own set of new challenges and complexities. The first alternative requires nothing less than the formation of an effective, credible and legitimate transnational governance structure, which participating governments are willing to bestow with at least some level of advisory and regulatory authority, and which outside governments are willing to recognise as an authoritative source of opinions and decisions. The second alternative avoids this particular pitfall, but invites the thorny debate on when a domestic variant ceases to be a transposition of REACH, and turns into a sui generis

[40]See ENDS Europe Daily, EU Chemicals Agency Could Go Bust By 2011.

[41]Consider the relevance of the 'undue delay' factor in the WTO GMO dispute, see Euractiv (2006) EU Accepts Trade Ruling on GMOs. http://www.euractiv.com/en/trade/eu-accepts-trade-ruling-gmos/article-159918.

approach to chemical risk regulation. A domestically differentiated version of REACH may perform as well as or even better than the EU example as a health and environmental protection instrument, but could forfeit the trade liberalisation and competitiveness benefits that the globalisation of norms seek to attain. Either way, there are no painless solutions.

13.4 Conclusion

The development of REACH into the new European standard for chemicals control was a long and intense process. Yet, internationally, it may be but the first step towards the emergence of a transnational approach to risk regulation, and perhaps risk governance. It is, admittedly, hardly earthshattering to claim that such development will create many new challenges and complexities. This chapter focused on one of these, namely, the fact that regulations are developed and operate within a specific institutional and social context, and that this context affects the normative and procedural regulatory choices made. This, in turn, has an impact on the Regulation likely effectiveness when adopted and implemented outside its original jurisdiction. What the chapter shows, is that a globalisation of REACH would trigger the emergence of a new variant of the familiar 'trade *versus* environment' tension. Countries contemplating accession to the REACH standard will have to navigate carefully between the twin perils of wholesale approximation and complete modification to build a risk regulation regime that facilitates international trade without foregoing the benefits of manageable and effective health and environmental risk control.

References

Case C-473/98, Toolex Alpha [2000] ECR I-9741.

Commission Communication to the Council and the EP – The European Chemicals Bureau [1993] OJ C1/3.

Directive 67/548/EEC on the approximation of laws, regulations and administrative provisions relating to the Classification, Packaging and Labelling of Dangerous Substances [1967] OJ 196/1.

Directive 76/769/EEC on the approximation of the laws, regulations and administrative provisions of the Member States relating to Restrictions on the Marketing and Use of Certain Dangerous Substances and Preparations [1976] OJ L 262/201.

Directive 93/67/EEC laying down the Principles for Assessment of Risks to Man and the Environment of Substances Notified in accordance with Council Directive 67/548/EEC [1993] OJ L227/9.

Regulation 793/93/EEC on the Evaluation and Control of the Risks of Existing Substances [1993] OJ L84/1.

Sixth Amendment to Dir. 67/548/EEC [1979] OJ L259/10.

Chalmers, Damian and Hadjiemmanuil, Christos and Monti, Giorgio and Tomkins, Adam (2006) European Union law: text and materials. Cambridge University Press, Cambridge.

Commission Communication on Implementing the Lisbon Programme: A Strategy for the Simplification of the Regulatory Environment COM(2005)535, 25 Oct. 2005.

Commission Working Document – Report on the operation of Directive 67/548/EEC, Directive 88/379/EEC, Regulation (EEC) 793/93, Directive 76/769/EEC SEC(1998)1986, 18 Nov. 1998.
Commission Communication on Updating and Simplifying the Community Acquis COM (2003)71, 11 Feb. 2003.
Council Decision 1999/468/EC laying down the procedures for the exercise of implementing powers conferred on the Commission [1999] OJ L184/23, Art. 5.
EUOBserver (2006) EU chemicals bill under fire from US-led coalition. http://www.euobserver.com.
Euractiv (2006) EU accepts trade ruling on GMOs. http://www.euractiv.com/en/trade/eu-accepts-trade-ruling-gmos/article-159918.
Euractiv (2008) US eyes REACH-style law for chemicals. Available via http://www.euractiv.com/en/environment/us-eyes-reach-style-law-chemicals/article-172968.
ENDS Europe Daily (2007) Switzerland mulls adoption of REACH-lite, ENDS Europe Daily. http://www.endseuropedaily.com/articles/index.cfm.
ENDS Europe Daily, EU chemicals agency could go bust by 2011. http://www.endseuropedaily.com/articles/index.cfm.
Finkel, Adam (1994) A second opinion on an environmental misdiagnosis: the risky prescriptions of breaking the vicious circle. *NYU Environmental Law Journal* 3: 341.
Fisher, Elizabeth (2008) The 'Perfect Storm' of REACH: charting regulatory controversy in the age of information, sustainable development and globalization. Journal of Risk Research 11: 541.
Golub, Jonathan (2000) Globalization, sovereignty and policy-making. In: Holden, B (ed.) *Global Democracy: Key Debates*. Routledge, London.
Heyvaert, Veerle (1999a) *The Regulation of Chemicals in the European Union*. Dissertation, European University Institute.
Heyvaert, Veerle (1999b) Reconceptualizing risk assessment. *RECIEL* 8: 138–141.
Heyvaert, Veerle (2001) Balancing trade and environment in the European Union: proportionality substituted? *Journal of Environmental Law* 13: 395–398.
Heyvaert, Veerle (2006) Guidance without constraint: assessing the impact of the precautionary principle on the European Community's chemicals policy. In: Etty, T (ed) *Yearbook of European Environmental Law*, vol 6. Oxford University Press, Oxford.
Heyvaert, Veerle (2008) The EU's chemicals policy: towards inclusive governance? In: Vos, Ellen (ed.) *European Risk Governance: Its Science, its Inclusiveness and its Effectiveness*, Connex, Mannheim.
Heyvaert, Veerle (2009) Globalising regulation: reaching beyond the borders of chemicals safety. *Journal of Law and Society* 36.
Hogue, Cheryl (2007) The future of U.S. chemical regulation. Chemical & Engineering News. Available via http://www.iisd.ca/process/chemical_management.htm.
Kjaer, Poul (2007) Rationality within REACH? On functional differentiation as the structural foundation of legitimacy in European chemicals regulation. EUI Law Working Papers, EUI, Florence. Available via http://cadmus.iue.it/dspace/handle/1814/6948.
Koch, Lars and Ashford, Nicholas (2006) Rethinking the role of information in chemicals policy: implications for TSCA and REACH. *Journal of Cleaner Production* 14: 32.
Lazer, David (2006) Global and domestic goverance: models of interdependence in regulatory policymaking, *European Law Journal* 12: 455–466.
Majone, Giandomenico (1995) Mutual trust, credible commitments and the evolution of the rules of the single market. EUI Working Paper RSC 95/1, EUI, Florence.
Majone, Giandomenico (2008) Unity in diversity: European integration and the enlargement process. *European Law Review* 457: 469.
Meunier, Sophie and Jacoby, Wade (2007) Europe and the management of globalization: defensive and offensive responses to globalization pressures. Paper presented at the conference on Europe and the management of globalization, Princeton University, 23 February 2007. Available via http://www.princeton.edu/~smeunier/conference_europeanization.htm).
Molyneux, Garcia Candido (2005) Current survey on chemicals. In: Etty, T (ed) *Yearbook of European Environmental Law*, vol 5. Oxford University Press, Oxford.

Moravscik, Andrew (1994) Why the European Community strengthens the state: domestic politics and international institutions. *Centre for European Studies Working Paper Series* 52. Centre for European Studies, Cambridge.

Q&A on the New Chemicals Policy, REACH, MEMO/06/488, 13 Dec. 2006. Available via http://europa.eu/rapid/pressReleasesAction.do?reference=MEMO/06/488.

Regulation (EC) No 1907/2006 of the European Parliament and of the Council of 18 December 2006 concerning the Registration, Evaluation, Authorisation and Registration of Chemicals (REACH), establishing a European Chemicals Agency, amending Directive 1999/45/EC and repealing Council Regulation (EEC) No 793/93 and Commission Regulation (EC) No 1488/94 as well as Council Directive 76/769/EEC and Commission Directives 91/155/EEC, 93/67/EEC, 93/105/EC and 2000/21/EC [2006] OJ L396/1 ('REACH').

Norwegian Pollution Control Authority (2007) REACH – a new chemicals regulation for the EU and EEA countries. Available via http://www.sft.no/publikasjoner/2326/ta2326.pdf.

Pesendorfer, Dieter (2006) EU environmental policy under pressure: chemicals policy change between antagonistic goals? *Environmental Politics* 15: 95.

Proposal for a Regulation of the European Parliament and of the Council concerning the Registration, Evaluation, Authorisation and Restriction of Chemicals (Reach), establishing a European Chemicals Agency and amending Directive 1999/45/EC and Regulation (EC) {on Persistent Organic Pollutants} {SEC(2003 1171} COM (2003) 644 , 29 Oct. 2003.

De Sadeleer (2003) Procedures for derogations from the principle of approximation of laws under article 95 EC. *Common Market Law Review* 40: 889.

Selin, Henrik (2007) Coalition politics and chemicals management in a regulatory ambitious Europe. *Global Environmental Politics* 7: 63.

Stewart, Richard (1981) Regulation, innovation and administrative law: a conceptual framework. *California Law Review* 1259.

Tovias, Alfred (2006) Exploring the 'Pros' and 'Cons' of Swiss and Norwegian models of relations with the European Union: what can Israel learn from the experience of these two countries? *Cooperation and Conflict* 41: 202.

Trachtman, Joel P. (2006) The world trading system, the international legal system, and multilevel choice. *European Law Journal* 12: 479.

Van Calster, Geert (2006) Regulating nanotechnology in the European Union. European *Environmental Law Review* 15: 242.

Vogel, David (1997) Trading up and governing across: transnational governance and environmental protection. *Journal of European Public Policy* 4: 556.

Vogel, David (2003) Regulation in Europe and the US. In: Somsen, H (ed) *Yearbook of European Environmental Law*, vol 3. Oxford University Press, Oxford.

Chapter 14
The Precautionary Principle in EU and US Chemicals Policy: A Comparison of Industrial Chemicals Legislation

Mikael Karlsson

Abstract In this chapter, the precautionary principle will be considered as the starting point for decision-making on chemicals in cases of scientific uncertainty. The principle will serve as the reference point for an analysis and a comparison of chemicals policies and, in particular, of legislation for industrial chemicals in the European Union and the United States of America. In the second section, the precautionary principle will be described on a general level and operationalised with respect to chemicals management. The third section will focus on EU precautionary and chemicals policy and, in particular, on the recently adopted REACH regulation. A similar analysis will be made of US policies in the fourth section, with a focus on the Toxic Substances Control Act. In the fifth and concluding section, the results from the analyses will be compared and discussed with the aim to identify measures that could improve the management of chemicals under uncertainty.

Keywords EU • Precautionary principle • REACH • TSCA • US

14.1 Chemicals and Complex Risks

Society would not look the same without man-made chemicals, being of utmost importance in medicine, industry and agriculture and for the daily welfare of citizens (European Commission 2009). At the same time, the production and use of many chemicals are causing severe health and environmental problems, including allergies, cancer and decline of biodiversity (see e.g. EEA 2007; EEA 1998); in fact, 70% of new chemical substances assessed under EU law have at least one property that is dangerous to the environment (European Commission 2003). Adverse effects result from the continued use of well-known hazardous substances and from new chemicals that are introduced without much control, not least chemicals found in

M. Karlsson (✉)
School of Life Sciences, Södertörn University, SE 141 89, Huddinge, Sweden
e-mail: mikael.karlsson@naturskyddsforeningen.se

J. Eriksson et al. (eds.), *Regulating Chemical Risks: European and Global Challenges*,
DOI 10.1007/978-90-481-9428-5_14, © Springer Science+Business Media B.V. 2010

various articles (i.e. goods). Comprehensive risk assessments have only been carried out for a few of the several tens of thousands of substances in use. Data about intrinsic properties and exposure conditions are lacking for most chemicals on the market (Allanou et al. 1999; Roe et al. 1997); it is even unclear how many substances are in use.[1] Results from toxicological and eco-toxicological studies are difficult to interpret and extrapolate from, and combination effects are seldom studied, meaning that the impact of exposure to mixtures of chemicals is more or less completely unknown (Cairns and Smith 1996). Furthermore, risks that would normally be considered as unacceptable are very difficult to detect for statistical reasons, even in well-designed epidemiological studies.[2] In addition, knowledge and data within the field have more often than not been heavily disputed.

Against this background, it is clear that scientific uncertainty will characterise knowledge about chemicals for the foreseeable future. A central challenge for chemicals management – and the object of this article – is, therefore, to answer the question of how to deal with uncertainty, particularly when the issues are considered controversial by various stakeholders.[3]

14.2 Core Elements of the Precautionary Principle

In spite of much debate about the precautionary principle and its more precise meaning, it is clear that precautionary decisions have been taken in environmental policy for a long time and that the precautionary principle is widely adopted today in policy and legislation, not least when it comes to chemicals policy (see overviews in e.g. de Sadeleer 2007a; Ashford 2007; Karlsson 2005; Sandin 2004; Lökke 2004; Tickner 2003a; O'Riordan et al. 2001; Applegate 2000). The political and academic criticism of the principle has been shown to be quite weak (Ahteensuu 2007; Gardiner 2006; Sandin et al. 2002). The principle is often described with reference to the Rio Declaration, Principle 15, which states that '... lack of full scientific certainty shall not be used as a reason for postponing cost-effective measures to prevent environmental degradation' in cases of threats of serious or irreversible environmental damage. This definition is quite weak, though, since it merely discredits lack of scientific proof as an argument for stalling preventive measures (UNCED 1993).

Much legislation and many binding international environmental agreements stipulate stricter and more action-oriented versions, which Sandin (1999) has summarised in the formula that 'if there is a threat, which is uncertain, then some kind of action is mandatory'. On the basis of this interpretation, I have previously elaborated five core elements of the precautionary principle in relation to chemicals

[1]The latter became evident when the pre-registration of chemicals under the EU REACH regulation came to include over 146,000 chemical substances, 20 times the number anticipated by the EU Commission (Dancet 2009).

[2]See further in Hansson and Rudén, Chapter 5 this volume.

[3]For a detailed discussion of management of complex environmental risks, see Karlsson (2005).

policy (Karlsson 2006).[4] As a starting point for my operationalisation, I reasoned that 'a threat' needs to respond to some level of seriousness, including effects on health or the environment that are severe, spatially or temporally dispersed, irreversible or non-linear. In the same manner, I consider 'uncertainty' to require some level of qualification, such as science-based suspicion, well-grounded practical experience or lay knowledge of threats. Obviously, any interpretation must be contextual.

Once the precautionary principle is invoked, the question of what kind of action is called for follows: Which additional measures need to be taken, in addition to traditional risk management? In the following, I will briefly outline the five core elements that I believe answer that question.

Group Classification: Generating more data is of key importance. In a scientific endeavour, though, a strong correlation relating cause to effect is required for making a claim, and scientists would rather miss true links than consider false incidents to be true. In precautionary policies, the opposite is justified, and unknown substances in a chemical group should be classified as the most hazardous known substance in that group, or as the 'worst-case', which is reasonably imaginable.

Management Based on Intrinsic Properties: The complex life cycles of many substances prevent proper estimations of exposure. Therefore, precautionary policies primarily focus on intrinsic hazardous properties, i.e. assume that critical exposure is the case. Since hazardousness (e.g. chronic toxicity) is difficult to estimate, also substances classified as 'persistent' and 'liable to bioaccumulate' should be treated as hazardous and as action targets.

Preventive Measures: Unless voluntary measures are proven sufficient, the precautionary approach requires anticipatory regulation. If a substance can be substituted with a suitable, better known and less hazardous substance or non-chemical option, then that should be the first priority. Secondly, partial or full restrictions can be implemented on substances that are precautionarily classified as being persistent, bioaccumulative, toxic or otherwise hazardous.

Maximin Decision-Makin: In the analysis of whether or not to require a specific preventive measure, it is common (in chemicals policy) to aim for maximised utility based on costs and benefits. When uncertainty about risks caused by a substance prevents that, precautionary decisions – based on reasoning and avoidance of obviously absurd requirements – favor measures that cause the least possible loss if the 'worst-case' turns out to be true.[5]

Reversed Burden of Proof: In each of the four core elements operationalised above, the precautionary principle favours placing the so-called 'burden of proof' on the operator, which thereby bears the responsibility for convincingly showing that stat-

[4]See also e.g. Ashford (2007), de Sadeleer (2007b), Tickner (2003b) and Applegate (2000) for interpretations and accounts of the precautionary principle that partly overlap with the operationalisation below.

[5]This follows the so-called 'maximin-rule' advocated by e.g. Rawls (1971); see also (Hansson 1997).

utes are being followed and, for example, that a regulatory decision on classification or a preventive measure is factually illegitimate or obviously unreasonable from a likely cost–benefit point of view.

These core elements of precautionary chemicals management will next be used as an analytical framework in relation to chemicals policy in general and legislation on industrial chemicals in particular in the EU and the US.

14.3 The Precautionary Principle and EU Chemicals Policy

In the following, the history of the precautionary principle within the EU and its member states will be briefly described, followed by a concise overview of EU chemicals policy and a more detailed description and analysis of the REACH regulation and its relation to the precautionary principle.

14.3.1 The Precautionary Principle in the EU

The precautionary principle is far from being a new component of environmental and health policies in Europe. Some of the core elements of the principle, as elaborated above, can be traced back in Swedish legislation centuries ago (Karlsson 2006), and the idea of reversing the burden of proof in environmental law in the Nordic countries goes back more than four decades (Sand 2000).[6] The strong standing of precautionary policies in, for instance, Germany and the Netherlands led to the inclusion of the principle in the EC Treaty in conjunction with the Maastricht Treaty (1992), followed by an interpretation by the European Commission (2000) in a so-called 'Communication', which was relatively well endorsed by EU heads of states and governments in 2000 (European Council 2000).

In the Communication, the Commission interprets the principle as being applicable when a hazard is identified but when a full scientific evaluation of the risk at hand cannot be completed. It requires that precautionary management is proportional, non-discriminatory, consistent with previous measures, based on an examination of potential benefits and costs and subject to review. However, these aspects are stated always to be relevant for risk management and do not help much in interpreting the specifics of the precautionary principle. On the contrary, some elements in the Communication can be seen to oppose common interpretations of the principle, for instances by relating the principle purely to risk management and not to risk assessment and by giving cost–benefit determinations a central role (Hansen et al. 2007; Karlsson 2005). However, the Communication acknowledges that the burden of proof may be reversed in some cases (see also the analysis by Rogers 2003a). Regardless of interpretation, though, it is clear that the precautionary principle nowadays is commonly applied in

[6]This is long before common use of the German concept 'Vorsorge Prinzip', a fact missed in much of the literature on the precautionary principle, which cites von Moltke (1988), Boehmer-Christiansen (1994) and others who claim that the German concept is the original one.

secondary EU law (see e.g. Krämer 2006) and in the practice of the EC courts (de Sadeleer 2007b), even though it has been claimed that the courts have recently increased demands on scientific justification for regulation (Stokes 2008).

14.3.2 The Development of Chemicals Policy in the EU

The chemicals policy of the European Union emerged in the 1960s, with a 1967 directive on classification, packaging and labelling of chemical substances, which had the purpose of harmonising market legislation (EEC 1967). Since then, EU chemicals legislation has been substantially developed and broadened, with several amendments to the early directives and new directives on restrictions (EEC 1976; EC 2003a), chemical preparations (EEC 1988), waste (EC 2003b) and specific types of chemicals, such as pharmaceuticals, cosmetics and pesticides.[7]

In 1979, the central classification directive introduced a differentiation between 'existing' – as of September 18, 1981 – and 'new' substances, the latter being thereafter only possible to list if certain basic data on the properties of the substance were reported. For the 100,106 substances categorised as 'existing', no new data were required, but following a 1993 regulation, 141 prioritised substances eventually became targets of risk assessments (EEC 1993). Due to a lack of data, high complexity and controversies, this process did not deliver what it promised for most chemicals (European Commission 1998).[8] Consequently, in 1998 and 1999, the EU ministers of environment expressed concerns and called for a stricter chemicals policy based on, among other aspects, the precautionary principle (see e.g. Environment Council 1999). Thereafter, the European Commission published a White Paper on a *Strategy for a Future Chemicals Policy* (European Commission 2001), which also emphasised the precautionary principle (see e.g. Rogers 2003b). The White Paper became highly contested, and the subsequent lengthy legislative process was probably the most controversial in the history of the EU (Fisher 2008; Selin 2007). In 2006, however, the REACH Regulation was finally adopted (EC 2006a). The regulation explicitly refers to the precautionary principle in the Preamble (9, 60), and in particular in the aim, stating that 'Its provisions are underpinned by the precautionary principle' (Article 1).[9]

14.3.3 The REACH Regulation

Before REACH, EU legislation on industrial chemicals was disparate, and different provisions applied to similar types of chemicals. The system did not provide

---

[7]See e.g. Krämer (2006) for an overview.

[8]Between 1993 and 2008 only 118 substances had gone through most of the risk assessment process, and only 56 substances had been dealt with completely (ECB 2008).

[9]REACH is divided into a preamble and titles, chapters and articles, which I will cite when appropriate.

sufficient data or sufficient protection for human health and the environment and did not charge companies with the main responsibility for chemicals management. The White Paper formulated seven objectives for the chemicals legislation reform: protection of health and the environment, enhancement of the competitiveness of the EU chemicals industry, a functioning internal market, international integration and conformity, increased non-animal testing and increased transparency (European Commission 2001). Implementation of the precautionary principle was mentioned as one of the cornerstones of the new legislation. The resulting REACH regulation replaces some 40 pieces of legislation and gradually harmonises the regulation of existing and new chemical substances, even though several groups of chemicals, such as pesticides and cosmetics, are still regulated separately. REACH will, thus, end the previous system that gave preferential treatment to existing substances and thereby created disincentives for the development of new substances.

In the following, the four cornerstones of REACH – registration, evaluation, authorisation and restrictions of chemical substances – as well as a number of other important aspects of REACH will be described and analysed.

14.3.3.1 Registration (Title II)

REACH charges each manufacturer and importer with responsibility to submit a registration to the new European Chemicals Agency (ECHA) for each substance, either on its own or in preparations or articles, that is manufactured or imported in a volume of at least 1 t per year and company (Article 6). For articles, registration for new uses is required for quantities over 1 t per producer or importer per year, but only if the substance in question either is intended to be released or meets the criteria for being placed on the candidate list for authorisation (see below) and is present in more than 0.1% by weight (Article 7).[10] With quite a number of statutory exemptions (including many polymers and intermediates, see Annexes IV and V), a substance not registered according to the provisions is not allowed to be manufactured or put on the market in the EU (Article 5). REACH thereby introduces the concept of 'no data, no market'.

A substance that was previously notified as 'new' is automatically considered to be registered under REACH (Article 24). For the group previously called 'existing substances', now called 'phase-in substances', data requirements and deadlines for registration primarily depend on the quantity of the substance in question, though to some extent also on intrinsic properties (Articles 10, 12–14, 23; Annexes VII–X).[11] The greater the quantity, the sooner the registration deadline and the higher the data requirements. The registration deadline for substances at or above 1,000 t, as well as for substances that, for example, are carcinogenic, mutagenic or toxic to reproduction (so-called CMRs) at or above 1 t, is 30 November, 2010, whereas it is 31 May 2013 for 100–1,000 t and 31 May 2018 for the span from 1 to 100 t.

[10] It is unclear whether the limit refers to the entire article, and thus often rendering registration exceptional, or to its various components.

[11] This is the case as long as pre-registration has taken place or is not required (Article 28).

The latter deadline is set nearly 20 years after the ministerial call for a new chemicals policy in the EU. Above 10 t, a 'Chemical Safety Report' with data on a relatively large set of parameters, including data on intrinsic properties, exposure scenarios and risk management measures, is to be included in the registration (Article 14, Annex 1). For substances in quantities of 1–10 t, a 'technical dossier' with more basic data (Article 10) is stated to be sufficient. Concerning so-called 'non-phase-in substances', i.e. basically those not being produced or marketed before REACH, the registration provisions entered into force on 1 June, 2008 (Article 141).

Compared to pre-REACH legislation on 'new chemicals', the data requirements have been lowered for all quantities. However, with regard to existing substances, for which no general data requirements were stipulated in previous law, REACH provides completely new requirements over time.[12] Furthermore, data requirements have been completely abolished for quantities below 1 t, thereby leaving the clear majority of all industrial chemicals outside of REACH registration (the previous notification limit was 10 kg). Also in the span 1–10 t, data requirements are very low.

The registration phase as such does not necessarily lead to reduced risks, and the quality of the data generated is not guaranteed. The ECHA is merely required to make a completeness check within three weeks (Article 20), after which the chemical in question can be used.

14.3.3.2 Evaluation (Title VI)

It is first in the evaluation phase that dossiers and substances are qualitatively investigated. In the 'dossier evaluation' part (Chapter 1), the ECHA checks how a small portion of the dossiers and reports received complies with the registration requirements, for instance concerning data and proposed risk management measures (Article 41). The ECHA also examines testing proposals, not least in order to prevent unnecessary animal testing, and decides on possible further measures (Article 40). Evaluation is not a prerequisite for starting manufacturing of substances, not even in cases where there are obvious needs for further testing. The registration and evaluation in REACH thus operate along quite independent tracks (EDF 2007a), as was the case under previous EU chemicals legislation. Most substances will not be evaluated at all – the ECHA is not obliged to carry out compliance checks on more than 5% of the registration dossiers for each tonnage band (Article 41).

The second form of evaluation involves substances that for one reason or another are considered or assumed to be problematic (Chapter 2). Both the ECHA and the Member State's Competent Authorities will continuously evaluate suspected substances, on the basis of a risk-oriented rolling action plan and criteria for prioritisation developed by the ECHA and member states (Article 44). If needed, further information can be requested from registrants (Article 46). If an evaluation indicates that further risk management measures are needed, it may result in the

[12]See further in Hansson and Rudén, Chapter 5 this volume, for details on data and test requirements.

application of the authorisation or restriction procedures under REACH, as well as measures under other laws.

14.3.3.3 Authorisation (Title VII)

Authorisation is focused on so-called 'substances of very high concern' (SVHCs), which is a new element as compared to previous chemicals legislation. These include substances that meet the criteria for being classified[13] as carcinogenic, mutagenic or toxic for reproduction (CMRs) in categories 1 and 2, substances that are persistent, bioaccumulative and toxic (PBT substances, according to Annex XIII) or very persistent and very bioaccumulative (vPvBs, as defined in Annex XIII) and other substances that cause equivalent concern, based on scientific evidence of probable serious adverse effects, for instance endocrine disrupters (Article 57). The Commission or member states may identify substances meeting these criteria, and the Agency is to – after a process that may involve a Member State Committee, interested parties and the Commission – establish a 'candidate list' for substances that may be included in Annex XIV (Article 59). After taking the opinion of the Member State Committee and the viewpoints of interested parties into account, the ECHA is to recommend, in line with its capacity, to the Commission a set of 'priority substances' to be included in Annex XIV, first in line being substances with PBT or vPvB properties, or with wide dispersive use or high volumes (Article 58).

The final decision rests with the Commission but is to be taken after Committee procedure (Article 133), commonly known as comitology, in this case with the 'Regulatory procedure with scrutiny' (EC 2006b), which basically means that the Commission has the unique power to draft proposals but that a member state committee, the Council or the European Parliament, given certain conditions, can block the proposals. In summary, these processes mean that there is no guarantee at all that a substance, which definitely meets the criteria in Article 57, will be listed.[14] In practice, the listing may depend on resources and viewpoints among member states. At present, the candidate list contains 15 substances, of which 7 have been placed on ECHA's first priority list (ECHA 2009), as compared with the 1,400 substances mentioned for authorisation in the White Paper (European Commission 2001).

Once a substance is included in Annex XIV and the so-called 'sunset date' has passed (Article 58),[15] authorisation must be sought by any manufacturer, importer or downstream user wanting to use the substance or to place it on the market on its own, in preparations (above certain concentrations) or for incorporation into articles

[13] The basis is the new EU regulation on classification (EC 2008), which implements the GHS (Globally Harmonised System of Classification and Labelling of Chemicals), see further in Bengtsson, this volume.

[14] See further de Sadeleer (2007c) on further details of authorisation and its relation to the restriction phase.

[15] There are no limitations on how early or how late the sunset date can be set. Furthermore, in spite of non-authorisation, a substance might still be permitted in articles after the sunset date (Article 69(2)).

(Article 56). Articles containing an Annex XIV substance may be imported to the EU without authorisation.[16] Authorisation is to be granted by the Commission on a case-by-case basis for a specified period and under conditions subject to review, if the risks to health and the environment arising from the listed intrinsic properties of a substance are 'adequately controlled' (Article 60, as defined in Annex I). However, the adequate control route is closed for a SVHC that has the properties PBT or vPvB or that is a CMR for which a threshold cannot be determined (Article 60). For such substances, or when control is not adequate, authorisation requires that socioeconomic benefits outweigh risks and that there are no substitutes available (Article 64). These aspects must be considered by a Committee for Risk Assessment and a Committee for Socioeconomic Analysis before the Commission can take a decision on an application, which must include a socio-economic analysis according to Annex XVI.

Concerning substitution, which is mentioned both in general and in relation to authorisations in the REACH Preamble (70, 72–75), a substitution plan is to be included in the application, but only if the applicant identifies a safer alternative (Article 62). There are options for third parties to present alternatives, but even if the ECHA committee considers them suitable, authorisation must still be granted if adequate control is considered to be in place (Article 60). In general, the burden of proof for substitution rests on the regulators, and substitution requirements are more or less restricted to some parts of the authorisation phase (Koch and Ashford 2006).

14.3.3.4 Restrictions (Title VIII)

The fourth cornerstone in REACH is the possibility of issuing restrictions. Restrictions apply to Community-wide situations when the production, market release or use of a substance (on its own, in preparations or in articles) entails an 'unacceptable risk' to the environment or to human health (Article 67–68).[17] 'Unacceptable risk' is not explicitly defined, but the provisions state that any restriction 'decision shall take into account socio-economic impact, including the availability of alternatives' (Article 68). In contrast to the authorisation title, under which the existence of alternatives might block authorisation, alternatives play the opposite role here, meaning that without alternatives, restrictions may be blocked.

The restriction procedure is complex and includes reviews by the Committee for Risk Assessment (during nine months) and the Committee for Socioeconomic Analysis (during 1 year), the publishing of draft recommendations and decisions and, eventually, a final decision by the Commission via comitology[18] (Articles 69–73). After the decision, the substance and the restrictive conditions placed on its

---

[16]Here, though, restrictions may apply.

[17]A substance in Annex XIV is not to be subject to new restrictions (Article 58), unless it entails risks from the presence of the substance in other articles (Article 58).

[18]Decisions are made in the same manner as for authorisation.

use are included in Annex XVII. For a CMR substance on its own, in mixtures or in other articles, which could be used by consumers, the process is simplified, and the Commission can take action much more easily (Article 68).

The restriction option has been considered to be a 'safety net' (European Commission 2007) if other provisions in REACH or action under other laws are deemed to be insufficient. However, the Commission, the ECHA or member states are charged with the responsibility for preparing a restriction decision and must file a dossier clarifying the motives and the most appropriate risk reduction measures, after which a complicated decision-making process follows. The principal aspects here are more or less the same as for substances and decisions under the previous EC chemicals legislation.[19]

14.3.3.5 Other Central Elements of REACH

In addition to the four cornerstones described above, REACH introduces a number of other relatively novel elements. Corresponding to central topics in the regulatory debate, several provisions in REACH promote data sharing during the registration phase, which is of importance for lowering industry's compliance costs and decreasing animal testing as far as possible (Title III).

Furthermore, REACH includes new or increased demands on the bidirectional flow of data and information in the supply chain (Title IV) and places demands on downstream users (Title V). For instance, REACH includes provisions on safety data sheets for substances on the candidate list (Article 31 and Annex II) and forces suppliers of articles to actively provide information for the safe use of the articles (Article 33). In addition, REACH entitles consumers to, without charge, request information within 45 days on the safe use of articles containing SVHCs in concentrations above 0.1 wt% (Article 33). It remains to be seen to what extent the increased flows of information will impact on the management of chemicals. However, companies that work with environmental management systems and companies located closer to consumers in commodity chains will probably seek more actively to decrease chemical-related risks.

When it comes to Confidential Business Information (CBI), REACH (Title XII) contains several provisions granting rights to companies, including protection of information regarding the full composition of preparations, the precise use and tonnage of substances and links between companies (Article 118).[20] On the other hand, Article 119 explicitly lists information that 'shall be made publicly available' by electronic means, for instance names and classifications of substances and safety-related data, unless the submitter can justify not doing so. Additional information from companies can also be made available upon request (see further in EDF 2007a).

[19]The substances previously restricted under EEC 1976 were transferred to Annex XVII on 1 June, 2009.

[20]However, in cases of urgent need for protection, this information can also be disclosed.

14.3.4 REACH and the Precautionary Principle

Some, though not all, parts of REACH have just entered into force. A detailed evaluation of the implementation and the impact in practice can clearly not be made at this point. However, there are already signs indicating that the administrative challenges connected to registration may have been underestimated and that the potential impact of authorisation may have been overestimated. For instance, there were seven substances on ECHA's first priority list in the authorisation phase – extremely few compared to earlier expectations. However, the following analysis will focus on the REACH system and the legal text, given previous experiences of EU chemicals law, in relation to the precautionary principle as interpreted above.

The first core element, 'group or worst-case classification' of unknown substances, could be implemented by interpreting uncertainty as if adverse effects exist until the opposite has been reasonably proven or by classifying on the basis of analyses of similarities between substances, for instance by using structural activity relationships. REACH does not incorporate such a classification concept but acknowledges the use of alternative testing methods.[21] However, REACH applies the idea of 'no data, no market', which is a form of worst-case classification in the sense that a non-registered and, therefore, more or less unknown substance is considered so problematic so as not to be permitted at all. In this way, and by placing the burden of proof on the company seeking registration, REACH provides strong incentives for generating data. However, substances in low quantities and those included in many articles are not included, data requirements are often limited and periods for transition to REACH are often long. Consequently, the registration phase under REACH can hardly be regarded as precautionary.

The second core element, that policies can be based on intrinsic properties, is clearly recognised in REACH. This is most visible in the fundamental requirement for authorisation for substances being vPvB, i.e. substances not necessarily being even suspected to be toxic or otherwise problematic for health or the environment. However, the complex processes for identifying and listing SVHCs for authorisation – combined with exemptions, the risk of very late deadlines and budget restrictions among regulators – may lead to similar implementation problems with REACH as with previous EU laws on risk assessment of existing substances (see also de Sadeleer 2007c). It is difficult to see, for instance, why member states that might wish to stall stricter policies would not be able to do so in the future as well. Furthermore, REACH allows authorisation of even the most troublesome chemicals if adequate control is assumed, albeit such control does not necessarily prevent exposure to humans and the environment. Nevertheless, the burden of proof in the

[21]Non-animal laboratory testing is preferable, but – in order to prevent large-scale risks outside laboratories – must be reliable, which requires scientific guidance, independent expert reviews, etc. (see EDF 2007a).

authorisation phase basically rests with chemical companies, which is a more or less unique precautionary measure in an international context.

Turning to preventive measures, the third core element of precaution, REACH offers a multitude of regulatory action possibilities. However, substitution requirements – a cornerstone in precautionary policies – will be the exception rather than the rule under REACH.[22] To be sure, possibilities for substitution will block the authorisation of some SVHCs, and some REACH elements promote substitution, in particular the increased flow of and access to information and transparency (Lahl 2007). Nonetheless, for most substances falling under REACH, no substitution requirements are provided at all. Some of the substances that will be authorised may comprise exceptions, however this is quite late in the regulatory chain (Koch and Ashford 2006), and even some of the most hazardous among these may be authorised in spite of existing substitutes, as long as adequately control is assumed. In addition, REACH provides no incentives for a company that applies for authorisation to consider seriously alternatives to what it actually wants to do. Finally, regarding restrictions, the burden of proof still rests with public agencies, with all the well-known implementation problems that are connected to a system of that kind (de Sadeleer 2007c). Thus, on this point, REACH will probably not be much more precautionary than previous legislation.

The fourth core element, the maximin principle, could partly be said to relate to the 'no data, no market-element', since this in principle prevents the marketing of a chemical rather than the opposite in case of uncertainty. However, this point refers to data gathering activities and not to decision-making on the basis of uncertain data. On the latter point, no provision in REACH explicitly recognises or implements the maximin principle, not even under the restriction title. On the contrary, traditional cost–benefit analysis plays a central role in the authorisation and restriction procedures.

Finally, REACH clearly reverses the burden of proof when it comes to registration and, to some extent, even during the authorisation phase, even though responsibility for placing a substance on the candidate list rests within the public domain. Concerning evaluation and restrictions, the burden of proof rests with agencies. This challenges the Commission, the ECHA and competent authorities of member states with heavy obligations, and there is an obvious risk for administrative overload, as under previous legislation.

In summary, REACH implements some of the core elements of the precautionary principle but only goes half way at best.

14.4 The Precautionary Principle in US Chemicals Policy

This section provides an overview of the precautionary principle and chemicals regulation in US policies and describes and analyses the Toxic Substances Control Act in more detail, in particular in relation to the precautionary principle.

[22]See Hansen et al. (2007) on previous proposals on substitution in the REACH regulatory process.

14.4.1 The Precautionary Principle in US Policies

Even though the precautionary principle is not explicitly mentioned in US federal law, fundamental aspects of the principle, such as taking anticipatory action under uncertainty, were reflected in much of the early US environmental legislation, including the 1970 Clean Air Act, the 1972 Clean Water Act and the 1973 Endangered Species Act[23] (Ashford 2007; Applegate 2000). Under the Clean Air Act, the concept 'precautionary' was, for instance, explicitly mentioned by courts in cases as early as in 1976 and 1979.[24] Concerning US legislation on chemicals substances, the initial ambition was to diverge from the otherwise typical US practice of taking action within the tort system first when damage has occurred and cause–effect relations are more or less proven. By establishing elements of precaution in law, legislators hoped to replace the reactive tort system with one that focused on prevention in the field of health and environment protection (Ashford 2007; Applegate 2000; Wagner 2000).

Over time, though, these elements have been increasingly countered both by amendments of statutes and by the development of other principles and ideas, such as risk acceptability and cost–benefit balancing, and the burden of proof is nearly always placed squarely on the regulator (Applegate 2000). In the field of chemicals policies, protection has been declining since the 1980s due to legislation on use cost–benefit analysis, review activities by the Office of Management and Budget, congressional review and replacement of members of the judiciary (Ashford 2007). In the following, the US legislation on chemicals in general and industrial chemicals in particular will be described and analysed.

14.4.2 History of Chemicals Law in the US

Chemicals policy in the US goes back quite some time. Today the laws are dispersed over several pieces of legislation. The 'Consumer Product Safety Act' regulates the general and chemical safety of thousands of common products, and the related 'Consumer Product Safety Improvement Act' of 2008 sets standards for children's products, including a ban on six phthalates. The 'Federal Hazardous Substances Act' requires labelling and authorises the Consumer Product Safety Commission to issue prohibitions when labelling is not considered sufficient for hazardous household products. Under the 1938 'Federal Food, Drug and Cosmetics Act', the US Food and Drug Administration assesses notifications of new substances in materials that come into contact with food, and the law holds industry responsible for the safety of ingredients in cosmetic products. The so-called 1958

[23]Endangered Species Act 16 U.S.C. §§1533–1539; Clean Water Act 33 U.S.C. §§1311–1317; Clean Air Act 42 U.S.C. §§7409–7412.

[24]Ethyl Corp v. EPA, 541 F2d 1 (D.C. Cir. 1976); Lead Industries Association v. EPA, 647 F2d 1139 (D.C. Cir. 1979).

'Delaney Clause' prohibited any carcinogenic additive in food, regardless of its potency, and was an early example of the application of a precautionary measure, even though the implications of the clause have been downplayed since then.

However, the cornerstone of US chemicals legislation is the Toxic Substances Control Act, which regulates most substances and most of the quantities, in particular industrial chemicals. In the following, the statutes and the more than 30 years of implementation of TSCA will be described and analysed. The investigation is limited to statutes with a general reach; the parts of TSCA that address specific substance groups will not be dealt with.[25]

14.4.3 The Toxic Substances Control Act

TSCA was enacted in 1976 and has basically not changed since then. The most important provisions in TSCA are Sections 4, 5, and 6 on testing, premanufacturing clearance and regulation of hazardous substances, respectively. TSCA differentiates between existing and new chemicals (Section 8(b)),[26] and it assigns authority to the US Environmental Protection Agency (EPA) to implement the various statutes. TSCA was partly based on the 'Report on Toxic Substances' by the US Council on Environmental Quality (CEQ 1971), which stated that lack of data and lack of government control caused problems needed to be managed by regulatory means. CEQ considered previous media-based environmental legislation to be insufficient and wanted a comprehensive life cycle-based approach, focusing on chemicals as such instead of on their presence in different environmental compartments. This view was supported by the Congress, which considered proactive policies to be both safer and cheaper than reactive ones, including the tort system (Applegate 2008). Congress wanted to shift the burden of proof, including placing the responsibility for generating data on manufacturers (see e.g. Section 2(b)(1)), and several legal techniques aimed at providing for this.

14.4.3.1 Testing of Chemical Substances and Mixtures (Section 4)

According to Section 4(a) of TSCA, the EPA is obliged to require testing of existing or new substances if necessary for filling data gaps in order to be able reasonably to determine if a substance or mixture '*may* present an unreasonable risk of injury to health or the environment [emphasis added]' or if it will be produced in substantial quantities and may reasonably either enter the environment in such quantities or may expose people in substantial or significant quantities.[27] According to the EPA, a Section 4 process can take between 2 and 10 years, and testing has

[25] These include PCBs (Title I, Section 6(e)), asbestos (Title II), radon (Title III) and lead (Title IV).

[26] I will use the numbering as TSCA was enacted (e.g. Section 8 corresponds to §2607 in the U.S.C.), and I will cite sections, subsections, paragraphs and subparagraphs, as appropriate.

[27] 'Substantial' and 'significant' are defined by EPA (EDF 2007a).

been required for some 200 substances since 1979 (OPPT 2008). There is an obvious Catch 22 here: the EPA needs data in order to determine 'unreasonable risk' but cannot require this data until such risks have been more or less demonstrated. Even the requirement of demonstrating that there is 'insufficient data' (Section 4(a)) can be extremely burdensome in practice EDF (2007a).

For these and other reasons – such as budgetary constraints and political priorities – the EPA is often forced to work on the basis of voluntary agreements with industry and other stakeholders. Perhaps the most well-known example is the 'High Production Volume Challenge Program' from 1998, initiated by the EPA, the Environmental Defense Fund, the American Petroleum Institute and the American Chemistry Council. The program focuses on 'sponsor' companies agreeing to generate basic hazard data for chemicals produced or imported at a volume of 1 million pounds[28] or more per year. Nearly 1,400 chemicals are in the program, with an additional 860 provided by coordinated international efforts (OPPT 2008). The program has been successful in the sense that industry actually accepted the challenge and that much previously unpublished data has been made public. However, obvious delays, incomplete submission of data and low quality data in the program as such (EDF 2007b) as well as non-inclusion of chemicals at lower volumes (GAO 2007) illustrate clear shortcomings in the voluntary approach, in particular if agencies do not have adequate resources.[29] It is not even certain that the program will yield data with the quality needed for the EPA to be able to determine if risks are unreasonable or not.

Simultaneously with aiming to finalise the HPV Challenge Program up to 2010, the EPA has initiated a new voluntary program, the Chemical Assessment and Management Program (ChAMP), launched in 2008 in response to the 2007 US–Canadian–Mexico Security and Prosperity Partnership. ChAMP focuses on data on both hazards and risks from some 6,750 chemicals produced or imported at volumes of 25,000 pounds or more per year and can obviously not be evaluated in detail, even though there is a lot to indicate that similar problems may occur within this program as with the HPVC program (EDF 2008).

Yet another initiative is the 'Voluntary Children's Chemical Evaluation Program', which has been running since 2000 as a pilot project and centres on 20 chemicals that children are exposed to.[30] In a recent evaluation, the program was commended by industry and criticized by civil society (GAO 2007).

14.4.3.2 Manufacturing and Processing Notices (Section 5)

Section 5(a) requires premanufacture notification (PMN) to be submitted by an operator at least 90 days before a new substance is produced or imported or before an existing substance is used in what has been determined by an EPA rule to be a

[28] 1 million pounds equals 454 (metric) ton.

[29] Ideas for managing industry data are presented by Applegate and Baer (2006), e.g. measures for increased transparency, penalising overuse of confidentiality claims and creating a registry of study results.

[30] The EPA originally proposed 23 chemicals.

'significant new use'.[31] TSCA requires data on, for instance, use, exposure and hazards, but only information that is already available, i.e. testing is not required even if key data are missing. The EPA states that 67% of PMNs include no test data and that 85% include no health data (OPPT 2008). The EPA, therefore, uses a number of other assessment options and screening tools, including models for structural activity relationships and voluntary agreements on testing (OPPT 2008). In addition, the EPA has published criteria for the identification of persistent and bioaccumulative substances, which can be used for further action under Section 5 (EDF 2007a).

If a substance '*may* present an unreasonable risk [emphasis added]' or will be produced in substantial quantities that may reasonably enter the environment or may cause significant or substantial human exposure, then the EPA, pending development of more information needed for evaluating effects, may limit or prohibit the use of the substance (Section 5(e)). If there is a reasonable basis to conclude that use of a substance '*will* present an unreasonable risk [emphasis added]', the EPA 'shall' take action with one or another form of restriction (Section 5(f)). Up until September 2006, this did not happen more than four times (OPPT 2008). Furthermore, TSCA stipulates that the EPA in some cases must publish a statement if action is not taken (Section 5(g)).

These PMN data, combined with regulatory opportunities, enable the EPA to work more effectively than under Section 4. The EPA can send signals on both avoidance of hazardous substances and guidance towards safer chemicals by issuing lists on categories of chemicals of concern and by informal communication and negotiation with submitters (Lowell 2003). Companies have withdrawn PMNs for some 1,700 chemicals once the EPA has indicated that it wanted to go further, and for over 1,300 chemicals, the EPA has placed demands on workplace controls (GAO 2007). However, less than 5% of PMNs go through a full risk assessment (EDF 2007a), and Applegate (2000) refers to the fact that the EPA took no formal action on 98% of PMN fillings in 1995. In addition, the 'new substances' correspond to only about 1% of the market.

14.4.3.3 Regulation of Hazardous Chemical Substances and Mixtures (Section 6)

With regard to existing substances, Section 6 authorises the EPA to regulate single chemicals in the 'least burdensome' way by nearly any type of measure in the life cycle, for instance by manufacturing restrictions, labelling or use requirements and regulations on concentration levels. The preconditions are that 'there is reasonable basis to conclude' that a substance or a mixture 'presents or *will* present an unreasonable risk of injury to health or the environment [emphasis added]'. The level of evidence required for regulatory action is clearly higher here than under Sections 4 and 5.

The central concept of 'unreasonable risk' is not defined in TSCA, but guidance can be found in the provision that the EPA can promulgate rules to protect against unreasonable risk 'after consideration of the effect on the national economy, small

[31]There are several exemptions for e.g. polymers, intermediates, R&D and low volumes, even though many granted exemptions are controlled (GAO 2007).

business, technological innovation, the environment, and public health' (Section 6(c)(1)(D)). For the sake of consistency, these factors should be included in the weighing of 'unreasonable' under other provisions in TSCA as well, in particular considering that the Congress stated that 'a determination that a risk … is unreasonable involves balancing the [risk] against the effect of … regulatory action on the availability to society of the benefits of the substance … taking into account the availability of substitutes … and other adverse effects which such proposed action may have on society.' (H. R. 1976). The EPA (1994) itself stated, in a proposed rule concerning lead fishing sinkers, that 'The unreasonable risk finding can be characterised as a judgment that the risk of health or environmental injury from the substance/mixture outweighs the burden to society of potential regulations'.

These requirements probably comprise key explanations as to why TSCA has not delivered the protection that was anticipated (Denison 2009; Applegate 2008; Ashford 2007). Since its enactment, only a handful of substances have been banned or restricted under Section 6 (OPPT 2008): certain CFCs (today superseded by air legislation), PCBs (statutory), dioxin in a specific case (superseded), certain uses of metalworking fluids (ban in place), asbestos (basically overturned) and hexavalent chromium (ban in place).

Furthermore, under judicial review, a court 'shall hold unlawful and set aside such rule if the court finds that the rule is not supported by *substantial evidence* in the rulemaking record [emphasis added]' (Section 19(c)), which causes problems in practice for the EPA, for example when attempting to prove that a requirement is the 'least burdensome'. This became evident after the EPA in 1989 had nearly completely banned asbestos, and some companies filed suit against the EPA by claiming, for instance, that there was a lack of substantial evidence on unreasonable risks. Supporting this claim by ruling that the EPA had not considered all of the evidence, not analysed the least burdensome option, not done a cost-efit analysis for all regulatory options, a US Court of Appeal returned most of the EPA rule for reconsideration (GAO 2007; Applegate 2000).[32] Considering the vast knowledge of the problems caused by asbestos and the fact that the EPA had worked on preparing the rule for about a decade, the case clearly illustrates the high level of evidence required. Since then, the EPA has not used Section 6 for restrictions (OPPT 2008).

On the other hand, a number of EPA-initiated voluntary agreements and programs have led to well-needed risk management measures in some important cases, for instance regarding fluorinated substances such as PFOA, detergents and certain brominated flame retardants (see further in EDF 2007a).

14.4.3.4 Reporting and Retention of Information (Section 8)

Section 8 regulates the division between 'new' and existing substances, the latter initially being all chemicals on the market before December 1979 and listed (according to Section 8(b)) in the 'TSCA Chemical Substance Inventory'. The approximately

[32] Corrosion Proof Fittings v. EPA, 947 F.2d 1201 (5th Cir. 1991). See also Harremoës et al. (2001).

62,000 substances that by then were included in the inventory were considered safe from a legal point of view, and they can still be used as long as restrictions under Section 6 are not adopted. This so-called 'grandfathering' means that most chemicals, at almost all quantities, are practically unregulated. Since the first list, nearly 21,000 'new' substances have been added in accordance with the PMN process, and the inventory, thus, encompassed nearly 83,000 substances in 2006 (OPPT 2008). As stated by the EPA, companies are requested to keep records of and report to the EPA on a number of basic chemical data (Section 8(a)) regularly included in the inventory.

Finally, a chemical company that obtains information that 'reasonably supports the conclusion' of 'substantial risk of injury to human health or the environment' caused by a chemical must inform the EPA as soon as possible (Article 8(e)). This function was intended as an early warning system, however as a consequence of a paucity of results, the EPA sent a letter to industry in 1990 about 'an amnesty program' for 3 years, giving companies a chance to submit any data they had failed to report, which many companies then did in the coming years (Lowell 2003).

14.4.3.5 Other Central Elements in TSCA

In addition to the provisions mentioned above, TSCA regulates a number of other issues. For instance, if a risk can be managed satisfactorily by other federal laws, TSCA prevents the EPA from taking action (Section 9(a)), unless it is in the public interest to regulate for protection (Section 9(b)). However, transfers of issues to other laws have been limited.

Section 14 in TSCA prevents information reported to the EPA from being disclosed to the general public if it is considered to involve trade secrets of different kinds; even the identities of companies and chemicals can be kept secret. Exceptions involve information concerning health and safety studies and information that the EPA considers necessary to disclose in order to protect public health or the environment (EDF 2007a). Taking 1990 as an example, chemical identification was claimed to involve confidential business information (CBI) in around 90% of PMNs, but it often turns out that the rate drops quite significantly after the PMN process has been completed or in the rare cases when the EPA challenges a CBI claim (OPPT 2008; EDF 2007a).

14.4.4 TSCA and the Precautionary Principle

As Applegate (2008) writes, the intentions behind TSCA were to put a legal system in place that was preventive and that placed responsibility for generating information on chemical companies. Despite 'pockets of precaution' (Wagner 2000), though, TSCA has not achieved this result. The separation between existing and new chemicals, a substance by substance approach, the omission of certain articles and the strong burden of proof placed on the EPA have led to a situation with an obvious lack of data and severe difficulties in imposing safety measures (Denison 2009; Sachs 2009; Applegate 2008; Ditz 2007; Ashford 2007; GAO 2005).

The failure is most obvious in relation to existing substances, with very little action taken. For new chemicals, the EPA has been somewhat more successful, but this category includes only a minority of the chemicals on the market. As a consequence, the EPA has often considered it necessary to enter into enforceable consensual agreements with industry that are far from being sufficiently effective.

Looking at the first core element of the precautionary principle – 'group or worst-case classification' – TSCA can hardly be seen to be stimulating the generation of data. Instead, it rewards the maintenance of unknown chemicals by producing 'incentives for scientific ignorance' (Wagner 2000). Besides the fact that the EPA sometimes uses, for instance, structural activity relationships when screening substances, neither policy nor law contains elements of group or worst-case classification. The burden of proof for making a claim about a specific property of a substance rests very strongly on the EPA.

The second precautionary element, basing policies on intrinsic properties, hardly plays any central role in TSCA. On the contrary, the provisions generally place a strong focus on the risk concept, so not even obvious toxicity is necessarily a factor leading to further regulatory action. Persistence and bioaccumulation have an even weaker position in the system and do not automatically play any legal role, even though the EPA uses such parameters, for instance in the screening of PMNs.

When it comes to the third element, preventive measures in the form of substitution or various types of restrictions are very difficult to implement under TSCA. The EPA is not completely free to take action (under e.g. Section 6), even when risks have been more or less completely proven, since a number of other provisions must be followed and since the burden of proof rests completely on the EPA. This is clearly illustrated by the passiveness of the EPA in relation to Section 6 action after the asbestos court case.

The fourth core element, the maximin principle, is not implemented at all. On the contrary, there is a strong reliance in TSCA and in the implementation of the act on formal risk assessments and cost–benefit analysis, which more or less completely blocks complementary approaches, such as decision-making aiming at reducing the likelihood of the worst case becoming a reality.

Finally, summarising the burden of proof in TSCA, it basically always rests on the EPA. Not even if the concept is divided into 'the burden of persuasion' and 'the burden of proving a causal link' (see further in Ashford 2007) does any component fall on chemical companies.

In summary, with few exceptions, TSCA is not at all in line with the core elements of the precautionary principle.

14.5 Discussion

The problems with chemicals management have in many ways been the same in most countries, including lack of data on chemicals, distinctions between new and existing chemicals, unconsolidated legislation, ineffective processes for risk assessment, burden of proof being placed on governments, limited prevention and weak incentives for substitution. Consequently, costs and administrative burdens have been high, and environmental and health effects have been severe – facts that are recognised by most parties today (European Commission 2009). Neglect of early warnings has led to several serious miscalculations during the past century (Harremoës et al. 2001).

It is quite clear that traditional risk assessment and risk management approaches are not working sufficiently well in the field of chemicals policy, in particular not in cases of high uncertainty. The traditional approach could hardly deal with the early chemical problems, characterised by evident impacts such as acute effects, and is even less effective in the present situation, with globalised flows of articles that contain hazardous chemicals and the resulting complex chemical cocktail, which may cause diffuse but significant adverse effects on human health and the environment.

The Toxic Substance Control Act is a nearly perfect realisation of the old approaches and is clearly outdated. The EPA has been somewhat successful with stimulating voluntary action, but that is far from sufficient, judging from the results in relation to expectations. Previous EU chemicals legislation was permeated by many of the same problems, and it is obvious that the designers of REACH have learned important lessons from the past.[33]

Comparing generation and interpretation of data, REACH potentially[34] requires between 22 and 54 tests, depending on the quantity, whereas TSCA requires zero tests for existing substances and potentially 14 – voluntary – tests for higher volumes of new substances (GAO 2007).[35] REACH no doubt requires more data and clearly places the responsibility for providing the data on chemicals companies and interprets lack of data as if a chemical is not desirable. Furthermore, data generated, even with respect to intrinsic properties, are more often used for further regulatory action under REACH than under TSCA, for example in the authorisation phase. REACH, however, suffers from the traditional approach in the restriction phase, though not as much as TSCA.

If TSCA is more protective than REACH on any single point at all, it is in relation to the evaluation of new substances, which in the US regularly takes place in a screening process before marketing, whereas marketing in the EU is more or less free once a registration is finalised, i.e. before the evaluation phase starts. TSCA is more protective on this point, but that is the exception that confirms the rule.

At present, the focus in the EU when it comes to chemicals law is on implementation of REACH. If there is political room for improvement, it should in my view focus on the following aspects[36]:

---

[33]Applegate (2008) focuses on the three decades between the CEQ and TSCA in 1971/1976, and the White Paper and REACH in 2001/2006, respectively, and analyzes the pairs as thesis and antithesis. Evidently, REACH is a reaction, but more to previous EU law than to TSCA, even if there were similarities between these.

[34]This depends on the situation; not all tests are always needed.

[35]The HPVC Program is based on potentially 18 tests. For details, see also Table 4 in EDF (2007a).

[36]See further important suggestions in e.g. Hansen et al. (2007).

- Fewer exemptions from REACH of specific categories of chemicals that are not regulated with the same degree of protection as REACH would require.
- Inclusion of chemicals in articles in a more comprehensive manner in REACH.
- Duty to register substances in lower quantities than 1 t per company and year.
- Increased data requirements for registration, in relation to all quantities.
- Extended time periods between registration and market introduction, enabling improved fast screening evaluation, possibly leading to further public management.
- Strict demands on general and early substitution; substitutes should be identified for substances in high quantities or with dangerous properties already in the registration phase.
- Fewer bottlenecks, lower barriers and decisive time limits for the process of identifying substances for the authorisation procedure.
- No authorisation when substitutes exist; statutory deadline on maximum time-limit for review of authorisation.
- Automatic phasing-out over time of CMRs, PTBs, vPvBs and chemicals with other hazardous intrinsic properties.
- Substantially lower burden of proof for public agencies when it comes to decisions on restrictions.
- Increased transparency regarding data provided by industry and agencies.

Whether or not the REACH statutes will change and develop in the coming years is, of course, difficult to say. However, it seems likely that continued administrative problems could reignite the debate, in particular once a new European Parliament and a new European Commission are in place during 2009.

Turning to the US situation, several agencies and scholars have for a long time pointed out shortcomings in TSCA[37] without any results. However, it seems plausible that US chemicals legislation could be amended in a fundamental manner in the coming years, due to both the new political landscape and the impact of REACH.[38] Considering the strong transatlantic trade relations, which includes exports from the US to the EU of chemical substances on their own and in various articles like electronics (Ackerman et al. 2006), REACH will clearly affect US companies that must comply with provisions concerning imports, thereby paving the way for regulatory reforms in the US (Scott 2009; Sachs 2009; Wirth 2007). At present, the debate over US chemicals policy is underway, and interest in REACH is quite apparent.[39]

On the federal level, a number of legislative initiatives are up for discussion in the US Congress, some being more inspired by REACH than others.[40] In May 2008, two identical bills called the 'Kids-Safe Chemicals Act' were sent to the Senate and to the

[37] See e.g. GAO (2005), Applegate (2000), Wagner (2000) and GAO (1994).

[38] EU and U.S. risk policies have influenced each other quite a lot over time; see Löfstedt and Vogel (2001).

[39] See e.g. the Report from the University of Pittsburgh' Seminar '*REACH: A New EU Approach to Chemicals Safety: Lessons for the United States*', on June 7–9, 2007, which gathered together several scholars and bureaucrats (see www.ucis.pitt.edu/euce/events/policyconf/07).

[40] This contrasts sharply with the critical view of REACH in the previous administration, see H. R. (2004).

House of Representatives.[41] The proposals are inspired by REACH and include a proposal for an additional part under TSCA, 'Title V – Child Safe Chemicals' as well as proposals for a number of amendments on existing sections in TSCA. The aim is to 'reduce exposure of children, workers and consumers to toxic substances', and the main elements include safety statements for manufacturers (i.e. partly reversed burden of proof), explicit safety standards and priority lists (including for prenatal exposure) and provisions in the direction of 'no data, no market', substitution and improved access to information (H. R. 2008; S 2008; see also Scott 2009 and Sachs 2009). A similar development is taking place on the state level, for instance in California, Maine and Washington (Scott 2009; Sachs 2009; Ditz 2007).

It is worth noting here that the initial main arguments against REACH – allegedly high costs and lower competitiveness – could not be successfully substantiated (see e.g. Sachs 2009; Selin 2007; Karlsson 2006) and would probably be even harder to make in the US since the development in the EU, a key export market, has already taken place. With REACH-inspired amendments in TSCA, a more level transatlantic playing field could develop, which would, in turn, have an impact on international chemicals law and chemicals legislation in other countries as well (Sachs 2009; Fisher 2008; Park et al. 2008).[42]

When it comes to developing US legislation in line with the precautionary principle, regulators must ask whether the best option would not be to toss TSCA into the garbage bin, without recycling the statutes. Developing a legal system based fundamentally on the core elements of the precautionary principle would be preferable, in my view. Proposals pointing in that direction have been developed by, for instance, the Swedish Committee on New Guidelines on Chemicals Policy (CNC 2000).

However, if the choice is made to develop TSCA, then a number of basic amendments are needed, partly in line with an improved version of REACH, as sketched above. In my view, a reform should include the following[43]:

- Inclusion of criteria for hazardousness, including toxic and non-toxic properties, guiding data requirements and further decisions on preventive measures
- Responsibility placed on companies to submit data, and lower barriers for the EPA to request additional information
- Treatment of existing chemical substances as, in principle, new substances; inclusion of the 'no data – no market' principle
- A comprehensive approach regarding chemicals in articles
- A general requirement to substitute hazardous chemicals when alternatives exist
- Duties for the EPA to decide on a span of preventive measures, even in case of uncertainty, with a burden to prove no need for regulation placed on companies

[41]Bill S. 3040, by Senators Lautenberg, Menendez, Whitehouse, Clinton and Kerry, and Bill H. R. 6100, by Reps. Solis and Waxman.

[42]See Heyvaert, this volume, who analyzes REACH as a blueprint for an international risk governance model.

[43]See also important proposals in e.g. Denison (2009) and Ditz (2007).

- A general requirement for substituting hazardous chemicals when alternatives exist
- Partial regulation of voluntary initiatives that fail to yield high quality data
- Increased disclosure of data on properties of chemicals

It remains to be seen how the debates in the US and its Congress develop and what the legal implications might be, but if TSCA were to develop gradually in accordance with the suggestions above, it could well mature into a legal system with the same protective functions as REACH, or preferably into something even better.

Evidently, managing uncertainty related to industrial chemicals is not an easy task. So far, the voluntary and regulatory approaches implemented around the world have not succeeded more than marginally. Implementing the core elements of the precautionary principle in EU and US legislation on industrial chemicals, as suggested and elaborated on in this article, would clearly enable better control of chemicals and, thereby, most likely lead to increased safety for human health and the environment.

References

Ackerman, F., Stanton, E. and Massey, R. (2006) *European Chemical Policy and the United States: The Impact of REACH*. Global Development and Environment Institute. Working Paper 6-06. Medford: Tuft University.

Ahteensuu, M. (2007) Defending the precautionary principle against three criticisms. *TRAMES* 11, 366–381.

Allanou, R., Hansen, B. G. and van der Bilt, Y. (1999) *Public availability of data on EU high production volume chemicals*. Ispra: European Commission Joint Research Centre, Institute for Health and Consumer Protection, European Chemical Bureau.

Applegate, J. S. (2000) The precautionary preference: An American Perspective on the Precautionary Principle. *Human and Ecological Risk Assessment* 6, 413–443.

Applegate, J. S. (2008) *Synthesizing TSCA and REACH: Practical Principles for Chemical Regulation Reform*. See http://ssrn.com/abstracts=1183942.

Applegate, J. S. and Baer, K. (2006) *Strategies for Closing the Chemical Data Gap*. Washington, DC: Center for Progressive Reform.

Ashford, N. (2007) The Legacy of the Precautionary Principle in US Law: The Rise of Cost–Benefit Analysis and Risk Assessment as Undermining Factors in Health, Safety and Environmental Protection. In: de Sadeleer, N. (ed.) *Implementing the precautionary principle. Approaches from the Nordic Countries, EU and USA*. London: Earthscan.

Boehmer-Christiansen, S. (1994) The precautionary principle in Germany – Enabling Government. In: O'Riordan, T. and Cameron. J (eds.) *Interpreting the precautionary principle*. London: Earthscan.

Cairns, J. and Smith, E. P. (1996) Uncertainties Associated with Extrapolating from Toxicologic Responses in Laboratory Systems to the Responses of Natural Systems. In: Lemons, J. (ed.) *Scientific Uncertainty and Environmental Problem Solving*. London: Blackwell Science.

CEQ (1971) *Toxic Substances*. Executive Office of the President: United States Council on Environmental Quality (CEQ).

CNC (2000) *Non-hazardous products – Proposals for implementation of new guidelines on chemicals policy*. Committee on New Guidelines on Chemicals Policy (CNC). SOU 2000:53. Stockholm: Fritzes.

Dancet, G. (2009). *The main challenges for ECHA and the chemical industry as a whole in the (near) future*. Speech at the Annual FECC Conference, Helsinki, 23/6 2009.

de Sadeleer, N. (ed.) (2007a) *Implementing the Precautionary Principle. Approaches from the Nordic Countries, EU and USA*. London: Earthscan.

de Sadeleer, N. (2007b) The precautionary principle in European Community Health and Environmental Law: Sword or Shield for the Nordic Countries? In: de Sadeleer, N. (ed.) *Implementing the precautionary principle. Approaches from the Nordic Countries, EU and USA*. London: Earthscan.

de Sadeleer, N. (2007c) The Impact of the Registration, Evaluation and Authorization of Chemicals (REACH) Regulation on the Regulatory Powers of the Nordic Countries. In: de Sadeleer, N. (ed.) *Implementing the precautionary principle. Approaches from the Nordic Countries, EU and USA*. London: Earthscan.

Denison, R. A. (2009) Ten Essential Elements in TSCA Reform. *Environmental Law Reporter* 39, 10020–10028.

Ditz, D. (2007) The States and the World: Twin Levers for Reform of U.S. Federal Law on Toxic Chemicals. *Sustainable Development Law & Policy*, Fall 2007.

EC (2003a) Directive 2002/95/EC of the European Parliament and of the Council of 27 January 2003 on the restriction of the use of certain hazardous substances in electrical and electronic equipment. *Official Journal of the European Union* L 37, 13.2.2003, 19–23.

EC (2003b) Directive 2002/96/EC of the European Parliament and of the Council of 27 January 2003 on waste electrical and electronic equipment (WEEE). *Official Journal of the Union* L 37, 13.2.2003, 24–38.

EC (2006a) Regulation (EC) No 1907/2006 of the European Parliament and of the Council of 18 December 2006 concerning the Registration, Evaluation, Authorisation and Restriction of Chemicals (REACH), establishing a European Chemicals Agency, amending Directive 1999/45/EC and repealing Council Regulation (EEC) No 793/93 and Commission Regulation (EC) No 1488/94 as well as Council Directive 76/769/EEC and Commission Directives 91/155/EEC, 93/67/EEC, 93/105/EC and 2000/21/EC. *Official Journal of the European Union* L 396, 30.12.2006, 1–849.

EC (2006b) Council Decision of 17 July 2006 amending Decision 1999/468/EC laying down the procedures for the exercise of implementing powers conferred on the Commission. *Official Journal of the European Union* L 200, 22.7.2006, 11–13.

EC (2008) Regulation (EC) No. 1272/2008 of the European Parliament and of the Council of 16 December 2008 on classification, labelling and packaging of substances and mixtures, amending and repealing Directives 67/548/EEC and 1999/45/EC, and amending Regulation (EC) No 1907/2006. *Official Journal of the European Union* L 353, 31.12.2008, 1–1355.

ECB (2008) *European Chemicals Bureau Newsletter*, No 1 2008. Ispra: European Commission Joint Research Centre, Institute for Health and Consumer Protection, European Chemical Bureau (ECB).

ECHA (2009) *Recommendation of the European Chemicals Agency (ECHA) of 1 June 2009 for the inclusion of substances in Annex XIV (the list of substances subject to regulation) of Regulation No 1907/2006*. Helsinki: European Chemicals Agency (ECHA).

EDF (2007a) *Not that innocent. A Comparative Analysis of Canadian, European Union and United States Policies on Industrial Chemicals*. Washington, DC: Environmental Defense (EDF).

EDF (2007b) *High Hopes, Low Marks. A Final Report Card on the High Production Volume Chemical Challenge*. Washington, DC: Environmental Defense Fund (EDF).

EDF (2008) *Environmental Defense Fund's Comments on ChAMP: EPA's Recent Commitments and Possible New Initiatives for Existing Chemicals*. Washington, DC: Environmental Defense Fund (EDF).

EEA (1998) *Chemicals in the European Environment: Low Doses, High Stakes?* The EEA and UNEP Annual Message 2 on the State of Europe's Environment. Copenhagen: European Environment Agency (EEA).

EEA (2007) *Europe's environment. The fourth assessment*. Chapter 2.5. Hazardous Chemicals. Copenhagen: European Environment Agency (EEA).

EEC (1967) Council Directive 67/548/EEC on the approximation of laws, regulations and administrative provisions relating to the classification, packaging and labelling of dangerous substances, as amended. *Official Journal of the European Communities* 196, 16.8.1967, 1.

EEC (1976) Council Directive 76/769/EEC on the approximation of the laws, regulations and administrative provisions of the Member States relating to restrictions on the marketing and use of certain dangerous substances and preparations. *Official Journal of the European Communities* L 262, 27.9.1976, 201.

EEC (1988) Council Directive 88/379/EEC on the approximation of the laws, regulations and administrative provisions relating to the classification, packaging and labelling of dangerous preparations. *Official Journal of the European Communities* L 187, 16.7.1988, 14.

EEC (1993) Council Regulation 793/93/EEC of 23 March 1993 on the evaluation and control of the risks of existing substances. *Official Journal of the European Communities* L 84, 1.

Environment Council (1999) *Council Conclusions on Community policy for chemicals products.* Minutes of the 2194th Council meeting (Environment) held in Luxembourg on 24 and 25 June 1999.

EPA (1994) *Lead Fishing Sinkers; Response to Citizen's Petition and Proposed Ban.* Federal Register 59, 11138.

European Commission (1998) *Report on the operation of community legislation in the chemicals sector.* COM (1998) 587/3.

European Commission (2000) *Communication from the Commission on the precautionary principle.* COM (2000) 1.

European Commission (2001) *Strategy for a future Chemicals Policy. White Paper.* COM (2001) 88.

European Commission (2003) *Extended Impact Assessment.* Commission Staff Working Paper. COM (2003)644 Final. SEC (2003)1171/3.

European Commission (2007) *REACH in brief.* Brussels: European Commission, Environment Directorate.

European Commission (2009) *Final Report of the High Level Group on the Competitiveness of the European Chemicals Industry.* Brussels: European Commission, DG Enterprise and Industry.

European Council (2000) *Presidency Conclusions. Nice European Council Meeting*, 7, 8 and 9 December 2000.

Fisher, E. (2008) The 'perfect storm' of REACH: charting regulatory controversy in the age of information, sustainable development, and globalization. *Journal of Risk Research* 11, 541–563.

GAO (1994) *Toxic Substances Control Act: Legislative Changes Could Make the Act More Effective.* Report to Congressional Requesters. GAO-05-458. Washington: United States Government Accountability Office.

GAO (2005) *Chemical Regulation: Options Exist to Improve EPA's Ability to Assess Health Risks and Manage its Chemical Review Program.* Report to Congressional Requesters. GAO-05-458. Washington: United States Government Accountability Office.

GAO (2007) *Chemical Regulation. Comparison of U.S. and Recently Enacted European Union Approaches to Protect Against the Risks of Toxic Chemicals.* Report to Congressional Requesters. GAO-07-825. Washington: United States Government Accountability Office.

Gardiner, S. M. (2006) A Core Precautionary Principle. *The Journal of Political Philosophy* 14, 33–60.

H. R. (1976) *House Report No. 94-1341*, pp 13–14. U.S. House of Representatives.

H. R. (2004) *A Special Interest Case Study: The Chemical Industry, the Bush Administration, and European Efforts to Regulate Chemicals.* U.S. House of Representatives, Committee on Government Reform, Minority Staff Special Investigations Division, Washington.

H. R. (2008) *A Bill to amend the Toxic Substances Control Act to reduce the exposure of children, workers, and consumers to toxic chemical substances.* U.S. House of Representatives, 110th Congress, 2D Session, May 20, 2008, H. R. 6100.

Hansen, S. F., Carlsen, L. and Tickner, J. (2007). Chemicals regulation and precaution: does REACH really incorporate the precautionary principle? *Environmental Science and Policy* 10, 395–404.

Hansson, S. O. (1997) The limits of precaution. *Foundations of Science* 2, 293–306.

Harremoës, R., Gee, D., MacGarvin, M., Stirling, A., Keys, J., Wynne, B. and Guedes Vaz, S. (2001) *Late lessons from early warnings: the precautionary principle 1896–2000.* Environmental issue report No 22. Copenhagen: European Environment Agency (EEA).

Karlsson, M. (2005) *Managing Complex Environmental Risks for Sustainable Development: Policies for Hazardous Chemicals and Genetically Modified Organisms.* Doctoral Thesis. Karlstad University Studies 2005:34. Karlstad: Karlstad University.

Karlsson, M. (2006) The Precautionary Principle, Swedish Chemicals Policy and Sustainable Development. *Journal of Risk Research* 9, 337–360.

Koch, L. and Ashford, N. A. (2006) Rethinking the role of information in chemicals policy: implications for TSCA and REACH. *Journal of Cleaner Production* 14, 31–46.

Krämer, L. (2006). *EC Environmental Law*. 6th edition. London: Sweet & Maxwell.

Lahl, U. (2007) REACH – assessment of the political agreement. *Environmental Law Network International Review* 1/2007, 34–38.

Löfstedt, R. E. and Vogel, D. (2001) The Changing Character of Regulation: A Comparison of Europe and the United States. *Risk Analysis* 21, 399–405.

Lökke, S. (2004) *The Precautionary Principle and Regulation of Chemicals*. Academic Thesis. Aalborg: Aalborg University.

Lowell (2003). *The Promise and Limits of the United States Toxic Substances Control Act*. Lowell Center for Sustainable Production. Lowell: University of Massachusetts.

Maastricht Treaty (1992) *Provisions Amending The Treaty Establishing The European Economic Community with a view to Establishing The European Community*. Title II, Article 130r. Maastricht, 7 February 1992.

O'Riordan, T., Cameron, J. and Jordan, A. (eds.) (2001) *Reinterpreting the precautionary principle*. London: Cameron May.

OPPT (2008) *Overview: Office of Pollution Prevention and Toxics Programs*. March 2008. U.S. EPA: Office of Pollution Prevention and Toxics Programs (OPPT).

Park, D., Song, M-Y., Lee, K-H., Yoon, D-C. and Cong, X. (2008) REACHing Asia: Recent Trends in Chemical Regulations of China, Japan and Korea. Working Paper. See http://ssrn.com/abstract=1121404.

Rawls, J. (1971) *A Theory of Justice*. London: Oxford University Press.

Roe, D., Pease, W., Florini, K. and Silbergeld, E. (1997) *Toxic Ignorance*. Washington: Environmental Defense Fund.

Rogers, M. D. (2003a) Risk analysis under uncertainty, the Precautionary Principle, and the new EU chemicals strategy. *Regulatory Pharmacology and Toxicology* 37, 370–381.

Rogers, M. D. (2003b) The European Commission's White Paper 'Strategy for a Future Chemicals Policy': A Review. *Risk Analysis* 23, 381–388.

S (2008). *A Bill to amend the Toxic Substances Control Act to reduce the exposure of children, workers, and consumers to toxic chemical substances*. U.S. House of Representatives, 110th Congress, 2D Session, May 20, 2008, S. 3040.

Sachs, N. M. (2009) Jumping the pond: Transnational Law and the Future of Chemical Regulation. *Vanderbilt Law Review*, Forthcoming. See http://ssrn.com/abstract=1413746.

Sand, P. H. (2000). The Precautionary Principle: A European Perspective. *Human and Ecological Risk Assessment* 6, 445–458.

Sandin, P. (1999) Dimensions of the precautionary principle. *Human and Ecological Risk Assessment* 5, 889–907.

Sandin, P. (2004) *Better Safe than Sorry. Applying Philosophical Methods to the Debate on Risk and the Precautionary Principle.* Thesis in Philosophy from the Royal Institute of Technology, No. 5, 2004. Stockholm: Royal Institute of Technology.

Sandin, P., Peterson, M., Hansson, S. O., Rudén, C. and Juthe, A. (2002) Five charges against the precautionary principle. *Journal of Risk Research* 5, 287–299.

Scott, J. (2009) From Brussels with Love: The Transatlantic Travels of European Law and the Chemistry of Regulatory Attraction. *American Journal of Comparative Law* 57:4.

Selin, H. (2007) Coalition Politics and Chemicals Management in a Regulatory Ambitious Europe. *Global Environmental Politics* 7, 63–93.

Stokes, E. (2008) The EC Courts' contribution to refining the parameters of precaution. *Journal of Risk Research* 11, 491–507.

Tickner, J. (ed.) (2003a) *Environmental Science and Preventive Public Policy*. Washington D.C.: Island Press.

Tickner, J. (2003b) Precautionary Assessment: A Framework for Integrating Science, Uncertainty, and Preventive Public Policy. In: Tickner, J. (ed) *Environmental Science and Preventive Public Policy*. Washington D.C.: Island Press.

UNCED (1993) *Report of the Conference of the United Nations on Environment and Development (UNCED)*, Rio de Janeiro, 3–14 June 1992, vol. 1–3. New York: United Nations.
von Moltke, K. (1988) The Vorsorge-Prinzip in West-German Policy. In: *Twelfth Report of the Royal Commission on Environmental Pollution*, Annex 3. London: H.M. Stationery Office.
Wagner, W. E. (2000) The Precautionary Principle and Chemicals Regulation in the U.S. *Human and Ecological Risk Assessment* 6, 459–477.
Wirth, D. A. (2007). *The EU's New Impact on U.S. Environmental Regulation.* Research Paper 144, November 9, 2007. Legal Studies Research Paper Series. Boston: Boston College Law School.

Chapter 15
Chemical Regulations in Central and Eastern Europe: The Pull of Transnational Markets and Associations

Liliana B. Andonova

Abstract This chapter considers the role of market integration and trans-European associations and coalitions, as potential explanatory factors that can account for the faster and more successful than anticipated horizontal diffusion of chemical standards. Market incentives are identified in mandatory and voluntary chemical safety standards not only in Europe but across the globe. However, the present analysis reveals that while market incentives can help explain the adjustment of interests of influential domestic actors in favour of harmonisation, they can hardly account for the overwhelming political support for the early adoption of EU chemical regulations despite the negative trade balance in chemicals in the region, competitiveness pressures faced by the majority of Central and East European chemical enterprises, and the high administrative cost of implementation. The chapter advances the argument that associations of the European chemical industry operating at the supranational and domestic levels have played a critical role for regulatory harmonisation in the new member states and their subsequent implementation. These industry associations spurred trans-European policy coalitions as a political force that pushed the harmonisation of chemical standards, thus serving as a transnational belt for integrating the new member states in the multi-level governance structure that characterises EU politics and regulations.

Keywords Chemicals • Industry • Markets • EU • REACH

L.B. Andonova (✉)
Graduate Institute of International and Development Studies,
Case postale 136 - 1211, Genève 21 - Suisse
e-mail: Liliana.Andonova@graduateinstitute.ch

J. Eriksson et al. (eds.), *Regulating Chemical Risks: European and Global Challenges*,
DOI 10.1007/978-90-481-9428-5_15,

15.1 Introduction

Chemicals regulations and labelling in the countries of Central East Europe were relatively weakly developed before their preparations for membership in the European Union (EU) and largely incompatible with EU standards (Regional Environmental Center 1996). Yet, within 10 years of pre-accession preparations these states fully adopted EU chemical safety regulations, which are considered to be among the most complex and advanced in the world. What explains the smooth diffusion of EU chemical safety regulations to the new member states despite the economic and administrative costs entailed?

The chapter is structured as follows. It first considers arguments of market integration and regulatory harmonisation in the context of EU enlargement and the market for chemical products. The chapter then analyses the role of EU chemical associations in translating market pressure into incentives and political support for regulatory reform. The section identifies several mechanisms through which European chemical associations facilitated regulatory harmonisation: capacity building for national counterparts in Central and Eastern Europe, promotion of norms and voluntary standards, and facilitating policy coalitions across institutions and levels of European governance. Through these mechanisms, the domain of chemicals safety regulation in Central and Eastern Europe, became integrated in the multi-level regulatory space of the EU much before and much more comprehensively than what might be anticipated on a purely market or conditionality logic. The chapter also provides a critical assessment of the corporatisation of chemical safety politics and regulations in the new member states by considering both the advantages and disadvantages of the dominant role of industry associations for the implementation and access to information on chemical safety.

15.2 The Market Logic of Harmonising Chemical Regulations

The market for chemicals in the EU and globally is highly integrated and dominated by large national or multinational companies. As worldwide competition is reported to have gotten fiercer, almost half of the 30 major world chemicals companies (14) had their headquarters in the EU in 2006. According to the European Chemical Industry Council (CEFIC), they held a 16% share of the chemicals sales around the globe, and they had a combined sales turnover of 526 billion €. Inside the EU, there were 27,000 enterprises. Only 4% of them employed more than 249 people – but they generated 70% of the total chemicals sales.[1]

The international market for chemicals, particularly in the Western hemisphere and in the EU, is also increasingly regulated to reflect concerns about the safety, environmental impacts, and management of chemical substances. As environmental movements

[1] http://www.cefic.be/factsandfigures/level02/profile_index.html.

gained strength and after several major chemical disasters, the industry has become the target of organised pressure for regulating its negative health and environmental externalities. Legislation on chemicals was first adopted by individual West European countries and the United States (US). The high market integration and interdependence of the industry along with the transnationalisation of societal pressures resulted in visible policy spillover effects manifested both horizontally across jurisdictions, as well as vertically in the strengthening of chemical legislation. One of the most prominent theories of regulatory harmonisation in the management of chemical substances in Europe and across the Atlantic emphasises the pull of large regulated markets and the incentives of exporters to support harmonisation as means to reduce transaction cost and competitive disadvantage (Vogel 1995). The adoption and exportation of voluntary standards such as the Responsible Care across the globe has been similarly motivated by concerns about market reputation, risks, and levelling the regulatory playing field (Garcia-Johnson 2000). The neofunctionalist theory of European integration similarly anticipated that economic integration on the old continent would create spillover pressures for further policy harmonisation (Haas 1958).

The logic of market integration and functional spillovers indeed played a visible role in the harmonisation of chemical safety standards in the European Community. The first of a number of directives aimed at protecting human health and the environment from harmful chemicals was adopted in 1967. It regulated the classification, packaging and labelling of more than 1,000 dangerous substances. It is notable that the goal of environmental protection was not included in the 1957 Treaty of Rome, establishing the European Economic Community. Most chemical safety standards were initially based on Article 100 of the Treaty, which explicitly links market integration and spillover regulatory pressures, and provides for the harmonisation of regulations directly related to the functioning of the Common Market.

The passage of the Toxic Substances Control Act in the United States in 1976 gave a further market-driven impetus to the harmonisation of chemical safety regulations within the European Community. Large chemical producers and national officials in Europe were concerned that the provisions of the US legislation on chemicals would limit the access of European chemical products to the lucrative US market. In response, European Community members sought to align their national chemical policies further and to consolidate a common regulatory framework to counterbalance US regulations (Vogel 1995). This prompted the rapid adoption of the sixth amendment of the 1967 Directive on chemicals, facilitated by trans-European coalitions of business, working in cooperation with the European Commission and national governments. This was a major step towards strengthening the chemical safety system in the European Community. It also gave the European Community a common voice and greater bargaining power in the negotiations with the US of a broader international system of chemicals control under the Organization for Economic Cooperation and Development (OECD).

The 1977 Seveso accident, which resulted in the contamination of a large area of Northern Italy with dioxin, further increased the societal pressure for safer production, use, and disposal of chemicals, and the activation of coalitions within the European Community. The authority to introduce environmental legislation was also strengthened by subsequent amendments of the Treaty of Rome, namely by the Single European Act (signed in 1986) which recognised environmental protection as a legitimate policy goal of European institutions, and by the Maastricht Treaty of the EU (signed in 1992) which introduced majority voting on environmental legislation, and co-decision procedures of the Council of Ministers and the European Parliament. As a consequence, EU institutions gradually opened space for non-market concerns to influence in a more profound and less functionalist way the making of EU chemical regulations.

The EU chemical safety legislation over time established a complex set of rules for assessing the risks associated with the use and production of chemicals, providing information on those risks and hazards, limiting their harmful effects, and reducing the risk of major industrial accidents. The entry into force in 2007 of Regulation (EC) No 1907/2006 on Registration, Evaluation, Authorization, and Restriction of Chemical Substances (REACH) further strengthened and broadened the requirements for testing, identification of properties, and registration of chemical substances. The adoption of the REACH regulation is significant not only because of its scope, but also because it represents a substantial and in many ways radical departure from the older pattern of functional coordination and incremental ratcheting up of chemical standards.

Following the end of the Cold War, Central and East European countries established a leading foreign policy objective for their reintegration in Europe, both economic and political within the framework of the European Union. The opening of European markets proceeded faster than political integration. The framework for EU enlargement to the East was established by the 1993 Copenhagen Council, with the following criteria for accession: stability of democratic institutions, functioning market economy, and ability to apply EU legislation. During the 1990s, the share of EU trade and investment increased rapidly, a trend that is clearly reflected in the development of the chemical sector in Central and Eastern Europe. Trade in chemicals between the EU and the Central and East European states roughly doubled between 1993 and 1997, and by 2003 grew threefold compared to its 1993 levels (Ghosh 2005). Investments in the chemical industry in the accession states also grew steadily, reaching an annual level of approximately 1.7 billion € in 1999 and 2003, and making the chemical industry in region the second most attractive for foreign investment after car manufacturing (Ghosh 2005).

As economic integration unfolded, the chemical industry of Central and East European countries entered highly cartelised and regulated regional and global markets. The chemical industry in the region was dominated by large, state owned, export-oriented chemical enterprises, many of which faced pressures for restructuring and for attracting foreign investment in the process of privatisation. At the same time the regulations for the management, safety, and labelling of chemicals in these countries were largely incompatible with those of the EU. A study by the Regional Environmental Center (1996) estimated a range of 21–33% compatibility between EU chemical standards and those of the Central and Eastern Europe candidates for accession. The re-orientation of increasing volume of chemical exports to the EU exerted pressure on chemical exporters in Central and Eastern Europe to adopt standards compatible with those of the EU. The chemical sector had also the reputation of one of the dirtiest industries in the region, further increasing incentives for outward looking enterprises to clean up their image (Andonova 2004). It is important to recognise in support of the market-focused functionalist logic of regulatory diffusion that pressure from subcontractors, EU chemical companies, and potential investors was felt by large chemical companies in the region before the preparations to meet EU accession conditionality began.

The environmental pressures and incentives associated with regional market and later on with political integration thus gradually re-shaped the environmental interests of key players in the chemical industry in EU accession countries. The process of East–West integration within the already highly regulated economic and political space of the EU, affected the incentives of domestic actors in a direction supportive of accession and regulatory harmonisation. Chemical enterprises in countries such as Poland and the Czech Republic, whose chemical sector reoriented the most towards EU markets in the course of the 1990s, played a leading role in establishing voluntary programs for eco-efficiency, chemical safety and industry organisations promoting such programs as mechanisms to improve reputation and bring performance standards closer to those of EU counterparts. There was thus a degree of anticipatory adaptation in the behaviour of leading exporters and their political interest in subsequent implementation of EU regulations (Andonova 2004).

The market incentives behind such change in behaviour and political interest in the chemical industry and beyond were succinctly summarised by the director of the Czech Business Council for Sustainable in an interview in 1997:

> EU markets are vital for export-oriented companies and for the Czech economy as a whole, and there is significant pressure by consumers and suppliers in those markets for improved environmental performance. It is in the interest of the country to accept EU legislation in such a way as to be acceptable for industry and to establish a stable legislative environment. That is why we insist that new laws should be, as much as possible, compatible with new EU legislation. The goal is to achieve harmonization in such a way as not to damage the competitiveness of Czech industry.[2]

But while market integration and the regulatory pressure associated with it provided a basic precondition for interest adjustment in support of horizontal policy spillover, they are far from sufficient in accounting for the politics, extent, and speed of regulatory adjustment. A largely market logic of regulatory spillover cannot tell us why large, EU-oriented exporters dominated regulatory politics over the interest of chemical companies that face high costs for regulatory adjustment compared to potential market benefits due to smaller size, restructuring, import competition, or orientation towards post-Soviet markets. Furthermore, certain economic characteristics of the chemical sector in Central and Eastern Europe such as the persistence of negative trade balance in chemicals with the EU (Deutsche Bank Research 2005), the exports of a significant share of chemical products from former Soviet markets, restructuring, and lower average productivity might indeed lead us to expect significant pressure for delays in the adoption of EU regulations, rather than early anticipatory adaptation. In order to better understand the aggregation of sector interest and the interplay of multiple sources of political pressure for

[2]Interview at the Czech Business Council for Sustainable Development, Prague, November 1997.

harmonisation of chemical regulations across old and new EU member states, it is critical therefore to examine in greater detail the political organisation of industry and its interaction with the multiple levels of EU and domestic institutions.

15.3 Industry Associations, Policy Coalitions, and Regulatory Diffusion

The process of EU integration brought the chemical industries of Central and Eastern European countries in closer contact not only with a highly regulated market space, but also with powerful industrial counterparts, highly organised at the EU level. The literature on the EU integration places a growing emphasis on the role of supranational actors and trans-European policy, elite, and expert coalitions in propelling and diffusing EU regulations, norms, and institutional change (Alter 2007; Burley and Mattli 1993; Hooghe and Marks 2001; Schmitter 2001). Such studies have redirected analytical attention back to the political and organisational aspects of the early neo-functionalist theory European integration advanced by Haas (1958), deepening the understanding of trans-European coalitions, their political roots and mechanisms of influence. Despite the strong state-centric focus on policy and normative adjustment under conditionality in the study on EU enlargement, a number of authors also draw attention to the role of transnational actors and multi-level politics in shaping the patterns of convergence or divergence across the expanded membership of the EU (Andonova 2004; Bruszt 2002; Zielonka and Mair 2002; Zielonka 2007).

EU chemical regulations exemplify a policy arena in which trans-European politics, organisations and coalitions, play a critical role both in the vertical deepening of chemical regulations (Selin 2007) as well as in their horizontal extension (Andonova 2004). CEFIC, the peak organisation of the EU chemical industry, served as the centre for policy coalitions that span all the levels of EU governance and promoted the diffusion of chemical safety standards in the context of the Eastern enlargement. CEFIC-centred coalitions influenced regulatory harmonisation in accession states through three main mechanisms: capacity building for partner organisations in accession countries; promoting normative consensus on voluntary and EU standards on chemical safety; and using technical assistance to facilitate expert coalitions across EU institutions, governmental agencies, and industry.

CEFIC is described by environmental advocacy groups as one of the most influential industrial lobby organisations operating in Brussels (Contiero 2006). The association was founded in 1972, precisely around the time the chemical industry was facing pressures for regulation and increasing incentives for harmonistion at the supranational level. Presently, CEFIC has a three-tier membership structure. National chemical industry associations of 22 countries in Europe are members of CEFIC, and 6 federations in Bulgaria, Croatia, Estonia, Latvia, Lithuania and Romania are listed as associate members. The corporate membership of the association includes over 40,000 large, medium and small chemical companies. In addition, CEFIC maintains a network of business members, associated companies, affiliated sector associations and partners.[3] Eleven members of CEFIC are also members of the European Roundtable of Industrialists (Contiero 2006). Linking to the global level, CEFIC is member of the International Council of Chemical Associations (ICCA), and co-ordinates on behalf of ICCA the promotion and implementation of Responsible Care in Europe. Historically, CEFIC works in close cooperation with the European Commission and particularly with DG Enterprise to influence the development of chemical safety and environmental legislation related to the chemical sector (Contiero 2006; Vogel 1995; Selin 2007).

CEFIC positioned itself to be a major player in the accession process much before accession preparations had started. As early as 1992, before Central and Eastern Europe countries had formally applied for EU membership, CEFIC began to establish relations with chemical enterprises and industry associations in post-communist countries, and particularly in Czechoslovakia, Hungary and Poland, which had taken more significant steps towards market reforms. In 1992, CEFIC initiated its first project for the diffusion of western chemical safety and environmental standards: Environmental Advisory Service for Technology Transfer (EASTT). The project sought to promote improved environmental management and energy efficiency practices in Central and Eastern Europe by facilitating contacts and training between the then state-owned Central and Eastern Europe chemical companies and West European companies (CEFIC 1992). Some of the Central and Eastern Europe companies, which first participated in such projects, became later the chief founders of national chemical industry federations and promoters of association with CEFIC and the adoption of voluntary environmental and safety standards under the Responsible Care Program (Andonova 2004).

After Central and Eastern Europe countries applied for EU membership in the period 1993–1995 and started accession preparations, CEFIC became the main counterpart of the European Commission in the horizontal diffusion of chemical safety policies. The organisation also maintained programs to help develop domestic chemical industry associations and promote their association with CEFIC and their inclusion in the process of policy harmonisation. Over the span of 5–10 years, a trans-European network of industry associations, and European Commission experts and experts in the Environmental Ministries of the candidate countries emerged and thickened. The network was centred on CEFIC, which sought to promote the smooth adoption of EU chemical safety regulations as well as Responsible Care standards in the new member states. The actors in the emerging trans-European coalition of epistemic and policy elites shared incentives to enlarge the EU regulatory space on chemicals eastwards, although each type of actor within these coalitions had somewhat different objectives.

At the supranational level, CEFIC and the European Commission became natural counterparts seeking to facilitate the transposition of chemical safety standards and market reforms in the sector. A position paper of CEFIC presented at the 1995 conference Competing in the New Europe: Strategies for the Central and Eastern European Chemical Industry, highlighted the growth in trade of chemicals between EU and Central and Eastern Europe, as Central and Eastern Europe exports to the EU rose by

---

[3]See http://www.cefic.be/.

26% to 1.9 billion ECU[4] between 1993 and 1994, and EU export increased by 24% to three billion ECU during the same period of time (De Bree 1995). CEFIC's position paper emphasised that in order to sustain this rate of economic integration and gradually achieve the political integration of Central and Eastern Europe countries into the EU, it was important to create free markets, transparent ownership structure, and a stable regulatory environment in the chemical industry of candidate countries (De Bree 1995). These conditions were of critical importance to CEFIC members, then largely of the EU-15 countries, to prevent competitive disadvantages and administrative or regulatory hurdles for trade and investment across Europe. The conditions for EU membership established by the Copenhagen Council of the EU (1993) and later promoted by the European Commission in the course of accession preparations were not much different: establishing free market economy, functioning democracy, and adoption of the full body of EU regulation. There was a visible convergence in the objectives of the European Commission and CEFIC with respect to EU enlargement and regulatory harmonisation.

At the sub-national level, CEFIC sought to engage chemical industry federations as central actors in this process, highlighting the adoption and implementation of these standards as a key condition for stable trans-European market. Collaboration with CEFIC provided a boost in the political position and technical skills of domestic industry federations, which were established in the early 1990s after the political changes in Central and Eastern Europe. Large, export-oriented companies played a key role in establishing, supporting and using national associations to promote their market and political interests (Andonova 2004). Project-based resources channelled trough CEFIC and West European chemical associations strengthened the position of industry associations in accession states to promote the adoption of EU standards as desirable for the sector. National associations were also able to extend some technical assistance and training to small and medium enterprises.

Government bureaucracies in the region, overburdened with a load of EU regulatory standards to understand, transpose, adopt, and implement, had little choice but to welcome the technical and political support from Brussels. In the case of chemical regulations, such support was not extended directly or only by the European Commission or through bilateral twinning programs. As a result of close collaboration between CEFIC and the European Commission on issues related to the transposition of the EU chemical regulations in the candidate countries, industry associations became a central player in capacity building and even legislative programs.

A coalition of industry associations, supranational and transnational experts and political elites thus emerged, lending political and technical support to harmonising chemical regulations. Such a coalition, as the discussion already implied, did not emerge, however, overnight or in any sense automatically. It was fostered by the active role of CEFIC, and by significant resources allocated by the European Commission for harmonisation projects in the sector, which were implemented by CEFIC in cooperation with national associations and relevant ministries in the accession states. Table 15.1 lists in chronological order these projects, their objectives, and participating actors.

[4]The European Currency Unit, or ECU, was the predecessor to the Euro and used as the unit of account of the European Community before being replaced by the Euro in 1999.

Table 15.1 EU Funded Projects on Chemical Safety in Central and Eastern Europe (http://www.cefic.be/Templates/shwStory.asp?NID=25&HID=289&PHID=288)

Years	Project	Objectives	Actors involved
1997–1998	Cefic/European Commission/Phare Project: Impact of the Commission's White Paper on the Chemical Industry in the Central and Eastern European Countries	Twinning between industry ecperts in the EU and CEECs; screening of national chemical legislation; identifying the changes required for approximation with EU laws; and anticipated effect on the chemical industry in the CEECs	EU Commission, CEFIC, national federations, public authorities
1999	CEFIC/TAIEX (Technical Assistance Information Exchange Office) Programme on Technical Assistance	Assisting CEEC Federations with regard to the transposition of the "acquis communautaire"	CEFIC, national federations, public authorities, DG Enlargement, DG Enterprise
1999–2000	Transposition Activities	Assisting candidate countries to align their chemicals legislation to that of the European Union, review state of transposition, seminars, workshops	
1999–2000	Industrial Forum on Enlargement	Involving representatives from the chemical industry from the EU, Bulgaria, Czech Republic, Estonia, Hungary, Malta, Latvia, Lithuania, Poland, Romania, Slovakia, Slovenia and Turkey in this initiative of the EU Industry Council	EU Commission, CEFIC, EU Industry Council, national federations and chemical enterprises
2000–2002	ChemFed	Strengthening the capacity of CEE Chemical Industry Federations, prepare for EU enlargment, promote voluntary programs such as Responsible Care	CEFIC, national federations of EU countries, CEE chemical federations, EU Commission (funding)
2000–2002	ChemLeg	Improving the capacity of CEEC Federations to provide regulatory services to assist business operators within the CEECs to cope with the requirements of the Community "acquis". Specifically, provide information on the requirements of the acquis, training and toolkits on implementation, training of trainers	CEFIC, national federations of EU countries, CEE chemical federations, EU Commission (funding)

(continued)

Table 15.1 (continued)

Years	Project	Objectives	Actors involved
2001–2005	ChemFed/ChemLeg 2	Assistance with implementation and compliance with the environmetnal and social acquis: information diffusion, training, prepare conditions for collaboration on registration of chemicals for the REACH (Registration, Evaluation, and Authorisation of Chemicals)	CEFIC, national federations of EU countries, CEE chemical federations, Eu Commission (funding)

Table 15.1 shows that considerable organisational and technical resources were channelled through EU funded projects into several strategic directions: the strengthening of national chemical industry associations; establishment of close working relations between national chemical industry associations, CEFIC and government authorities; and technical support for the transposition and implementation of chemical safety regulations. Information on the financial support for these projects is not readily available. However, the ICIS information reporting service for the chemical and oil industry, estimated that just the ChemFed and ChemLeg programs alone amounted to 3 million € in technical assistance provided by the European Commission and implemented in cooperation with CEFIC (ICIS Chemical Business 2000).

The significance of these projects and the trans-European coalitions that emerged around them for the horizontal diffusion of chemical safety standards is evident in several sets of outcomes. First, in the span of 10 years, chemical industry associations across Central and Eastern Europe became important and powerful interlocutors of national governments, supporting under the oversight of CEFIC the adoption of EU legislation. A second and rather direct outcome was a smoother and faster than anticipated drafting and adoption of chemical legislation compatible with the EU body of regulations. One of the interesting outcomes of the 1997–1998 PHARE project managed by CEFIC was, for example, the drafting of national legislation jointly by industry and government experts to advance the process of harmonisation. The influence of supranational actors and transnational cooperation was so pervasive that in some accession countries draft chemicals legislation was prepared first in English and was later translated into the local languages (Andonova 2004). CEFIC and the European Commission had become important mediators between industry and political decisions on chemical safety legislation in the national context.

Finally, as a consequence of trans-European collaboration and policy coalitions, voluntary standards on safety and the environment diffused throughout the new member states. One of the main objectives of CEFIC was to promote not only EU legislation but also Responsible Care as another mechanism to level the regulatory playing field in Europe, to promote a culture of safety and corporate responsibility, and to protect the reputation of the European chemical industry. Table 15.2 presents the growths of Responsible Care participation by chemical enterprises in Central and Eastern Europe.

Table 15.2 Responsible Care in Central and Eastern Europe

Country	Year adopted	RC companies
Hungary	1995	44 (90% of its members)
Poland	1992	37
Czech Republic	1995	55
Slovakia	1996	20
Slovenia	2000	14 (60% of industry)
Estonia	2002	9
Latvia	2002	9
Lithuania	2002	N/A
Bulgaria	2002	19 (30% of its members)
Romania	N/A	N/A

Doktor (2002) for year of adoption; data on the number of RC companies is from the web sites of national chemical industry associations, accessed in July 2007.

A 2002 presentation by Frantisek Doktor, Director for Central Europe and Regulatory Affairs of CEFIC, highlighted that the establishment of an ever denser network of industry and policy experts across new and old member states within the European Community was one of the most important outcomes of the ChemFed and ChemLeg programs.[5] ChemFed, which targeted primarily the development of national federations, facilitated the establishment of '60 local networks and working groups initiated and developed involving over 1,000 experts' across Central and Eastern Europe (Doktor 2002). These networks were furthermore linked to the supranational and national levels of policy by establishing working relations with CEFIC, the European Commissions, and national authorities. In essence, a thick transmission belt of industry and policy elites was created to facilitate the horizontal diffusion of chemical safety policies and voluntary standards.

The ChemLeg program, in turn, focused on closer cooperation between policy makers, industry experts and national federations to facilitate the implementation of these standards. The main indicators of achievement were the enlargement of CEFIC with new full and associate CEFIC members representing nine of the ten Central and Eastern Europe states that subsequently joined the European Union; the creation of an influential regional network of chemical industry; and 'active involvement in European advocacy' on the part of federations from Central and Eastern Europe (Doktor 2002). The political dynamics of trans-European industry mobilisation, organisation at the domestic and European level, and alliances with supranational and domestic policy elites and experts developed in full swing and in support of the horizontal diffusion of EU chemical safety policies and norms.

Following the accession of eight Central and Eastern Europe countries to the EU in 2004 and of Bulgaria and Romania in 2007, the chemical industry associations continued their close cooperation with CEFIC on several fronts. The trans-European network of chemical associations continued to promote the Responsible Care program, resulting in a growing number of enterprises across the region that adopted and implemented

[5] http://ec.europa.eu/enterprise/chemicals/conferences/enlargement/enlarge_2002_en.htm.

its principles. On a parallel track, however, the chemical industry associations of the new member states also cooperated with CEFIC in resisting the adoption of REACH and exerting pressure for its modification. While not as influential as the industry associations of members with large chemical sectors and enterprises such as Germany or the United Kingdom, the associations of the new member states were instrumental in emphasising the additional burden to already strained small and medium enterprises in the sector of the new member states (Angerera et al. 2008).

Following the adoption of REACH, the chemical industry associations in the new member states, with technical support from Brussels, remain a central institutional mechanism in translating the implications of European regulations and mechanisms for implementation of different sub-sectors and types of member companies (Pelovski 2006). With respect to REACH implementation, however, there has been considerably less programmatic and capacity building support which links industry associations, supranational institutions and government regulators, compared to the pre-accession period. This pattern reflects the considerably greater divergence in the objectives of industry associations and the policy coalition which involved advocacy organisations, DG Environment of the European Commission, key states, and the European Parliament which successfully pushed for the adoption of REACH (Selin 2007).

15.4 Corporatisation of Chemical Safety and Implications

The role of environmental advocacy organisations in the diffusion of chemical safety and environmental policies from the EU to the Central and Eastern Europe states has been surprisingly limited. The vacuum of societal activism on the issue illustrates once again the critical meditating role of actor associations in linking domestic and EU regulatory and normative arenas.

Domestically, a limited number of environmental organisations in Central and East European countries focused on issues of chemical safety. The database of the Regional Environmental Center Directory on Environmental Non-Governmental Organizations in Central and Eastern Europe does not even list chemicals or chemical safety as a category of NGO 'priorities'.[6] The majority of organisations listed focus primarily on issues related to democratisation such as environmental impact assessment, access to information and environmental education; as well as substantive issues in the realms of biodiversity, air pollution, water management and climate change. The limited societal attention to chemical safety in the new member states is somewhat surprising, given the negative reputation of poor chemical safety management under communism and the negative environmental externalities of chemical enterprises.

At the EU level, environmental NGOs were also ambivalent and fragmented in their interest in accession and the diffusion of chemical safety policies, compared to the highly centralised and highly involved position of CEFIC. EU environmental NGOs developed an interest in enlargement relatively late, and for the most part

[6] http://www.rec.org/rec/databases/ngodirectory/ngofind.html.

maintained a critical perspective on integration, emphasising the negative impacts of consumerism, infrastructure development, agricultural subsidisation, and trade that were to come in the same package with EU accession.

With respect to the regulation of chemicals, European NGOs did develop a complex network and increasingly powerful coalitions spanning domestic interests, supranational and national governmental elites interested in the tightening of regulation and the adoption of REACH. This coalition, however, sought first and foremost a *vertical strengthening of regulations*, e.g. the deepening of regulations within the EU policy space. Relatively little attention and resources were allocated to processes related to the widening of the EU policy space. As a result of weak organisation of domestic interests, and limited incentive on the part of the European NGOs and Commission experts in looking for or creating domestic societal counterparts in the process of chemical policy harmonisation, no transnational advocacy coalition emerged. Nor did EU NGOs seek to strengthen the capacity and networks in the East that span supranational and national authorities on chemical safety matters.

The consequences of the absence of East–West advocacy networks working on chemical safety were several. The diffusion of norms that seek more ambitious regulation of chemicals and improved access to information was significantly constrained in the new member states. Since industrial associations dominated the policy agenda, they were successful in framing the agenda entirely around the technical requirements of the approximation of existing EU regulations. The second outcome was the very limited engagement of public opinion in issues related to chemical substances and chemical safety. Indeed, the coalition of EU and national industrialists and experts encouraged the de-facto corporatisation of chemical safety politics and regulations. This approach was highly successful in achieving the immediate objectives of the adoption of EU chemical legislation and voluntary international standards. However, the lack of counterbalancing public opinion and advocacy mobilisation implies that the implementation of chemical safety standards and policies is left without societal oversight if failures were to occur. To the extent that non-governmental organisations are involved in discussion on the implementation of chemical regulations and REACH in the new member states, they generally participate on the invitation of industry associations. There is limited capacity and indeed limited advocacy interest in the new member states to serve as a watchdog for chemical risk and the implementation of chemical regulations.

Information on enterprise level compliance with regulations and safety standards is limited. A report of Pricewaterhouse Coopers (2004) published on the eve of the first round of the eastwards enlargement of the EU concludes that:

> […] many companies in Central and Eastern Europe are still a long way from meeting the European Commission's current social and environmental regulations. Moreover, the need for 'cleaning up' often makes any restructuring efforts very costly, and can even prove an insuperable obstacle unless public funds are available. All such companies must put review or audit arrangements in place, and develop management systems to deal with particular areas of performance […].

Several years later, information on the level, obstacles or achievement to implementation remains limited. The speed and level of regulatory harmonisation for managing chemical risk in the new member states has been impressive. Micro-level information on the actual impact of regulation remains, for the general public, elusive and at best incomplete. The case study of chemical safety reveals that in the absence of careful balancing on the part of European institutions and capacity building programs around which networks emerge, policy agendas can be skewed in direction of interests that are already entrenched at the supranational and domestic level, marginalising potential counterbalancing political forces.

15.5 Conclusion

The chapter developed the argument that the extension of EU chemical safety regulations to the new member states did not follow a simple logic of market-driven functional diffusion. Powerful industry association operating at the supranational and domestic levels, in alliance with supranational institutions and government experts, provided the political link between markets and regulations. Clearly not all areas of EU regulations are likely to follow the same pattern of diffusion. Even in the context of chemical politics, advocacy coalition spanning East and West failed to materialise, resulting in greater emphasis on formal harmonisation and industry training, and limited attention to access to information and independent assessment.

This study illuminates both the enduring as well as the conditional logic of the neofunctionalist arguments of regulatory diffusion in the European Union, emphasising the critical role of trans-European and domestic associations and alliances in this process. The empirical material presented in the chapter also reveals that programs for technical assistance can play a critical role through the power of financial resources and expertise in influencing the direction, participation, and issue orientation of trans-European policy coalitions. It suggests that as integration and practical implementation of EU policies proceed after the accession, such trans-European networks and coalitions could play an important role to smoothen the bumpy road of horizontal integration. The Eastern enlargement indeed has added and would continue to contribute to the multi-level nature of European regulations and governance.

References

Alter, Karen (2007) Jurist social movements in Europe: the role of euro-law associations in European integration (1953–1975). *EUSA Review* 20: 6–12.

Andonova, Liliana B. (2004) *Transnational Politics of the Environment: The European Union and Environmental Policy in Central and Eastern Europe*. MIT Press, Cambridge.

Angerera, Gerhard, Nordbeckb, Ralf and Sartoriusa, Christian (2008) Impacts on industry of Europe's emerging chemicals policy REACH. *Journal of Environmental Management* 86: 636–647.

Bruszt, Laszlo (2002) Markets and Eastern enlargement: diverging convergence? *West European Politics* 25: 121–41.

Burley, Anne-Marie and Mattli, Walter (1993) Europe before the court: a political theory of legal integration. *International Organization* 47: 41–76.

CEFIC (1992) Central & Eastern Europe. The challenges, opportunities and risks for the European chemical industry. Available via www.cefic.be. Cited July 1998.

ICIS Chemical Business (2000) EU beckons for East Europe. http://www.icis.com/Articles/2000/12/11/128284/eu-beckons-for-east-europe.html. Cited in May 2008.

Contiero, Marco (2006) Toxic lobby. How the chemical industry is trying to kill REACH. Report of Greenpeace International. http://www.greenpeace.org/raw/content/international/press/reports/toxic-lobby-how-the-chemical.pdf. Cited in May 2008.

De Bree, Simon (1995) Global industry perspective. Paper presented at the conference on competing in the new Europe: strategies for the Central and Eastern European chemical industry.

Deutsche Bank Research (2005) Western chemical companies move into Eastern Europe. EU Monitor. http://www.dbresearch.de/PROD/DBR_INTERNET_DE-PROD/PROD0000000000183437.pdf. Cited in March 2009.

Doktor, Frantisek (2002) Outcome of the ChemFed and ChemLeg projects. Presentation prepared for the conference on making the enlarged internal market for chemicals a reality, DG Enterprise, 21–22 November 2002.

Garcia-Johnson, R. (2000) *Exporting Environmentalism. U.S. Multinational Chemical Corporations in Brazil and Mexico*. MIT Press, Cambridge.

Ghosh, Probal (2005) Growth prospects for the European chemical industry in Central and Eastern Europe. Frost & Sullivan Market Insight. http://www.frost.com/prod/servlet/market-insight-print.pag?docid=32643008. Cited in August 2008.

Haas, Ernst B. (1958) *The Uniting of Europe: Political, Social and Economic Forces*. Stanford University Press, California.

Hooghe, L. and Marks, G. (2001) *Multi-Level Governance and European Integration*. Rowman and Littlefield, Lanham.

Pelovski, Yontcho (2006) Bulgarian chemical industry from economic and social point of view. Presentation at the chemicals policy, REACH and health and safety in the candidate countries conference, Bucharest, Romania, 29–30 May 2006. http://www.emcef.org/euproj.asp?job=re

Pricewaterhouse Coopers (2004) Looking forward: the impact of EU enlargement on the chemical industry. http://www.pwc.com/extweb/pwcpublications.nsf/docid/C811A9F2AC541F8E85256E93003D3847. Cited in October 2008.

Regional Environmental Center (1996) *Approximation of European Union Environmental Legislation. Case studies of Bulgaria, the Czech Republic, Estonia, Hungary, Latvia, Lithuania, Poland, Romania, Slovak Republic and Slovenia*. REC, Budapest.

Schmitter, Philippe C. (2001) The impact of Europeanization and globalization on national patterns of business interest intermediation (unpublished manuscript). European University Institute, Florence.

Selin, Henrik (2007) Coalition politics and chemicals management in a regulatory ambitious Europe. *Global Environmental Politics* 7: 63–93.

Vogel, David (1995) *Trading Up: Consumer and Environmental Regulation in a Global Economy*. Harvard University Press, Cambridge.

Zielonka, Jan (2007) Plurilateral governance in the enlarged European Union. *Journal of Common Market Studies* 45: 187–209.

Zielonka, Jan and Mair, Peter (2002) Introduction: diversity and adaption in the enlarged European Union. *West European Politics* 25: 1–18.

Chapter 16
Capacity Building for Chemicals Control: Legislation, Institutions, Public–Private Relationships

Bengt Bucht

Abstract The risk management of chemicals is of concern for many areas of society. In most countries this is generally regulated in several pieces of legislation, such as legislation on the environment, waste, work environment, consumer safety, rescue services, transport, trade, pesticides, cosmetics, additives in food and feedstuff, etc. Often different ministries are responsible for various aspects of the problem. This chapter focuses on the specific product and trade orientated type of preventive risk management, referred to below as chemicals control. This part of the system concentrates on measures for risk management early on in the supply chain, before or at the time industrial and consumer chemicals are introduced on the market. Chemicals control is in this sense horizontal in nature, comprising the first steps in all risk management activities, may they be for protection of workers, consumers, public health or the environment.

Keywords Chemicals • Control • Legislation • Public–private • Risk mangement

16.1 Introduction

Control of pharmaceuticals, food and feed additives and narcotics, which is generally regulated in separate pieces of legislation, is not dealt with in this chapter. And traditional risk management, aiming at reducing exposures as regulated in legislation on the environment, on workers protection, on transport, etc., is only dealt with in very general terms.

The discussion and reflections below on the legislative and institutional infrastructure for control of chemicals are based on personal experiences from developments in Swedish and European chemicals control, from long-term Swedish participation in international co-operation on chemicals control, and from

---

B. Bucht (✉)
Swedish Chemicals Agency, P.O. Box 2, SE-172 13, Sundbyberg, Sweden
e-mail: bengt.bucht@kemi.se

J. Eriksson et al. (eds.), *Regulating Chemical Risks: European and Global Challenges*, DOI 10.1007/978-90-481-9428-5_16, © Springer Science+Business Media B.V. 2010

Swedish co-operation projects in a number of developing countries and countries with economies in transition. The views that are put forward reflect the general views on chemicals control as expressed in international agreements and in various national programmes on chemicals control. Practical views on how conventions and other agreements may be implemented and enforced are also presented.

Distinct and clearly formulated legislation on chemicals is, together with efficient implementation and enforcement, a necessary prerequisite for effective chemicals risk management. Two recent publications from the Swedish Chemicals Agency, KemI, analyse and discuss in more depth the need for an appropriate national infrastructure for chemicals control, such as legislation and institutional capacity and capability (KemI PM 1/07 and KemI PM 4/08). KemI PM 4/08 includes an example of a full-text basic law on chemicals.

The word chemical is used here in general as a common term for chemical substances and mixtures of substances. For clarity, a distinction between substances and mixtures is made in some cases.

16.2 Background

16.2.1 Chemicals Control – An Internationally Prioritised Issue

In industrialised countries and in international organisations, management of chemical risks is a highly prioritised issue in programmes on protection of the environment, public health and workers health. In the EU a new chemicals policy was introduced in 2006 followed by the new chemicals legislation: Regulation on Registration, Evaluation and Authorisation of Chemicals (REACH). A number of international agreements, such as the United Nation's programme for a Strategic Approach to International Chemicals Management (SAICM), the Stockholm Convention on Persistent Organic Pollutants (POPs), the Rotterdam Convention on the Prior Informed Consent (PIC), Procedure for Certain Hazardous Chemicals and Pesticides in International Trade, the Montreal Protocol on substances that deplete the Ozone Layer and the Globally Harmonized System of Classification and Labelling of Chemicals (GHS), are focused either very strongly or completely (GHS, PIC) on risk management measures to be taken by suppliers before or at the time chemicals are introduced on the market. These international agreements, therefore, provide important starting points for the development of national control of the marketing of chemicals.

In developing countries and countries with economies in transition, risks due to the use of pesticides in agriculture may be well known, though often not dealt with in an appropriate way. However, in spite of increasing interest, there still is a low awareness of risks due to the increasing use of other chemicals in industrial production, in agriculture and in households. Information on risks and safe use of chemicals is often lacking or insufficient. Legislation on chemicals with regulations regarding, for example, warning labelling is lacking or inadequate, and when legislation exists, implementation and enforcement is often weak due to the inadequate organisation and resources of governmental institutions.

16.2.2 Basic Elements of Chemicals Risk Management

Chemicals risk management has four main goals:

1. *To obtain knowledge of the intrinsic hazardous properties of chemicals*. This is accomplished by toxicity testing or by other means of data generation and compilation, evaluation and assessment of these data. International test standards are available for and applicable to this part of the process.
2. *To disseminate information on hazardous properties of chemicals placed on the market with the purpose of promoting safe use*. This is accomplished by hazard warning labelling, safety data sheets and other types of information on risks and safe use. International standards are also applicable to this part of the process.
3. *To make informed choices of chemicals in order to avoid hazards*. There are comparatively few chemicals for which this choice is governed by international or national restrictions on the use of individual chemical substances or by regimes for pre-market approval of pesticides and other chemicals of particular concern.
4. *To organise a safe use of chemicals*. To some extent, public institutions regulate the use of particularly hazardous chemicals in more detail. In other cases, regulations are limited to general requirements for employing a cautionary approach.

Chemicals control (i.e. risk management steps before or at the time of the marketing of chemicals) focuses mainly on the three first goals, while traditional risk management for protection of workers and the environment and corresponding legislation focus on the fourth.

16.3 Legislation on Product and Trade-orientated Risk Management

The various measures for product orientated risk management of chemicals early in the supply chain (Fig. 16.1) before or at the time chemicals are introduced on the market, referred to here as chemicals control, aim at improved control of the flow of chemicals to the market, including appropriate information to users on risks and safety measures.

The main steps in this product and trade orientated, preventive and precautionary risk management are:

- Identification and assessment of inherent hazardous properties of, and possible risks with, chemicals
- Information to users on hazards, risks and safe use
- Restrictions on sale and use of chemicals that are determined to cause unacceptable risks

These measures for the control of the flow of chemicals are essential for managing risks wherever they occur, and primary responsibility for these measures should rest with enterprises importing or manufacturing chemicals, i.e. those enterprises that introduce chemicals on the market, the primary suppliers.

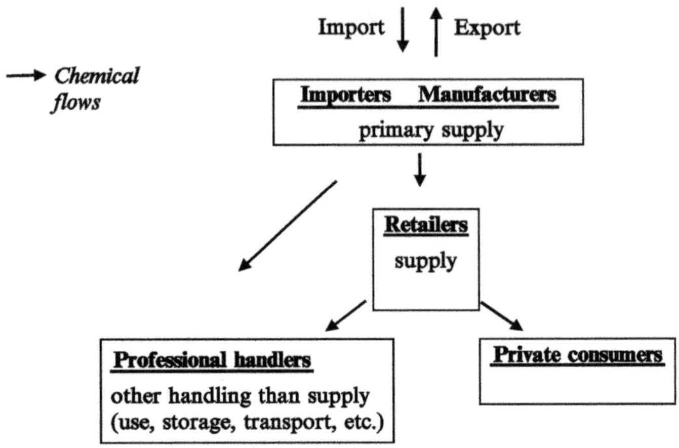

Fig. 16.1 Supply chain – the basis for risk management

Often an effective way of reducing risk is by substituting hazardous chemicals with more harmless ones. To enable substitution, the users of chemicals and products further down the supply chain need to have appropriate and sufficient information on the technical and hazardous properties associated with the chemicals they purchase. Such information is necessary in order to assess potential risks at the point of intended use and to compare different alternatives from a risk reducing perspective. Therefore, suppliers of chemicals, especially manufacturers and importers, need to identify and assess all of the hazardous properties and foreseeable risks – to human health, to the environment and to property due to fire or explosion – in order to be able:

1. To decide whether or not to market a chemical
2. To disseminate full information on hazards and possible risks to users of chemicals that are marketed

There are advantages in regulating these rather few early measures for risk elimination and reduction through comprehensive horizontal legislation on chemicals.

16.4 Rationale for Separate General Legislation on Chemicals

There are some key reasons for regulating the flow of chemicals to the market independently of other legislation. One reason is the need for preventive measures early in the supply chain, covering protection of the environment, workers, consumers and property. Preventive measures, such as risk and safety information (labelling and SDS), as well as bans and restrictions on trade with very hazardous chemicals support, simplify and make more cost-efficient risk management for exposure and emission control at later stages (transport, use, waste handling, etc.).

National measures, such as classification, labelling, SDS and bans and restrictions may, if very country specific, have negative effects on the export and import of chemicals. Almost by definition, national measures in these areas lead to trade barriers, which is why they must be carefully designed in order to comply with international trade agreements that countries may have ratified. A high degree of international harmonisation of chemicals legislation is most important, a consideration that also justifies having legislation separate from traditional legislation on, for example, the environment, work environment, etc.

In order to facilitate compliance with the legislation and efficient enforcement and monitoring of compliance, the various pieces of legislation regulating responsibilities and tasks of suppliers of chemicals should be easy for all of the actors – enterprises as well as authorities – to find and to understand. Horizontal trade and product orientated legislation should, therefore, be as coherent and condensed as possible. Enterprises that market chemicals have to be able to assess and classify them with respect to available data for all types of hazards and to provide them with information (labels and SDS) taking into account all kinds of risks and use. Furthermore, they have to comply with bans and restrictions, if relevant. Regulating the obligations of suppliers through numerous laws and regulations issued by several ministries and authorities would greatly complicate the legislation as well as its implementation and enforcement. This would be disadvantageous and costly for governmental institutions as well as enterprises. This is a third reason for dealing with it separately.

Furthermore, to ensure efficient national implementation of international agreements, implementation should be carried out in a coherent and co-ordinated manner. Implementation of trade orientated parts of SAICM, the Stockholm and Rotterdam conventions, GHS and the Montreal Protocol within the framework of a single act of legislation on chemicals would contribute to making such legislation more transparent and efficient. This might also be the case with respect to other international agreements on chemicals, such as the ones regulating trade.

It might be advantageous in practice to formulate general, comprehensive legislation for chemicals control in terms of an independent chemicals law. This solution would make the legislation clearly separate from other legislation on chemicals risk management (Fig. 16.2). It would provide a greater focus on the importance of preventive risk management before or at the time of the marketing of chemicals to the benefit of risk management further down the supply chain. General chemicals legislation, however, might also be organised as a separate, specific part of some existing legislation, such as an environmental protection law. In the latter case, it would be important

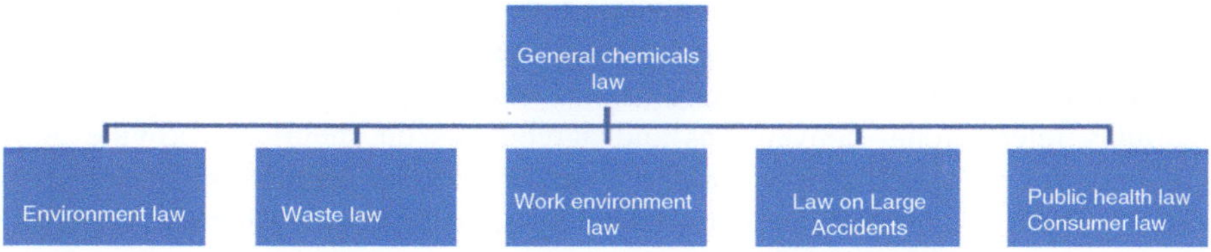

Fig. 16.2 General chemicals law and examples of sector specific legislation

to make clear that this separate part was to regulate chemicals on the basis of a broader risk perspective rather than on the basis of protection of the environment alone.

16.5 Design of Legislation on Chemicals

There are some important issues that should be dealt with through national legislation on chemicals.

(a) Responsibilities and obligations of enterprises with respect to, for example, providing information on hazards, risks and appropriate use
(b) Limitations (bans, restrictions, authorisation requirements)
(c) Registering, licensing, and reporting
(d) Confidentiality
(e) Implementation and delegation; responsibilities, obligations and rights of governmental institutions
(f) Sanctions

Primary legislation on chemicals, to which secondary legislation may be attached, is essential for effective regulation of chemical hazards and risks. This primary legislation would make it possible to clarify general responsibilities, obligations and tasks of both enterprises and governmental institutions. Such a law would set the framework for more detailed, secondary legislation, further specifying responsibilities, obligations and tasks. This could be done as ministerial regulation or at an even lower level. Specific regulations required in relation to, for example, classification and labelling are too scientific, technical and detailed and need to be revised too frequently for them to be decided upon by parliaments, in many cases even by Governments.

In secondary legislation issued by the national government or at a lower organisational level, specifics regarding, for example, testing and assessment of and information on chemicals could be regulated. Among the most important regulations to be elaborated are those concerned with classification, labelling and safety data sheets needed to guide the crucial flow of hazard, risk and safety information from suppliers to users and others that handle chemicals. High quality in the flow of this information is essential for efficient management of chemical risks. Bans and restrictions on supply and use of chemicals could also be dealt with through secondary legislation. In addition, supportive legislation might involve specifics with regard to authorisation, licensing and registering in order to stipulate the specific obligations of legal and private persons and the procedures and organisational processes and routines needed.

Many international conventions and agreements on chemicals provide good guidance for the specific design and content of national legislation. The internationally agreed upon Globally Harmonized System of Classification and Labelling of Chemicals (GHS) should, preferably without any substantive changes, be considered as the accepted national standard for classification, labelling and safety data sheets. Any unilateral revision of this international standard will make national chemicals control more complex and resource demanding for authorities and enterprises. Deviations from the standard may, furthermore, lead to unnecessary trade barriers.

The Stockholm and Rotterdam conventions on persistent organic pollutants and on prior informed consent for certain chemicals in international trade provide a good foundation for national legislation on bans and restrictions and for information in conjunction with the export and import of chemicals, respectively. The OECD has issued several documents on chemicals control, e.g. on testing and on the exchange of data. The Food and Agriculture Organization of the United Nations, FAO, has issued Guidelines for Legislation on the Control of Pesticides, which regulate the authorisation of pesticides (plant protection products).

Government institutions for managing and in other respects taking responsibility for legislation could be specified in secondary legislation. In cases in which decision making is not delegated to the Government, responsibilities must be established in primary legislation.

The publication KemI PM 4/08 outlines in more detail how chemicals legislation with basic legislation and secondary legislation can be designed.

16.6 Responsibility of Enterprises

One of the main aims of modern legislation on chemicals control is to stipulate and specify the responsibilities and obligations of enterprises. Enterprises and other actors handling chemicals, including private consumers, not only own their chemicals but also own the potential problems associated with the use of them. Therefore, they have to take responsibility for ensuring that use does not affect human beings and the environment in an unacceptable way. Every actor in the supply chain, such as the producer, the importer, the retailer, the user, the waste handler, has its specific responsibilities.

Figure 16.3 illustrates in a simplified form the main responsibilities of various actors in the supply chain as commonly expressed in legislation. According to specific chemicals legislation regulating the supply of chemicals, producers and importers, being the key actors as primary suppliers, are assigned primary responsibility for the implementation of the regulations. They are required to provide appropriate information about their chemical products that are to be placed on the market regarding composition (chemical substances) and hazardous properties and to provide customers/users with adequate information about risks and safe use. The *users* are required to search for data, to take risks into account in their choice of chemicals and to ensure safe use to avoid unacceptable exposures and risks. It is obvious that extensive co-operation between the actors in the product chain is essential in order to achieve a desired result.

Type of legislation	Primary target groups in supply chain	Typical responsibilities regulated
Legislation on chemicals ('chemicals legislation')	Suppliers:	Responsibilities at the time of marketing (supply):
	Exporters Importers Producers = primary suppliers	**Measures to make possible safe use and other handling – Risk management/prevention** - Data retrieval (testing, literature, ..) - *Hazard* assessment, classification - Information to customers through e.g. labelling, SDS - Packaging in connection with supply
	Retailers	- General bans and restrictions in relation to supply and use - Substitution with less hazardous chemicals - Licenses imp/exp/ trade - Authorization (pesticides, biocides, others) - Registration of chemicals
Legislation on: - the environment - work environment - major accidents	Handlers: Users and other handlers (including suppliers with regard to responsibility for own handling)	Responsibilities at time of handling (use, transport, storage, etc.): Measures to ensure safe handling at each specific enterprise - Risk management/reduction - Data retrieval (primarily from suppliers) - *Risk* assessment - Specific bans and restrictions on use - Substitution with less hazardous chemicals - Information to workers, instructions, training - Safe handling - Exposure control - Emission and waste control - Licenses/permits for production etc.

Fig. 16.3 Chemicals control. Supply chain risk management – two main types of legislation

16.7 Capability and Capacity of Enterprises

16.7.1 Organisation and Expertise

In countries with a less developed chemicals control (developing countries, newly industrialised countries, countries with economies in transition), many if not most enterprises handling chemicals (producers, importers, traders and users) lack the capability and the capacity needed for implementing efficient chemicals control. Normally, this is a main reason why chemicals placed on the market are not assessed, classified, labelled, provided with SDSs or used in an appropriate way to avoid risks.

Enterprises need a certain internal organisation for chemicals control. A clear allocation of responsibility and efficient routines must be established within the company for control of the purchase of chemicals, data retrieval, hazard and risk assessment, classification, labelling, SDS, work instructions, training, exposure and emission control, etc. Obviously, smooth and efficient co-operation between actors in the supply chain simplifies the work of enterprises (c.f. Fig. 16.4). Enterprises may meet demands on expertise by making use of the skills of their chemical suppliers, by hiring their own experts or by hiring external expertise as consultants. Normally, a combination of these alternatives is used.

To improve capacity building, enterprises, in particular small and medium-sized ones (SMEs), will often benefit from being members of business associations. In addition to providing a channel to and from governmental institutions, business associations may provide assistance to enterprises in the form of information on regulations and requirements, advice on problem solving, assistance in training, etc. Unfortunately, such organisations, if they exist at all, are often not very well developed in many countries.

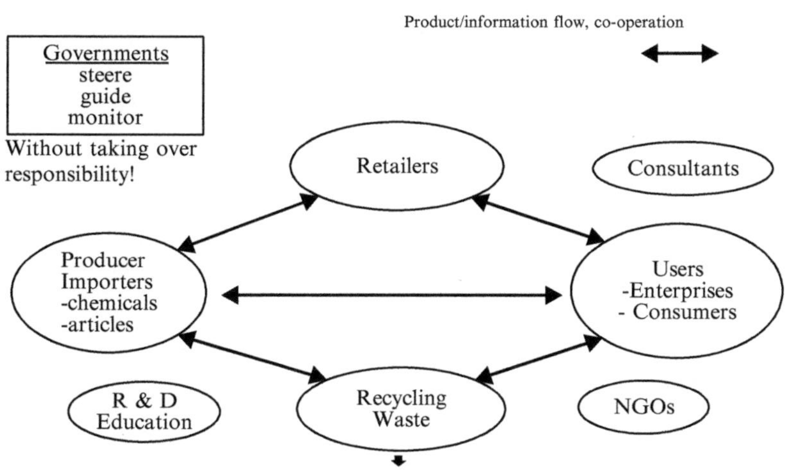

Fig. 16.4 Chemicals control. Decentralised risk control – shared responsibilities of actors in the supply chain

16.7.2 Good Chemicals Control Promotes Business

Enterprises acting on the international markets are facing increasing demands for safe, clean products, for clean production and for good information that can be passed on to their customers. In order to protect the health of their own workers, their clients and the environment, customers in industrialised countries are increasingly demanding that their suppliers apply stringent rules and routines for chemicals control. They, furthermore, increasingly demand that the international standards for environmental management systems are complied with. It is obvious that enterprises in countries with inadequate chemicals control – at the national level and in enterprises – will face growing problems due to competition in international markets.

16.8 Role of Public Institutions

The main role of governments and authorities is to steer the activities of actors in the supply chain, to force them to take responsibility, through legislation, through general information on chemical risks and by defining the responsibilities of enterprises, enforcing legislation and monitoring compliance with legislation. This is to be done without public institutions assuming responsibilities from the actors in the supply chain and from those handling chemicals – 'the problem owners'. Therefore, the legislation must clearly differentiate between the responsibilities of the actors and the obligations of government and authorities.

To some extent, public institutions may be involved in hazard and risk assessments and decide on risk management, such as for (harmonised) classification and labelling of chemical *substances*, for authorisation of biocides and of plant protection products and for specific regulatory activities like bans or restrictions on the use of certain very hazardous chemicals.

However, even for these groups of chemicals, primary responsibility must lie with the enterprises. Public institutions should not be made responsible for supplying specific information on chemical *mixtures*, including their contents, toxicological data, hazard assessment, classification, labelling and SDS, nor for detailed advice on how to ensure safe use. Due to the very comprehensive, varying and ever changing use of chemicals, no governments or authorities in any country have the capacity and capability required for such tasks. And, most important, if public institutions assist with tasks that are the responsibility of enterprises, the latter will not develop their own capacity in these regards.

16.9 NGOs

NGOs, especially trade unions and environmental organisations, have a very important role to play in chemicals control, primarily on issues relating to raising general awareness and maintaining pressure on enterprises as well as on governments and authorities by functioning as a 'watchdog'. However, NGOs should carefully design their actions so that they do not assume responsibilities from either authorities or enterprises. They should rather take measures aiming at, for example, strengthened legislation, improving implementation and enforcement and improving performance of enterprises in performing chemicals control work in the supply chain.

16.10 Capability and Capacity of Public Institutions

The establishment of a basic infrastructure for chemicals control, i.e. primary and secondary legislation and a sound governmental institutional set-up with the responsibilities of different institutions clearly defined, is the first step in developing chemicals control and a prerequisite for cost-efficient management of chemical risks.

As management of chemical risks is a horizontal issue that concerns health and safety of consumers, the general public and workers as well as the protection of the environment and property, several governmental institutions have an interest in this issue. Accordingly, the need for integration is significant. Integration is needed in order to achieve coherence, concentration, co-ordination, co-operation, continuity and cost-efficiency ('the six C's') of efforts made in order to achieve the ultimate goal, good risk management of chemicals.

16.11 Organisation of Public Institutions

The management of chemical risks presupposes activities at three main levels:

Policy level: preparatory and executive legislative actions, international co-operation on policy issues, co-ordination/co-operation between ministries
Management: supportive legislative work, daily scientific/technical expert implementation *level* work, co-ordination/co-operation between implementing institutions
Enforcement: enforcement and supervision, co-operation/co-ordination between *level* institutions for enforcement and supervision

It is quite common in many countries for responsibility for existing pieces of chemicals legislation to be unclear and dispersed among a number of ministries and other institutions. The more responsibility is dispersed, the more overlapping of work that can be expected, resulting in public institutions having to spend more resources on inter institutional co-ordination. A division of responsibilities and lack of clarity regarding responsibilities make it difficult for enterprises to identify the responsible institutions and contribute in general to making implementation of legislation unnecessarily complicated and costly.

16.11.1 *Policy Level*

16.11.1.1 Allocation of Responsibilities at the Ministerial Level

It is preferable to designate one ministry together with subordinate institutions as being responsible for legislation regulating the marketing of chemicals, including overall responsibility for national co-ordination and international contacts (Fig. 16.5).

When designating which ministry is to have primary responsibility for chemicals control, familiarity with legislation of relevance for the management of chemical risks is of importance. Therefore, countries often confer responsibility for chemicals legislation on ministries of the environment or ministries of health. Ministries of the environment (or the like) may be preferable due to the fact that issues of risk assessment and risk management in general, as well as for chemicals, are frequently dealt with in these ministries (air, water and soil pollution, waste problems). Furthermore, modern chemicals control at an international level, in addition to focusing on health problems, focuses very much on environmental risks and environmentally mediated health risks, such as problems with POPs.

As management of legislation on consumer and industrial chemicals, biocides and plant protection products has much in common, countries may find it advantageous to aspire to a high degree of co-ordination between these areas.

Modern legislation on chemicals is highly scientific/technical and regularly revised, a process often requiring prompt response and action. An adequate delegation of power from the parliament and the government to institutions with the ability to act quickly and with appropriate expertise is a prerequisite for meeting these demands. Therefore, ministries and even authorities should be appropriately empowered to make decisions concerning secondary legislation.

Irrespective of how primary responsibility for chemicals legislation is allocated, several ministries will need to contribute, which requires good co-ordination and

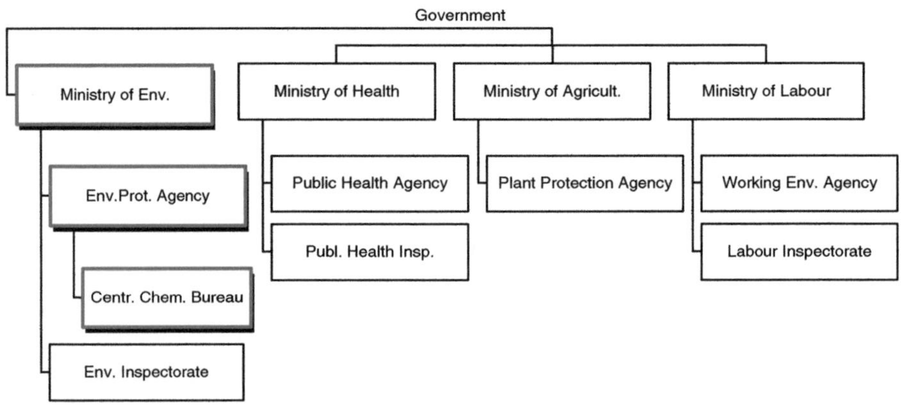

Fig. 16.5 Organisation of public institutions for chemicals control. An example

co-operation between ministries. Every ministry involved should have the capacity to deal with legislative and other policy issues concerning chemicals in co-operation with other ministries.

16.11.1.2 Co-ordination of Institutions

The ministry designated for management of specific chemicals legislation could be supported in its work by a special co-ordinating body connected to the ministry with representation from other ministries (and authorities) concerned. The role of this special body should be advisory and consultative only. It would be of great value for discussion and consultation on general matters, such as policies and strategies concerning legislation and organisation, monitoring compliance as well as co-ordination of national positions and contributions in international work on chemicals. Co-operation on and co-ordination of collection, dissemination, retrieval and storage of data are examples of other issues that could be discussed. When establishing a co-ordinating body, countries could make use of experiences from the establishment of National Profiles based on the UNITAR/IOMC National Profile Guidance Document (1996).

16.11.2 Implementation and Management Level

As part of the legislative work on chemicals is highly scientific, technical and resource demanding, ministries in charge need support with regard to technical and scientific issues from subordinate institutions. The daily work of implementation and management of the various parts of chemicals legislation is even more demanding in terms of scientific and technical capability and capacity. There is a need to organise support to ministries and to organise the work of authorities.

Especially in countries with scarce resources, it is advantageous to concentrate support to ministries and daily management of the chemicals legislation to a special managing institution for chemicals control. Preparation of the technical aspects of proposals for regulations on chemicals and administration of systems for classification and labelling and for authorisation of biocides and plant protection products, etc. largely requires the same kind of expertise (lawyers, chemists, toxicologists, and others) and the same type of routines and methods, irrespective of the type of chemical or risk. It is, therefore, logical and practical to amalgamate existing units and activities in ministries and at the level of authorities and thereby facilitate an efficient use of resources and co-ordination.

A special managing institution (Central Chemicals Bureau in Fig. 16.5) for chemicals control could be connected to an existing governmental agency with scientific and technical tasks and qualifications in the area of chemicals risk assessment and management.

If a special institution is not established, appropriate resources should be made available for management of legislation, e.g. through efficient networking between the managing institutions under the ministries in question. A solution of this kind involving the dispersion of responsibility will, however, most probably be less cost-effective than a special institution and will be complicated to manage in terms of achieving the required co-ordination and co-operation.

Qualified scientific assessments of hazards and risks can be made by scientific institutions at universities or institutes contracted to provide assistance. As hazard and risk assessments of, for example, industrial chemicals, consumer chemicals, biocides and plant protection products have much in common, it would be cost efficient to co-ordinate and concentrate scientific support in order to avoid a thinning out of resources and expertise.

16.11.3 Enforcement Level

16.11.3.1 Organisation of Supervision Is Vital

As noted above, modern legislation on chemicals allocates considerable responsibility to enterprises handling chemicals. Except for chemicals of special concern, such as plant protections products and biocides, current legislative systems applied internationally include very few elements of central steering of details.

Therefore, it is of utmost importance that countries have efficient means for monitoring compliance with the legislation among enterprises and efficient systems of sanctions in cases of violations. Supervision carried out as post-market control is, when combined with sanctions in cases of non-compliance, the most cost-efficient and, in practise, the only possible means for authorities to ensure that enterprises comply with the regulations. Properly planned and executed random post-market monitoring contributes to ensuring fair competition among enterprises, all running the risk of sanctions for non-compliance.

The organisation of responsibility for supervision of chemicals legislation should be based on the three main levels in the supply chain: producers/importers, retailers

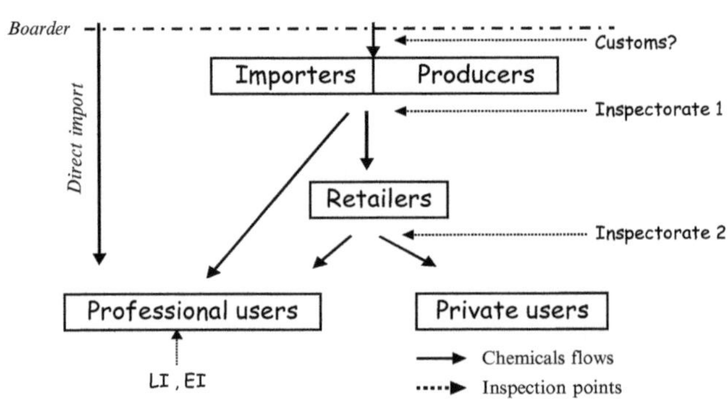

Fig. 16.6 Organisation of supervision/enforcement. An example LI = Labour Inspectorate, EI = Environment Inspectorate

and users, taking into account the differences in their respective responsibilities and, therefore, the different types of methodology and expertise needed by inspectors (Fig. 16.6).

Supervision of producers and importers as primary suppliers of chemicals is a task requiring specific skills and qualifications on the part of inspectors, e.g. in toxicology and hazard assessment. To supervise retail sales is a simpler type of control. Supervision of users is, again, more complex, requiring still other types of inspector skills and qualifications for assessment of exposure and of risks, control of technical safety measures, etc.

The task of monitoring compliance with chemicals legislation should be carried out by a relatively small number of special authorities. It is often appropriate to make use of existing inspectorates to the extent possible, provided they have the capability, capacity and specific expertise and methodology needed. Normally, traditional labour inspectors or environmental inspectors do not have the skills required for a chemicals inspection of primary suppliers. Regardless of the choice of inspectorates to supervise the various actors, appropriate statutes should ensure that enterprises do not run the risk of being checked by more than one inspectorate in relation to compliance with specific chemicals regulations. The responsibilities of inspectorates must be clearly separated.

As other public institutions, inspectorates should not provide specific advice to enterprises on how to solve a problem. They should not assume responsibility for managing chemicals from the enterprises.

The possible role of customs in chemicals control is often overestimated due to a lack of understanding of the need for skilled, highly qualified chemical inspectors and the often time consuming task of verifying possible non-compliance with chemicals legislation. In most cases, as already mentioned, the monitoring of compliance with regulations on chemicals, as with those on classification, labelling, bans, etc., should be done as post-market surveillance through random inspections of suppliers. Customs may, however, have a role in some specific cases, such as in checking that licenses for imports or other documents are in order, e.g. for the implementation and enforcement of the Rotterdam Convention on banned and severely restricted chemicals. In addition, customs may alert chemical inspectorates in cases of suspected violations of legislation.

16.12 Mechanisms of Financing

16.12.1 Costs and Gains

16.12.1.1 Public Sector

The costs for governments of establishing the institutional infrastructure needed for efficient chemicals control may be considerable. The highest costs in most cases are related to the need for ensuring adequate personnel capacity, both in terms of numbers and of qualifications, and to other expenses for the overall tasks involving legislative work, such as scientific and technical assessments, surveillance and supervision. Other costs are related to data retrieval and processing and to other kinds of technical support.

Smaller countries may require resources at the central level amounting to 10–20 persons, excluding resources for inspection. By concentrating responsibility for chemicals control issues to a few institutions and by improved co-ordination and co-operation, it should, as discussed above, be possible to achieve a more cost-efficient use of resources, reducing the need for additional resources. Even developing countries may have available resources close to the level required, although perhaps dispersed among several ministries and/or authorities that work relatively isolated from each other.

The gains for society of improved chemicals control may be substantial, though not easy to quantify in monetary terms. They include, for example, reduced costs for health care as a result of fewer accidents with chemicals, fewer acute health effects caused by poisonings, skin corrosion or burns, reduced risk for chronic effects such as allergies, cancer, etc. Furthermore, improved chemicals control will lead to a reduction of costs for remediation of environmental damage and of other costs following from emissions, e.g. water and soil pollution due to accidents or misuse of chemicals.

16.12.1.2 Enterprises

The costs for enterprises in trade and industry of organising good chemicals control may in the short-term perspective seem high. Enterprises need to organise their internal administrative systems for chemicals control, recruit personnel, including experts for running the systems, educate existing personnel, establish routines for retrieval, assessment and dissemination of data and for risk management, etc. The services and expertise needed may, however, in part be purchased from consultants, which may lower the costs, especially for small and medium-sized enterprises, as will support they may receive from their associations.

Good chemicals control has, however, considerable positive economic effects for enterprises. It is increasingly important for the competitiveness of enterprises. Furthermore, investments in preventive chemicals control leading to the use of less hazardous chemicals and improved information on risks and safe use will have paybacks in the form of a reduced need for costly risk reduction measures for control of exposure, emissions and waste. In addition, better control of chemicals very often results in more cost-effective processes with reduced use of chemicals and less hazardous waste. By applying the concepts of Clean Products and Clean Production as aspects of improved chemicals control, costs for initial investments in many cases may have paybacks within just a few years time.

16.12.1.3 Possible Alternatives for Financing Work of Public Institutions

The work of public institutions on chemicals control may be financed in the traditional manner through the government budget and existing regular taxes. However, there may be alternatives available, such as levying special taxes on enterprises or their products or introducing special fees. Many countries already exact fees to pay for national authorisation systems for plant protection products. Other systems that may be financed by fees are licensing regimes and authorisation or other administrative systems for biocides or other chemicals of concern. Inspectorates may be at least partly financed by fees on inspections. In such cases, where there is a clearly-defined activity on the part of an authority in relation to a specific enterprise, it is possible to quantify and to put a price on the efforts of that authority.

16.13 Conclusion

The management of chemical risks is increasingly becoming the focus of national and international attention. As chemicals are distributed internationally through trade as well as through the long-range transport of pollutants, no country can any longer manage its own risks. Efforts to achieve a sustainable use of chemicals are being increasingly made through international co-operation. In order to do this, countries need a national infrastructure, legislation and institutions on a scale proportionate to the level of chemicals produced, imported, used and exported.

When adapting to modern legislation and modern systems for chemicals control, many countries will face challenges. They will have to develop appropriate and well-balanced legislative systems required for efficient chemicals control. The legislation must clearly stipulate separate the responsibilities of governmental institutions and of enterprises in trade and industry. Primary responsibility for avoiding chemicals risks should be placed on enterprises.

Most probably, the major challenge will be to establish an appropriate institutional set-up with the capacity and capability needed to manage and enforce the legislation. The greatest demands for capability and capacity arise with the tasks connected to day-to-day management of legislation at the agency level and to monitoring of compliance with the legislation. Both tasks require special expertise.

In order to get the priorities right, governments have to analyse in depth the implications of modern chemicals control for their legislation as well as their institutional capabilities and capacities to manage appropriate legislation and establish programs for necessary improvements and reinforcements. Enterprises and their federations have to carry out corresponding analyses.

The horizontal nature of chemicals control is highly demanding with regard to efficient and smooth co-ordination of the work of public institutions at all levels. Even in countries with scarce resources, there may be good opportunities for meeting new challenges without having to allocate a very high level of additional resources by simply reallocating existing resources, by clarifying responsibilities and by establishing efficient co-ordination of activities of public institutions and co-operation between public institutions. To some extent, the establishment of new structures may be necessary for efficient and cost-effective management of the tasks faced. The costs for the institutions could, at least in part, be financed by fees or earmarked taxes.

References

OECD Chemicals Programme http://www.oecd.org

EU (2007) Regulation (EC) No 1907/2006 of the European Parliament and of the Council concerning the Registration, Evaluation, Authorisation and Restriction of Chemicals, REACH. http://eur-lex.europa.eu

FAO (1989) *Guidelines for Legislation on the Control of Pesticides*. Food and Agriculture Organization of the United Nations. http://www.fao.org

KemI (2007) *Capacity Building for Chemicals Control*. KemI PM 1/07, Swedish Chemicals Agency, Sundbyberg, Sweden. http://www.kemi.se

KemI (2008) *Legislation for risk management at marketing of chemicals*. KemI PM 4/08, Swedish Chemicals Agency, Sundbyberg, Sweden. http://www.kemi.se

UN (1987) *The Montreal Protocol on Substances that Deplete the Ozone Layer*. http://ozone.unep.org

UN (1998) *Rotterdam Convention on the Prior Informed Consent (PIC) Procedure for Certain Hazardous Chemicals and Pesticides in International Trade* , Rev. 2005. http://www.pic.int

UN (1998) *UNECE Convention on Access to Information, Public Participation in Decision-making and Access to Justice in Environmental Matters*, Aarhus Convention. http://wwww.unece.org

UN (2001) *Stockholm Convention on Persistent Organic Pollutants* (POPs). http://www.pops.int

UN (2003) Gl*obally Harmonized System of Classification and Labelling of Chemicals* (GHS). Rev. 1 2005. http://www.unece.org

UN (2006) *Strategic Approach to International Chemicals Management* (SAICM). http://www.chem.unep.ch/saicm

UNITAR 1996. Preparing a National Profile to Assess the Management of Chemicals: Guidance document. United Nations Institute for Training and Research. http://www2.unitar.org./cwm/publications/cw/np/np_gd (nov2003).pdf

Chapter 17
Scientific Committees and EU Policy: The Case of SCHER

Johan Eriksson, Mikael Karlsson, and Marta Reuter

Abstract This chapter analyses the science–policy interface in EU chemicals policy, with a particular focus on the relationship between Risk Assessment and Risk Management. This is achieved through a case study of SCHER – the scientific committee responsible for assessing chemical risks in the EU. Thus this chapter also makes a contribution to the study of 'committee governance', and the politics of expertise in the EU more generally. This study has shown that, by and large, SCHER seems to be able to maintain a traditional role as scientific peer-reviewer, with some, though seldom any direct or significant, impact on policy decisions made by the Commission. Views on risk assessment and particularly on risk management vary among the committee members, with some voicing industry-friendly ideas and others supporting 'green' visions, including the precautionary principle. However, SCHER almost always reaches consensus on its opinions. An unexpected result, however, is how managing DG Sanco officers tried to control the publication of this study, which illustrates a political fear of policy studies such as the present one.

Keywords Experts • Policy • Risk • SCHER • Science

J. Eriksson (✉)
Södertörn University, SE 141 89 Huddinge, Sweden
e-mail: johan.eriksson@sh.se

M. Karlsson
Naturskyddsföreningen, P.O. Box 4625, Åsögatan 115, SE 116 91
Stockholm, Sweden
e-mail: Mikael.karlsson@snf.se

M. Reuter
Department of Political Science, Stockholm University, SE 106 91
Stockholm, Sweden
e-mail: Marta.reuter@statsvet.su.se

J. Eriksson et al. (eds.), *Regulating Chemical Risks: European and Global Challenges*,
DOI 10.1007/978-90-481-9428-5_17,

17.1 Introduction

The policy system of the European Union (EU) relies heavily on various forms of expert committees, including scientific committees, for assessing and preparing input into the decision-making process. Indeed, scholars studying EU policy talk about 'committee governance' as a key feature of the integration process (Christiansen and Kirchner 2000; Christiansen and Piattoni 2004; Dunlop and James 2007; Gornitzka and Sverdrup 2008; Joerges and Vos 1999; Rhinard 2003; Van Schendelen 1998; Weale 2000).[1] Moreover, health and environmental policy, which are the focus of chemical regulations, are traditionally 'science-heavy' domains, and the EU is no exception.

Given the institutionalisation of scientific committees in the EU, how are such committees managing the science–policy interface? In the EU context, it can be assumed that members of scientific committees experience political pressure from various actors – including the EU institutions, national governments and agencies and external actors, such as industry, the environmental movement and the news media. Are members of scientific committees trying to maintain a traditional and distinctive role as independent scientists with little or no interest in shaping policy? If so, is such a position feasible, given the highly politicised context of EU committees? In contrast, are committee members accommodating or even advocating particular interests or ideologies? How are scientific committees managing internal conflicts – to the degree that they exist – and in what ways are internal conflicts based on scientific concerns or on political concerns? How are they managing contact with stakeholders?

We are addressing these questions through a case study of the Commission's scientific committee, which, more than others, has chemical risks on its agenda – the Scientific Committee on Health and Environmental Risks (SCHER).[2] This study is based partly on official documents produced by the Commission and SCHER (including protocols and 'rules of procedure') and partly on interviews

[1]There is no clear definition of what an 'expert group' or 'expert committee' within the EU is (see e.g. Larsson, 2003). In this article we focus on expert committees (i.e. groups) that are set up by the European Commission for the preparatory phase in the EU governance system, which is the most common type of EU committee (as distinguished from committees under the Council and Parliament for the adoption phase and committees under the Commission for the implementation phase, i.e. for the Comitology procedure). Depending on the purpose, committees in the preparatory phase can be composed of people from, for example, member states, civil society, industry, and academia. See the further explanation and the register of expert groups at the European Commission (2009a).

[2]Commission Decision 2004/210/EC, which was amended in 2007 (Decision 2007/263/EC). SCHER is one of three scientific committees that have been set up by the Commission. The other two are the Scientific Committee on Consumer Safety (at the time of our study, called the Scientific Committee on Consumer Products) and the Scientific Committee on Emerging and Newly Identified Health Risks. All three committees are regulated by the same 'rules of procedure'.

with people in key positions within and related to SCHER.[3] We are focusing on SCHER as set up in the late 1990s until the reorganisation and appointment of new members in 2009.

In the following, we will first briefly discuss past research on the role of expert committees in policy processes (specifically those of the EU) and more generally on the science–policy interface. The four subsequent sections present the case study of SCHER, which specifically discusses the mandate and composition of SCHER, the working process, external contacts and the role and influence of this committee in the EU system.

17.2 Perspectives on the Science–Policy Interface

Previous theory and research on the role of expert committees in the EU show that they can have an impact on EU policy (Christiansen and Kirchner 2000: 6; Christiansen and Piattoni 2004; Dunlop and James 2007; Gornitzka and Sverdrup 2008; Joerges and Vos 1999; Rhinard 2003; Van Schendelen 1998). While the Commission, the Council and the Parliament hold the ultimate powers of decision-making, expert committees may 'nudge' policy proposals in a certain direction or 'put the brakes on existing proposals' (Rhinard 2003: 4). The EU bureaucracy, which increasingly has to deal with many technical and scientific issues, does not have in-house expertise but must rely heavily on outside sources of knowledge and advice. This is true not least for the Commission, which has developed a system of various expert committees – including scientific committees – with members from, for instance, enterprises, environmental organisations, national agencies and the academic research community in Europe. The role, rules and composition of these committees varies greatly. Some expert committees, such as SCHER, are institutionalised and exist for several years, while others are created ad-hoc for a limited period of time or for a specific piece of legislation.

While the political influence of expert committees has been observed, it is difficult to assess how great this influence actually is – a question that we will soon return to with regard to SCHER. Moreover, and of particular interest for the

[3]We have interviewed three committee members: Chairman Helmut Greim, Vice-chairman Bo Jansson and the sole member from the new members states from the former Eastern Bloc (i.e. the Polish representative Hanke Wojciech), two persons working at DG Sanco, in the unit responsible for risk assessment and directly involved in the work of SCHER and one person who, as an external expert, has participated in meetings with SCHER (and who wished to remain anonymous). The interviews were conducted during the period October 2008 to March 2009. The interviews with the three committee members were digitally voice recorded; the other interviews were not. Each interview took about 1–1.5 h. All of the interviews were semi-structured, and the questions were open-ended. The themes that guided the interviews are also used to structure this chapter: the mandate and composition of SCHER, the working process, coping with external pressure and the role and influence of SCHER.

scientific committees, it is rarely clarified whether such influence is based on scientific knowledge or on more or less overtly political considerations. Many studies of the 'politics of expertise' do not problematise the matter of what kind of advice experts provide but simply categorise the involvement of scientific committees as 'expert governance' or as 'venues of science', as opposed to 'political venues' (Timmermans and Scholten 2006). By contrast, some scholars are ardent critics of 'expert governance', which they consider undemocratic and, therefore, illegitimate (Black 2001; Rhinard 2003; Weiler 1999; cf. Weale 2000).

In our view, it is somewhat too naïve to assume that scientific experts only communicate scientific facts and too exaggerated to assume that they are simply pursuing their own or legitimating others' political interests and values disguised as 'scientific truths'. Why not allow such questions to be answered empirically rather than by assumption? It may be more realistic to expect experts involved in policy processes to provide *combinations* of fact-based and value-based advice (Fischer 1990; Jasanoff 1990; Radaelli 1999). The classical and diametrically opposed perspectives on the science–policy interface – the technocratic/truth-oriented perspective versus the interest-based/power-oriented perspective – are hypothetically ideal-types rather than accurate generalisations of how scientific committees actually work. We will investigate this proposition empirically, informed by a perspective that is realistic but not disillusioned: expert committees can certainly provide scientific facts and knowledge that is more or less instrumental for policy, but they can also provide concepts and ideas that enhance understanding and play a symbolic role, legitimating prevailing policies or reform proposals (Beyer 1997: 17; Amara et al. 2004: 75–77).

Our arguments will be based partly on the difficulties of maintaining a strictly scientific role in such a politicised policy system as that of the EU and partly on the nature of risk assessment, which, in our view, has implications for risk management. Also, as shown by numerous empirical studies of public policy, if experts are to be influential in policy making, some form of advocacy is needed (Sabatier and Jenkins-Smith 1993).

17.3 Mandate and Composition of SCHER

SCHER has existed since 2004, having replaced a previous committee that had a similar area of responsibility.[4] The severe criticism and perceived failure of the EU in managing the BSE crisis and Belgian dioxin scare were important reasons behind reorganising the scientific advisory system (Byrne 2003; Jansson 2008; Rhinard 2003). The Commission, however, considers scientific committees to be significant for the quality of risk management and policy decisions. As stated in the 'Rules of Procedure' for the scientific committees: 'Sound and timely scientific advice is an essential requirement for Commission proposals, decisions and policy relating to

[4]The Scientific Committee on Toxicity, Ecotoxicity and the Environment, which started in 1997.

consumer safety, public health and the environment' (European Commission 2009b). To ensure that decisions are based on the best available scientific knowledge is not the only motive, however. The Commission also seeks to ensure public trust and legitimacy among stakeholders. In the words of Robert Madelin, Director General for Health and Consumer Protection:

> The Commission in general and DG SANCO in particular attach high importance to the use of sound science to underpin its work and strive therefore to ensure that its scientific advice is of the highest quality. But, scientific excellence is not enough. To be effective, the advice must enjoy the confidence of all stakeholders, consumers and industry alike. (Madelin 2004)

The willingness to open up for dialogue and, thus, for the politicisation of policy advice follows from the Commission's broader approach to engage civil society and stakeholders (Madelin 2004). This might be a response to the distrust of EU policies and institutions witnessed in polls and in the series of negative outcomes in national referenda that have blocked major reforms.

The primary task of SCHER is to review Risk Assessment Reports established according to the regulation on evaluation and control of risks of existing substances (EEC 1993). These reports evaluate the risks of prioritised substances, and they are drafted by member state rapporteurs and decided on through comitology procedures. The job of SCHER is to review the scientific quality of these reports, particularly with regard to how they specify risks to public health, consumers and the environment. The committee put together its review in an 'opinion', which basically clarifies what it considers to be the health and environmental risks of the substance under study. Opinions also discuss the level and conditions of scientific uncertainty as well as whether and, if so, what further research is called for. The Commission may also ask SCHER for advice on other specific issues, even if there is no RAR to review.

Generally speaking, the Commission has the initiative. There is, however, a formal opening for committee initiatives, although our interviewees claim that this has hardly ever been utilised and that some committee members do not even know about this possibililty. The 'Rules of Procedure' state the following:

> The Scientific Committees shall draw the Commission's attention to a specific or emerging problem falling within their remit, which they consider may pose an actual or potential risk to consumer safety, public health or the environment by adopting and addressing to the Commission's memoranda or position statements. (European Commission 2009b: Ch. 9.10.1)

There are other, more informal ways in which experts and policy makers interact, however. While power and authority tends to be unequally distributed in policy making systems, there are no water-tight boundaries between levels and participants. Ideas and suggestions for what initiatives the Commission should take can be informally communicated through subordinate bodies, such as the scientific committees. Likewise, by interacting with the Commission, committee members may become increasingly socialised into the EU system of governance and, thus, more or less consciously come to shape their ideas and advice in order to fit the expectations and needs of the Commission.

There is a regular exchange of ideas and communication between committee members and Commission bureaucrats, partly through the regular meetings in which both participate and partly through informal communication that take place in corridors, at cafés, over lunches and dinners and via e-mail and other means of communication. While the Commission still makes the decisions and may choose to ignore or reject the opinions of scientific committees, both groups can have an influence on the mindsets of each other. Still, simply by virtue of their role as the EU's official scientific experts, the committee members have a salient position in the policy system. The Commission may choose to ignore or reject the expert opinions, but this requires either that the scientific base is particularly weak or that the Commission explicitly adheres to non-scientific arguments (such as preserving jobs or market shares).

Commission officials and the basic documents[5] regulating SCHER and other scientific committees clearly show an awareness of the risks that scientific expertise and, thus, risk assessment can be influenced by various external interests and values. Therefore, paragraphs on 'independence' and 'transparency' and a mandatory and annually renewed declaration of commitment to the principles of 'excellence and independence' are signed by every member every year (European Commission 2009b). Yet, at the same time, the written rules demonstrate belief in the possibility of maintaining scientific objectivity and integrity, as long as the risk of politicisation is carefully managed. Through the rules of procedure, the Commission seeks to maintain a clear distinction between risk assessment and risk management, which corresponds to the distinction between science and politics. The DG Sanco interviewees, who were given the opportunity to read a draft of this chapter, wrote to us that 'the exchanges between risk managers and risk assessors exist in all systems, are necessary and do not per se imply a confusion of roles.'[6] We will soon return to the question of whether such distinctions and rules of procedure are realistic.

During the period under scrutiny here, all 17 members of SCHER were scientists, most of them university professors, and a few affiliated with independent research institutes. The official criterion for selecting members is scientific excellence – that only the best and most knowledgeable scientists are invited. The first stage in selecting members is an open call to which interested individuals can apply. The Commission screens applications and rejects anyone with a clear 'double role', such as scientists working directly for the industry. Committee members also emphasise 'experience' of being a member of previous committees or of similar advisory bodies. Indeed, 40% of the members of the 2004–2009 scientific committees were members of earlier committees (Madelin 2004). Members are officially appointed by the Commission through a selection board composed of representatives of the Commission services, which in practice relies partly on suggestions and advice from existing members and on personal contacts. This board is chaired by an external senior scientist. This way of selecting members is not unique for SCHER or other expert communities within the EU but is a common practice in many merit-based advisory and decision-making bodies. Nevertheless, this entails

---

[5]European Commission 2004: The Scientific Committees on … Rules of procedure. SCs/01/04 final. Directorate C – Public Health and Risk Assessment.

[6]Comments by DG Sanco Unit C7 on a previous draft of this chapter; e-mail sent to us on 17 July, 2009.

significant power for existing members and individuals that have established contacts with the committee and with the Commission. Responding to our interpretation of the selection process, the DG Sanco representatives wished to emphasise that the scientific chairman of the board guaranteed 'quality and neutrality of the process'.[7] In our reading, this simply underlines the wish of EU Commission officers to maintain the image of strict objectivity and absence of any kind bias, a wish that may be politically motivated but an image that may be somewhat naïve.

As mentioned, the official criteria for selecting committee members are based strictly on merits and expertise rather than on gender, geographical affiliation or anything similar. It may nevertheless be of interest to note the gendered and geographical background of the committee. Of the 17 members, only 2 were women, confirming observations about the very strong male dominance in EU committees and the EU policy-making process more generally. Moreover, the two DG Sanco representatives we interviewed confirmed that the principle of geographical representation has played an informal role when committee members have been selected. Three members came from Germany (including the Chair), three from Denmark, three from Italy, two from Finland, one from Poland, two from Belgium, one from the Netherlands, one from Sweden (the Vice-chair) and one from Spain. Notably, Northern Europe was strongly represented, and the new member states from the former Eastern bloc were clearly underrepresented – by the one Polish member. As noted by the Vice-Chair Bo Jansson, members do not represent their countries, but they do carry on 'some of the politics from back home' (Jansson 2008), implying that there is rarely any direct pressure but that differences between national political cultures might have an indirect effect.

A Commission secretariat (in DG Sanco) runs administration and helps coordinate and supervise the three scientific committees. This secretariat has an important function as a link between the scientists and the Commission. The secretariat is also a point of contact for external stakeholders. In addition, an inter-committee coordination group has been set up in order to manage overlapping issues and to avoid contradictory advice.

SCHER, like the other scientific committees, can and does set up working groups and assigns external advisors and experts for particular issues and tasks. Working groups are always led by a committee member, but non-member experts can also be invited (European Commission 2009b). External experts are invited whenever the committee expresses a need for specific and relevant knowledge that is not found within the committee. For example, Vice-chair Jansson noted that SCHER lacked significant expertise for exposure analysis as well as for epidemiology. Given the growing specialisation and compartmentalisation of scientific research, the need for external expertise on particular issues is likely to increase. While SCHER consists of 17 members, covering several areas of expertise of relevance for assessing health and environmental risks, it goes without saying that the committee cannot cover all relevant research fields. Invited non-members include not only university scientists, however, but even bureaucrats and advisors involved in risk assessment and risk management on a national level.

[7]Comments by DG Sanco Unit C7 on a previous draft of this chapter; e-mail sent to us on 17 July, 2009.

17.4 The Working Process: Managing Facts and Values

As noted, the 'Rules of Procedure' make a very clear distinction between risk assessment and risk management – which the Commission equates with 'science' and 'politics', respectively. How, then, are such distinctions perceived and managed in practice by SCHER?

On the one hand, all of our interviewees confirmed the official ideal-type distinction between risk assessment and risk management: They were of the opinion that SCHER devoted itself to what it was supposed to, that is risk assessment. SCHER reviews scientific Risk Assessment Reports and provides science-based answers to questions from policy-makers. On the other hand, there are several indications of why it can be difficult in practice to maintain a clear distinction between risk assessment and risk management – which basically equates science versus politics and facts versus values. We will illustrate this below.

First, our own experience of communicating with DG Sanco officers (Unit C7) – two of our interviewees – revealed an unexpected attempt to politically control research, namely the very study presented in this chapter. In response to a draft of the present chapter, the DG Sanco officers not only – and contrary to our explicit request – sent the draft to at least one other EU officer, the head of the responsible unit in DG Sanco, and to one of our other interviewees (the Chair) but also threatened to take action to prevent the publication of this chapter if the 'revisions' suggested were not made. For the sake of clarity and scrutiny, we are quoting the officers' e-mail in full without editing it (except for typos and the added italics) – sent to us on 17 July 2009:

> Thank you for the opportunity to revise the draft document 'Science Committees and EU Policy: the SCHER case'. Please find attached hereby the draft which includes our corrections and comments. Please note that we have also discussed with Prof H Greim and Prof. B Jansson the parts of the interview attributed to them and agreed with them on the changes. Unfortunately, parts of the draft, give the impression of a prejudicial document, reporting some of the comments of the persons interviewed in a somewhat distorted or out of context way, referring to some off-the-record remarks, drawing conclusions and making statements which do not accurately reflect the real situation and the outcome of at least some parts of the interviews. Therefore, *we regret not to be in a position to agree with the publication of the document and request you to include the indicated comments and corrections related to the interviews with us and the experts mentioned above*. We also request to be given the opportunity to review the revised draft before publication. *In the event that you should decide to proceed with publication without our and the interviewed scientists' consent we will be put in the unfortunate situation to inform the editor of our objections to the publication of the document*. Please do not hesitate to contact us again whether you need further clarification on the changes to be made.

This response from the officers within DG Sanco is not only a violation of established principles of academic integrity but also shows how some EU Commission officers try to control how leading members of a scientific committee respond to an academic study.[8] While this does not necessarily say anything in general about the

[8]We asked Jansson and Greim if they had been contacted by the DG Sanco officer, and they confirmed that Greim had been contacted but that there had been no direct contact between DG Sanco and Jansson regarding these interviews.

Commission or the EU at large, it is a disturbing illustration of how bureaucrats in an EU institution dealing with policy and risk management are seeking to control what scientists formally responsible for risk assessment say. Whether the committee members agreed with these bureaucrats is, in fact, irrelevant. That the officers contacted, attempted to edit and, thus, attempted to exercise political control over how our interviews with the two top committee members should be interpreted is a violation of academic integrity and an attempt to violate our own integrity as well as that of the Chair and the Vice-chair.

Second, the Vice-chair complained about the principle that experts should not provide any advice on risk management, particularly as they believe that the official risk managers, within the Commission and elsewhere, do not always understand the scientific advice. The Vice-chair observed a wide gap between assessment and management, which he found very unfortunate, as this entails a risk for political misinterpretation of scientific advice. In particular, he emphasised the difficulties of defining and communicating uncertainty in risk assessments. He also noted that 'during coffee breaks' committee members might express their views about policies and regulations to the EU bureaucrats, but they are not formally charged with management. The DG Sanco officers interviewed suggested in their response to a draft of this article that the complaint is 'saying that the separation between "risk assessment" and "risk management" is meticulously maintained'.[9] However, as will be shown below regarding brominated flame retardants, even formal SCHER opinions can contain suggestions on risk management measures.

According to the Chair, Helmut Greim, the higher up in the hierarchy – and, thus, the farther away from direct contact with the scientific community – the less informed bureaucrats and decision-makers seem to be. Greim also emphasised the need for improved communication, specifically to avoid policy makers misinterpreting scientific advice.

Problems of communication have also been addressed by top-level decision makers, such as David Byrne and Robert Madelin. In their perspective, however, the problem is not so much a lack of interest among policy makers but mainly the ways in which scientists disseminate knowledge. In the words of Commissioner Byrne (2003, cf. Madelin 2004):

> As a lawyer, it has not always been easy to grasp the more subtle scientific arguments or to interpret phrases which are designed to give conclusions without compromising scientific accuracy. But I am lucky. I have a team of competent colleagues who come to my rescue, although, even they have occasionally struggled to draw a unique conclusion from the advice.

Third, some committee members occasionally expressed their views about political decisions. For example, the Chair expressed his disagreement with the decision taken by the European Parliament (EP) to ban animal testing of cosmetics (Greim 2008). This is a classic clash of norms, which, at least on the surface, is about consumer safety versus animal rights. However, this also comes with the politically assigned mandate of SCHER, which concerns public health and the environment but not explicitly

[9]Comments by DG Sanco Unit C7 on a previous draft of this chapter; e-mail sent to us on 17 July, 2009.

animal rights. From a scientific point of view, it can be argued that animal testing is a superior method of assessing risks of substances and products. However, the normative implication is still undeniable – consumer safety is prioritised over animal rights. According to Vice-chair Jansson, the EP ban on animal testing of cosmetics entails the necessity of human tests, which some EP members were allegedly unaware of. The point here is not to deny the right of scientists to express political views; on the contrary, we are doubtful as to whether it is possible or even valuable for scientists to remain silent on essentially political issues, especially if they are as involved in the policy system as SCHER members are. What may be important, though, is that scientists strive to be as precise as possible in explaining the grounds they have for taking a certain position, i.e. to be explicit on value issues.

Fourth, our interviews revealed some political views over which there is a great deal of controversy within the EU, as well as within SCHER. Some interviewees did not seem entirely cognisant of the essentially political nature of these views. An example of this is the much debated ‘precautionary principle’ (see Chapter 14), which states that even if there is absence of full scientific data or consensus on whether a substance or product might cause severe or irreversible harm to the public or to the environment, measures should be taken, including placing the burden of proof for safety on those advocating the use of these substances or products. Importantly, we consider the precautionary principle as well as its contrary alternative essentially to involve both risk management and risk assessment (cf. Ashford 2007; Karlsson 2005; de Sadeleer 2002). It is interesting to note here that the aforementioned DG Sanco officers, in response to a draft of this chapter, asked why we discussed differences between the attitudes of committee members regarding risk management, since this goes beyond the formal mandate of SCHER.[10] This illustrates a concern with maintaining the image of scientific committees as completely disconnected from ‘politics’ and ‘policy’, a statement that may be politically warranted but that is misleading and unrealistic, given results in the existing body of research on policy processes in general and science–policy interfaces in particular.

One committee member (Hanke 2008) argued that the precautionary principle is advocated by NGOs but not by the scientific community. In his view, science ‘operates on the basis of facts’, not on the basis of this kind of ‘moral principles’. Notably, however, his argument as to why the precautionary principle should be avoided is distinctively political; he claimed that precautionary principle has ‘economic implications’ and that industry should not be prohibited from using compounds simply because they are not thought to be safe. His view is in stark contrast to the views expressed by the Commission as well as by many scientists, which advocate using the precautionary principle (cf. Chapter 14). According to the Chair as well as the Vice-chair, the precautionary principle should be applied whenever there is insufficient data on risks, although they emphasise that this is a principle for management rather than for assessment. The internal disagreements regarding the precautionary principle were also observed by a non-member, who participated in several meetings with SCHER as external expert.

---

[10]Comments by DG Sanco Unit C7 on a previous draft of this chapter; e-mail sent to us on 17 July, 2009.

Moreover, some opinions issued by SCHER could, in our view, be interpreted as going beyond risk assessment, for example some conclusions reached on deca-BDE (a type of brominated flame retardant), which SCHER, after having criticised the Risk Assessment Report it reviewed, suggested should be the object of further risk reduction measures:

> The previous scientific committee (CSTEE) said that the uncertainties in the fate of DeBDE warrant risk reduction measures. Today there is further evidence for degradation of this substance to potentially harmful compounds and SCHER also strongly recommends further risk reduction (conclusion iii). Alternatives with properties similar to those of DeBDE should not be used until proven environmentally safe.[11]

According to the Opinion Vice-chair Bo Jansson played a key role in formulating SCHER's opinion in this case. SCHER argued, for example, that 'breakdown products' of deca-BDE were more harmful than the substance itself and, therefore, recommended risk reduction. This met with protests from stakeholders,[12] and the institutions swung back and forth between allowing and banning deca-BDEs. The issue is still unresolved, currently being on the agenda of, e.g., the European Court of Justice (cf. Eriksson et al. 2010).

The DG Sanco interviewees, in a comment to a draft of this article,[13] claimed that the opinion cited was 'perfectly within [the] remit of SCHER' and that the 'fact that the Commission may not have followed the SCHER conclusions speaks for … the clear separation of "risk assessment" and "risk management" ' However, both of the recommendations in the quotation above – on risk reduction and alternatives – clearly concern risk management, not risk assessment. Furthermore, the recommendations do not answer the three questions to SCHER that were formulated in the 'Terms of reference' in the opinion.

At the end of the day, contending views within the committee – whether of a scientific or a political nature – are not visible in reports and opinions from the committee. The 'Rules of Procedure' are cognisant of the possibility of diverging opinions both within a committee and between the committee and other bodies in the policy system. Voting rules and other procedures are carefully laid down for such situations. All of our interviews indicate a strong consensus-oriented process, however. There is a possibility for members to issue a minority statement if the committee's decision goes against their will, however this is apparently extremely unusual, if it ever takes place. This consensus orientation could be explained partly by an effort to make the voice of the committee as strong as possible, which it obviously is if consensus is reached and, thus, a clearer statement can be made, and partly by the very strong Chairmanship in this particular committee (External expert 2009). As noted above, there is also an element of political control, indicated by our own experience of how

[11]SCHER. 2005. *Opinion on "Update of the risk assessment of bis(pentabromophenyl) ether (decabromodiphenyl ether)"*. Adopted 18 March, 2005. DG SANCO. European Commission.

[12]See e.g. critical comments from the Bromine Science and Environmental Forum, the bromine producers' lobby organisation, on www.bsef.com/science/scientific-studies-4/deca-bde-2/.

[13]Comments by DG Sanco Unit C7 on a previous draft of this chapter; e-mail sent to us on 17 July, 2009.

DG Sanco officers tried to control the contents and publication of the study presented here. Fundamental disagreements are more likely voiced in the early stages of the discussion of a particular issue. However, eventually, after a series of meetings, an opinion is to be formulated, which typically works towards consensus.

17.5 Coping with External Pressure

The Commission is most certainly aware of the risks of politicisation of scientific expertise, particularly with regard to the perceived failures of expert governance in connection with the BSE crisis and the dioxin scare (Jansson 2008). In the words of Commissioner David Byrne in an address to the then newly elected members of the scientific committees:

> Although you will not be called to defend Commission policy, and should not even if you are invited to, you need to be aware that your names are in the public domain. You will be asked to give the Commission your advice on matters where there are often powerful commercial and social pressures and you may be the subject of direct or indirect lobbying. If you find yourself asked to comment on matters which arise because of your membership, it is important that you also take account of the need to protect the integrity of your scientific committee. I therefore stress the importance of rules of procedure which have been extensively revised to cover matters relating to independence, transparency and relations to stakeholders. (Byrne 2003)

Committee members have to sign a general declaration of independence, and all but one declared they were independent. The exception was the Dutch member Jan Linders, who, in his mandatory online 'Declaration of interests', noted that he did 'contract work for industry', which apparently was not considered a sufficiently great obstacle to become a committee member. In addition to the general declaration of independence, members also have to state any 'conflict of interests' at the beginning of each meeting. In contrast with the general declarations of independence, it is noteworthy that at more or less every meeting there are a few members that declare a particular 'conflict of interests'. It is not always the same members that declare a 'conflict of interest', as this seems to differ depending on the issue on the agenda for the particular meeting.

There seems to have been conflicting views within SCHER as to whether a connection to the chemicals industry is problematic or not. The Chair argued that this was not really a problem, as long as all connections with the industry or any other stakeholder are clearly declared. He went even further, claiming that the fact that a member has done work for the industry can be seen as a sign that this person is a 'real expert', since the industry allegedly only wants to work with the best scientists. The Chair, thus, expresses a fundamental trust in the scientific integrity and ability of committee members to keep their various roles apart. However, both the Chair and, more clearly, the Vice-chair noted that external stakeholders representing industry had sometimes been invited but that they very often did not provide scientific arguments. According to the Vice-chair, industry representatives were often not telling the entire truth – a comment made twice during the same interview (Jansson 2008).

All interviewees confirm that EU bureaucrats – in particular representatives of DG Sanco, DG Enterprise, DG environment and the European Chemicals Bureau (ECB) – usually participated in meetings. The Chair noted that these 'receivers' occasionally disagreed with the committee's advice, to which he responded that 'we are an independent group' and that they would just have to agree to disagree. None of the interviewees thought that there was ever any direct pressure from the Commission, or any other actors within the EU institutions, to influence SCHER's opinions.

Indeed, the interviewees all replied that there was not much external pressure at all.[14] Several interviewees also noted that if members received any communication from external pressure groups, this was usually passed on to the secretariat, which typically did not send much external information back to the members. The DG Sanco representatives argued that committee members are 'clever persons' that are concerned about their integrity and that know that what they say and recommend can be used – and misused. The Commission's rules on integrity, independence and transparency are supposed to minimise this problem, but this obviously cannot be fully guaranteed (DG Sanco 2009). It was observed, however, that since all of the committee members have other jobs, they could simply quit if they felt that the Commission would try to influence their assessments in a particular direction.

However, as noted, the way the aforementioned DG Sanco officers tried to control our presentation of what the Chair and the Vice-chair said during our interviews, including threatening to take action against publishing of the present study, indicates a readiness to control and, thus, violate the integrity of scientific studies. We cannot tell on the basis of this study whether similar attempts on the part of DG Sanco are made to influence SCHER members in their work within the SCHER.

One interviewee also observed that there was a tendency for countries with a significant chemicals industry to tend to be more eager to receive assignments for official risk assessments (Jansson 2008).

17.6 The Political Role and Impact of SCHER

What, then, is the impact or wider political role of SCHER? Has it in any discernable way shaped the Commission's risk regulation policies? Generally speaking, our study has indicated that SCHER achieves what it is primarily supposed to do – that is, reviewing the scientific quality of risk assessments. By its very nature, it is difficult to say anything conclusive about what kind of policy impact this entails. More often than not, SCHER's opinions grant a lot of leeway to the Commission, particularly in those many cases when data on risks is either unavailable or plagued with severe uncertainties. In those cases, the impact of SCHER has often been a call for more and better data, which has not in any clear way translated into policy.

[14]This is strikingly different as compared to the pressure put on DG Sanco.

This function, the ability to prolong a process by calling for more and better data, is what one interviewee (External expert 2009) calls the 'watchdog' function of SCHER.

In cases where data is considered to be more reliable and conclusive, such as on tin-organic compounds, SCHER has reached conclusions that have entailed a clearer 'green' or 'red' light for the Commission, which has usually followed suit (External expert 2009; DG Sanco 2009). The DG Sanco interviewees also maintained that SCHER's opinions had a significant impact on, e.g., the Commission's application of the so-called 'restriction directive'. Thus, the better the data, the more influential the opinions of scientific committees such as SCHER.

The opinions issued by SCHER are basically scientific peer reviews, similar to a longer referee report for a piece submitted to a scientific journal. Just like any peer reviewer would do, SCHER comments on the data used in risk assessments, especially on the lack or reliability of data as well as on methods, approaches and interpretations. Uncertainty of exposure and effects and factors conditioning uncertainty are discussed, often resulting in statements such as 'firm conclusions cannot be reached'.

As shown above, EU bureaucrats and policy-makers find these kinds of statements difficult to interpret, and committee members agree that it is extremely difficult to express uncertainty in a way that is both accurate and intelligible to laymen. In our view, it is usually difficult to discern any explicit recommendations for policy-making in SCHER's reports, although implicitly it might matter whether indications of risks are described in terms of, for example, 'possible risk' or 'uncertain risk'. Nevertheless, the Commission finds SCHER's opinions useful, as they are seen as legitimating Commission policy, even when the opinions explicitly state that no firm conclusions can be reached. The very existence of SCHER and the other scientific committees gives scientific credibility to Commission policy, whether or not the experts offer instrumental advice. This is a symbolic function (Amara et al. 2004; Beyer 1997) of expert committees that should not be underestimated.

Moreover, informal contacts between SCHER and the Commission certainly exist (it would be surprising if they did not), and they may potentially entail influence beyond the formal mandate of SCHER. Such contacts are, however, extremely difficult to observe, let alone evaluate in terms of policy impact. The interviews we made indicate that, taken together, formal and informal channels have hardly yielded any substantial policy impact beyond the mere provision of scientifically-based reviews of scientific knowledge. This is corroborated by the general perception expressed by our interviewees that there is very little feedback from the Commission regarding how they are using SCHER's opinions, a problem that is perceived as having gotten worse over the years.

In addition, impact is difficult to discern due to the typical 'lag' that characterises policy processes in general, to which the EU system is no exception. As noted by the Vice-chair of SCHER, once SCHER has delivered its report to the Commission, there may be negotiations between governments and with stakeholders, which can be prolonged and entrenched. Likewise, scientific research and assessment takes time. The DG Sanco interviewees said that they were generally satisfied with SCHER's work but that they would like to see processes speeded up in general.

Several of our interviewees indicated that SCHER and the scientific committee system at large are changing. In 2009, a new SCHER committee was set up, with several new members and with revised 'Rules of Procedure'. It was argued that SCHER is moving on to deal with broader issues (DG Sanco 2009), which in the Vice-chair's perception seems to imply a return to the broader focus that the committee had at its inception in 1997. SCHER will probably still do ad-hoc work (DG Sanco 2009), however, and there are also discussions about developing a 'quick response' function (Jansson 2008) with the scientific committees being able to reply to urgent questions from the Commission within a few days. There is also a chance that SCHER and the other committees will receive a clearer mandate for 'self-tasking', which, if realised, would definitely increase the agenda-setting powers of the scientific committees.

17.7 Conclusion

This study has shown that, by and large, SCHER seems to be able to maintain a traditional role as scientific peer-reviewer, with some, though seldom any direct or significant, impact on policy decisions made by the Commission. We have not been able to discern any political campaigning with SCHER as an explicit partner in advocacy coalitions. The interviews all noted that external contacts were very limited, particularly with the industry and the green movement, respectively, mainly because they were seen as providing normative arguments rather than scientific information. Views on risk assessment and particularly on risk management vary among the committee members, with some voicing industry-friendly ideas and others supporting 'green' visions, including the precautionary principle. These differences, however, do not seem to have impacted on SCHER's reports in general. SCHER almost always reaches consensus on its opinions, and the mechanism behind this seems to be that they apply strict (and traditional) scientific criteria when reviewing risk assessments. Above all, our impression is that SCHER's members seem to be concerned about maintaining credibility within the scientific community. While often concerned about policy decisions, the members were careful not to let their normative positions be seen to guide conclusions in their reports. Indeed, the normative-political tensions with SCHER seem to have prevented SCHER from becoming an advocate for particular policies. At the end of the day, focusing on scientific criteria and peer review made consensus possible.

An unexpected and much more disturbing result, however, is how managing DG Sanco officers tried to control the publication of this study, indeed the very words we have written here. As noted, contrary to our explicit request (and in violation of a fundamental principle in scientific publishing), these DG Sanco officers not only disseminated a draft of this article to at least one other EU officer and some of the other interviewees but also tried to control how we have reported the results of the other interviews, indeed, threatening to take action to prevent this study from being published unless the 'corrections' suggested were made. This smacks of undemocratic control of science. The reaction illustrates a political fear of academic studies, which may be seen as a critique of the Commission's management of scientific expertise and which signals a threat to such fundamental values as academic integrity and freedom of expression.

References

Amara, N., Ouimet, M. and Landry, R. (2004) 'New evidence on instrumental, conceptual and symbolical utilization of university research in government agencies', *Science Communication* 26(1).

Ashford, N. (2007) 'The Legacy of the Precautionary Principle in US Law: The Rise of Cost-Benefit Analysis and Risk Assessment as Undermining Factors in Health, Safety and Environmental Protection'. In de Sadeleer, N. (ed.) *Implementing the precautionary principle. Approaches from the Nordic Countries, EU and USA*. London: Earthscan.

Beyer, J. S. (1997) 'Research Utilization: Bridging the Gap Between Communities', *Journal of Management Inquiry* 6: 17–22.

Black, I. (2001) 'People losing faith in 'anonymous' EU', *The Guardian*, July 18, 2001.

Byrne, D. (2003) "Irrational Fears or Legitimate Concerns" – Risk Perception in Perspective. Presentation given at the *Risk Perception: Science, Public Debate and Policy Making Conference*. Brussels, 4 December 2003. European Commissioner for Health and Consumer Protection. Available at: http://europa.eu/rapid/pressReleasesAction.do?reference=SPEECH/03/593&format=HTML&aged=0&language=EN&guiLanguage=en

Christiansen, T. & E. Kirchner, eds (2000) *Committee Governance in the European Union*. Manchester: Manchester University Press.

Christiansen. T. & S. Piattoni, eds (2004) *Informal Governance in the European Union*. Cheltenham: Edward Elgar.

de Sadeleer, N. (2002) *Environmental Principles: From Slogans to Legal Rules*. Oxford: Oxford University Press.

Dunlop, C. A. & O. James (2007) 'Principal-agent modelling and learning: the European Commission, experts and agricultural hormones', *Public Policy and Administration* 22(4): 403–422.

EEC (1993) Council Regulation (EEC) No 793/93 of 23 March 1993 on the evaluation and control of the risks of existing substances, *Official Journal* L. 84, 5.4.93, 1.

Eriksson, J., Karlsson M. and Reuter M. (2010) 'Technocracy, politicization, and non-involvement: politics of expertise in the European regulation of chemicals', *Review of Policy Research*, 27(2): 167–185.

European Commission (2009a) Register of Experts Groups. Brussels: European Commission, Secretariat General. Available: http://ec.europa.eu/transparency/regexpert/

European Commission (2009b) *Rules of Procedure. The Scientific Committees on: Consumer Safety (SCSS), Health and Environmental Risks (SCHER), Emerging and Newly Identified Risks (SCENIHR)*. Brussels: European Commission, Health and Consumers Directorate-General. Available: http://ec.europa.eu/health/ph_risk/documents/rules_procedure_en.pdf

External expert (2009), which has been working in connection with SCHER. Personal interview, March 2009.

Fischer, F. (1990) *Technocracy and the Politics of Expertise*. Newbury Park: Sage.

Gornitzka A. and U. Sverdrup (2008) 'Who Consults? The Configuration of Expert Groups in the European Union', *West European Politics* 31(4): 725–750.

Greim, H. (2008) Chair of SCHER. Personal interview, November 2008, Munich.

Hanke, W. (2008) Member of SCHER. Personal interview, November 2008, Warszaw.
Jansson, B. (2008) Vice-chair of SCHER. Personal interview, October 2008. Stockholm.
Jasanoff, S. (1990) *The Fifth Branch: Science Advisors as Policymakers.* Cambridge, MA: Harvard University Press.
Joerges, C. and E. Vos, eds (1999) *EU Committees: Social Regulation, Law and Politics.* Oxford: Hart Publishing.
Karlsson, M. (2005) *Managing Environmental Risks for Sustainable Development: Policies for Hazardous Chemicals and Genetically Modified Organisms.* Doctoral Thesis. Karlstad: Karlstad University Press.
Larsson, T. (2003) *Precooking in the European Union. The World of Expert Groups.* Ds 2003:16. A Report to the Expert Group of Public Finance (ESO). Stockholm: Fritzes.
Madelin, Robert (2004) 'The Importance of Scientific Advice in the Community Decision Making Process'. Director General for Health and Consumer Protection, the European Commission. Presentation given at the inaugural joint meeting of the members of the non-food scientific committees, Brussels, 7 September 2004.
Radaelli, C.M. (1999) 'The public policy of the European Union: Whither politics of expertise?' *Journal of European Public Policy* 6(5): 757–74.
Rhinard, M. (2003) 'Committees in the European Union: An Empirical and Normative Assessment.' Paper prepared for the EUSA 8th International Biennial Conference, Nashville, Tennessee, March 27–29, 2003.
Sabatier, P. and Jenkins-Smith, H-C. (eds) (1993) *Policy Change and Learning: An Advocacy Coalition Approach*, Boulder: Westview Press.
DG Sanco (2009) Two representatives of DG Sanco, European Commission. Personal interview, March 2009, Brussels.
Timmermans, A. and Scholten, P. (2006) 'The Political Flow of Wisdom: Science Institutions as Policy Venues in The Netherlands', *Journal of European Public Policy* 13(7): 1104–18.
Van Schendelen, M. (1998) *EU Committees as Influential Policymakers.* Aldershot: Ashgate.
Weale, A. (2000) 'Government by Committee Lost the People's Confidence?' In *Elitism, Populism, and European Politics*, edited by J. Hayward. Oxford. Clarendon Press.
Weiler, J. H. H. (1999) 'Epilogue: 'Comitology' as Revolution – Infranationalism, Constitutionalism and Democracy'. In *EU Committees: Social Regulation, Law and Politics.*, edited by C. Joerges and E. Vos. Oxford: Hart Publishing.

Chapter 18
Implementing Chemical Regulation: The Role of Inspectors

Vicki Johansson

Abstract Inspection is one of the methods commonly used by public authorities to monitor compliance and goal achievement. But can inspections improve goal achievement and, if so, under what conditions? In this chapter the role of inspectors in the implementation of chemical regulation will be discussed. The theoretical focus is on three conditions that must be met if inspections are to be able to contribute to goal achievement. Since a comparison between inspection groups improves the possibility for highlighting general characteristics of inspection as well as specific characteristics of chemical inspection, the results of an interview study conducted with Swedish chemical and labour safety inspectors serve as an empirical starting point for the discussion.

Keywords Chemicals • Compliance • Inspectors • Monitoring • Sweden

18.1 Introduction

Environmental and work environment problems are two issues that engage policy makers at the national and European as well as international levels. These are issues that are of utmost importance for the lives and health of human beings now and in the future. But the environment can only improve if patterns of behaviour and consumption are changed at both an individual and an organisational level.

In recent decades regulation as a part of governance has increased in magnitude and importance (Braithwaite 2006). The reasons behind this development can be discussed, but, among other hypothesis, it has been suggested that the complexity of the welfare state in combination with decreased financial resources has advanced regulatory enforcement at the expense of reform policies (Kleinman Mark 2002).

V. Johansson (✉)
School of Public Administration, University of Gothenburg, P.O. Box 100, SE 405 30, Göteborg, Sweden
e-mail: vicki.johansson@spa.gu.se

J. Eriksson et al. (eds.), *Regulating Chemical Risks: European and Global Challenges*,
DOI 10.1007/978-90-481-9428-5_18,

There has also been a shift in governmental use of policy tools from input to output centred instruments (Pierre and Peters 2000), with evaluations and inspections being examples of two policy instruments that have grown in importance (Johansson 2006). It has also been suggested that these changes have been fostered by the expansion of the European Union, mainly because regulations are the EU's most common policy instrument (Majone 1996). Regulations dominate within the area of environmental policy in Europe, even if economic and communicative policy instruments also are used (Mac Neil et al. 2002).

Regulations are the oldest and most frequently used policy instruments that states adopt in order to stimulate behavioural change and thereby achieve desired political outcomes (Schneider and Ingram 1990; Vedung 1998). Regulations, such as laws, decrees and directives, place limits on what is permissible at the same time as some types of behaviour are stigmatised (de Bruin and Hufen 1998; Schneider and Ingram 1990; Vedung 1998).

A regulation tool can only work if people respect and comply with the rules. The growing individualism in western societies has fostered a situation in which people to an increasing extent ignore rules that they dislike or they do not feel obliged to comply with (Woodside 1998). The legitimacy of rules in force tends to be undermined when the distance between norms and rules grows (Hydén 2002). From the perspective of political steering, this kind of development is highly problematic, since it makes it more difficult to achieve political objectives. In order to counteract such a development, states make use of both preventive and reactive policy instruments. Sanctions, such as fines, prohibitions and deprivation of freedom, are negative incentives that are supposed to discourage people from transgressing rules (Schneider and Ingram 1990). In areas where the risk of being detected when breaking a rule is low, the government often gives state authorities the task of monitoring compliance with regulations. Such policy areas are diverse and include, among others, work environment, social services, social security, gender equality, nuclear power, road safety and national heritage. Inspection is one of the methods commonly used by public authorities to monitor compliance and goal achievement. In Sweden there are 230 inspection laws, and no less than 380 public authorities on the national, regional and local levels have mandates as supervision authorities. One central assignment for these authorities is to check that activities comply with existing regulations (SOU 2002:14).

At inspections, authorities control whether public and private organisations are complying with existing regulations. If this is found not to be the case, the authorities are obliged to take action. The requirements that authorities make in such situations must have legal support in formal rules. Inspections, as evaluations and audits, can be understood as a reactive policy instrument used by governments in order to investigate if pro-active policy instruments have expected outcomes and to promote goal achievement.

But can inspections improve goal achievement and, if so, under what conditions? In this chapter the role of inspectors in the implementation of chemical regulation will be discussed. The theoretical focus is on three conditions that must be met if inspections are to be able to contribute to goal achievement. Since a comparison between inspection groups improves the possibility for highlighting general characteristics of inspection as well as specific characteristics of chemical inspection, the results of an interview study conducted with Swedish chemical and labour safety inspectors in 2004 will serve as an empirical starting point for the discussion.

18.2 The Swedish Chemicals Inspectorate and the Swedish Work Environment Authority and Their Inspectors

Public administrators are not a homogeneous group working in similar settings, a point not always noted in the literature (Hood 2002). They can be found on all levels of government – from the national to the local level - and on all levels within a bureaucracy – from higher positions of management to case-work at the street level. Some are primarily involved in the preparation of proposals, while the main objective of others is to implement policy on a day-to-day basis. Some operate under regulatory frameworks that make their decisions rather predictable, while others do not. Some are trained in public administration, others as professionals or experts. Some perform their official duties on their own, while others would hardly be able to perform a single task without close and continuous interaction with other actors, public as well as private. The contextual setting, including the regulatory framework, the organisational structure and the national political administrative system, will influence how public administrators perform their duties: what tasks they are supposed to carry out and how these tasks are performed in practice. In order to be able to analyse administrative practice within inspection authorities, it is, therefore, necessary to understand the kinds of public bureaucrats that inspectors are and the settings within which they find themselves.

In the Swedish political system authorities on the local, regional and national levels are commissioned by the government to monitor the compliance of public and private organisations with regulations. In Swedish, this governmental activity is called 'tillsyn', which can be translated into English as inspection or supervision. 'Tillsyn' literally means to 'see to' or look after something in a practical sense, so that things are being done in a secure and safe manner. It includes not only inspections but also other activities, such as providing advice and guidance, evaluating, mapping, granting permits, and issuing decrees. Inspectors are usually involved in all of these activities; they do not only carry out inspections. Nonetheless, in what follows the term inspection will be used, even though that term is to narrow in a Swedish context.

Some kind of supervision of labour safety and working conditions has been carried out by public authorities on the local, regional and central levels for almost 200 years. Today such activity is a governmental responsibility (SOU 2002:14). The Swedish Work Environment Authority (SWEA) was established in 2001 through the amalgamation of the Labour Inspectorate and the National Board of Occupational Safety and Health. SWEA's objective is to reduce the risks of ill-health and accidents in the workplace and to improve the work environment (in terms of social, mental and physical conditions). Regulations stipulating how a good work environment should be designed can be found in the Work Environment Act. SWEA's main task is to check by means of inspections that private and public organisations comply with the regulations. The authority is also required to issue detailed binding regulations in accordance with the law.

In 1998 the Swedish parliament, Riksdagen, enacted an environmental code in which the majority of the rules within the environmental area today can be found. Legislation had been increasing during the 1900s but was not based on modern knowledge of what creates environmental problems or how these kinds of problems ought to be dealt with. At the same time, the legislation was hard to grasp, since regulations were to be found in many different laws (Setterlid 2000).

Of the 318 Swedish enforcement authorities within the environmental policy area, seven are on the state level, 21 on the regional level, and 290 on the local level. Broadly stated, the municipalities are responsible for environmental enforcement within their own territories, the county administrative boards for organisations and enterprises whose effects on the environment impinges on more than one municipality, while the Swedish Environmental Agency (SEA) is the largest environmental enforcement authority on the national level. SEA's main tasks are to co-ordinate and to promote environmental work on both a national and international level in accordance with 16 environmental quality objectives decided on by the Swedish Riksdag. Objectives such as 'clean air', 'a safe radiation environment' and 'a rich diversity of plant and animal life' are supposed to guide Sweden towards a sustainable society. In order to fulfil its tasks, SEA is commissioned to support and advise the county administrative boards and the municipalities in their enforcement activities (Setterlid 2002). In certain environmental areas the competence required is regarded as so specialised that it is considered neither possible nor desirable to delegate the entire enforcement responsibility to county administrative boards and municipalities or to give primary responsibility to SEA. Chemical control is such an area. The Swedish Chemicals Inspectorate (SCI) is the state government authority that is to promote the environmental quality objective: 'a non-toxic environment'. The responsibilities of the Inspectorate include co-ordination of co-operation between central agencies, county administrative boards and municipalities and evaluation of goal achievement for the objective of 'a non-toxic environment'.

The Swedish Work Environment Authority (SWEA) and the Swedish Chemicals Inspectorate (SCI) are, thus, two enforcement authorities that work with environmental issues, though in different respects. The Swedish Work Environment Authority focuses on factors that have an impact on the mental, social and physical health and well-being of employees, while the Swedish Chemicals Inspectorate focuses on factors that influence toxic levels in nature and thereby both directly and indirectly the health and well-being of human beings as well. While SWEA's activity is dominated by inspections of workplaces in public and private organisations, inspections play only a minor role in the activities of SCI. See Table 18.1 for further differences between the authorities.

Chemical inspectors, in addition to carrying out inspections, are responsible for providing support to and guide enforcement authorities on the local and regional

Table 18.1 Differential characteristics between the Swedish Work Environment Authority (SWEA), and the Swedish Chemicals Inspectorate (SCI) (Johansson 2006: 87)

SWEA	SCI
Core activity	Non core activity
Old authority (amalgamation)	New authority (partitioned)
Many inspectors (ca 400 inspectors who in total carry out 30,000 inspections a year)	Few inspectors (ca 15 inspectors who in total carry out 300 inspections a year)
Many objects of inspection (every work-place in Sweden)	Few objects of inspection (ca 2500)
District organisation	Central organisation
Financed by taxes	Financed by fees
Different educational level among inspectors	University educated (natural sciences) inspectors
Extensive internal education	Limited internal education

levels, i.e. county authorities and municipalities as well as the organisations they supervise. SCI, for example, invites local and regional inspectors to participate in their inspections, arranges conferences and educational programmes and, on the basis of inspection results from different enforcement projects, prepares reports and brochures that can be used as information material. A majority of the chemical inspectors also have an oversight function that involves either keeping in contact with a number of different authorities or following developments within a certain chemical branch. The chemical inspectors have, in other words, a number of different tasks that cannot be described merely as inspection or supervision in a strict sense.

Similarly, work environment inspectors have many different tasks. As with chemical inspectors, they disseminate information about their activities and participate in groups having specific tasks. One assignment can, for example, be to oversee and to follow developments within a particular area of work environment or to contribute to the development of new inspection methods on both the local and central levels.

The preventive and evaluative tasks that the inspectors have, which in a Swedish context are defined as promoting, counselling or supportive enforcing, will not be further discussed in this article. It is, however, important to keep in mind that Swedish inspectors generally have both supportive and evaluative tasks, even if the occupational title emphasises the control functions.

18.3 The Discretion of Inspectors

Inspectors are public officials that work on what is sometimes referred to as the frontline, which means that they work directly with people in their roles as citizens, clients and representatives of different organisations and operational activities. This group of public employees is positioned, so to speak, 'in between' civil society and the state. They are both to be compliant and to engender compliance.

In order to be compliant, the inspectors must be sensitive to the political and administrative management and to the objectives that are supposed to guide the operational activities. They must understand what they are supposed to do, they must have a desire to fulfil their mission and they must have the capacity to do so (Van meter and van Horn 1975; Sannerstedt 2001). How they relate to, interpret and respond to political signals will affect how they prioritise between different objectives and will thereby indirectly affect the degree of goal achievement in the policy area in which they work. A high degree of goal achievement requires further that the inspectors can engender compliance with regulations among those citizens or groups of citizens they are to supervise. Inspectors, thus, find themselves in a middle position where their work is judged both from above (the political and administrative management) and from below (operational activities and citizens). Therefore, they have much in common with what in the public administration literature have been called street-level bureaucrats (Lipsky 1980).

During the past 3 decades, street-level bureaucrats have been the object of countless studies within the social sciences. It has been concluded that these officials – given their discretion and relative autonomy in relation to organisational authority – can be understood as a specific kind of policy maker. The presence of visionary, complex and contradictory goals in combination with limited resources shape a setting in which street-level bureaucrats need to prioritise between tasks and to routinise their day-to-day work. This, together with the difficulties in controlling and directing street-level bureaucrats, both enables and constrains them in making policy (Lipsky 1980; May and Winter 1999; Johansson 2006; Meier and O'Tool 2007). The extent to which it is possible to circumvent the discretion of street-level bureaucrats and to influence their priorities through political and administrative measures is an important research area, one which has been carefully investigated (Jewell and Glaser 2006; May and Winter 2007). In general, the results of this research do not reveal any clear-cut conclusions. However, the literature identifies a number of organisational factors that are likely to influence the priorities made and the styles adopted by street-level bureaucrats. Inspectors usually work in policy areas with visionary, complex and contradictory goals and limited recourses, and they utilise discretion when carrying out inspections (Fineman 1998; Hutter 2001; May and Wood 2003; Johansson 2006; Winter et al. 2007).

Existing regulations within a policy area can be understood as a specification of goals. Regulations are, however, always open to interpretation. The extent to which this is the case depends on policy area and on the level of complexity. *Vagueness in regulations* creates leeway for various interpretations and may increase the likelihood that rules are applied differently by different inspectors, a common problem within street-level bureaucracies. In combination with limited resources, these circumstances may force inspectors both to interpret goals and regulations and to *give different priorities* to them. Limited resources in this context means that, regardless of the amount of financial recourses actually allocated to inspection, it will always be possible to make improvements. At the same time, improvements will often create demands for further improvements. This has been defined as the paradox of public service production (Birgersson 1975). Characteristic for inspectors is, thus, that their tasks, in practice, never come to an end. Therefore, inspectors, as other street-level bureaucrats, tend to build in a time margin in their daily work by creating a time buffer. This buffer is used to deal with unforeseen events, tasks that are complicated and tasks that officials themselves regard as important, to accomplish a less stressful work situation and/or to reduce working hours. In order to build up a buffer and thereby facilitate the opportunity for greater discretion, the inspector will try to minimise the time use for various tasks and will argue for the management that the time needed to fulfil a task is greater than it in reality is (Lipsky 1980).

Various techniques are used to build up a buffer, but a common trait is that the bureaucrat develops *routines that simplify decision making*. The routines aim at shortening and streamlining the decision process. The primary methods used involve limiting (restricting) and standardising decision making as well as contacts with inspection objects. Street-level bureaucrats adjust their strategies to the policy instruments that management uses in order to control their work and, therefore, street-level routines and patterns of decision making tend to change over time. The manner in which routines are modelled and used will affect policy outcomes and thereby the degree of goal achievement.

Management, political as well as administrative, is usually aware of the fact that street-level bureaucrats, as a consequence of their discretion, can produce political outcomes that deviate from those that have been decided on and that they, thus, have the power to change and to make policy. Therefore, management adopts strategies aimed at controlling and steering the priorities made by street level bureaucrats and the time they use to fulfil different tasks. Common management measures are, for example, formulating goals, conducting educational programmes, rebuilding organisational structures and using monitoring systems. The choice of policy instruments on the part of management affects strategies and routines that street-level bureaucrats develop and can develop.

One conclusion that is possible to draw from this discussion is that inspectors have, in principle, discretion in relation to three important elements that can affect the efficiency and legitimacy of inspections, these are the degree of vagueness in regulations, the selection of inspection themes and objects and the choice of compliance strategies used by inspectors in order to persuade inspection objects to follow relevant regulations.

18.4 Vagueness in the Law (Regulations)

The main task of public inspectors is to check whether companies, municipalities, organisations and citizens are acting in accordance with existing regulations (SOU 2004:100).

A basic condition that must be met in order for inspection to be an effective policy tool is that the rules in relation to which the inspectors operate are based on 'correct' knowledge about how human activities affect the environment. If the causal logic inherent in the rules is invalid, the environment will not improve, even if every rule is followed down to the last detail as a result of the work of the inspectors. The main task of inspectors is to check for compliance with existing regulations. They are not supposed to take a personal position on the causal logic inherent in the rules. They can, of course, refer concerns back to the legislators if they believe that existing rules are based on faulty premises or that compliance with existing regulations does not yield expected effects. However, a basic, if not sufficient, condition that must be met if inspection is to be able to function efficiently is that the causal logic in environmental legal provisions is adequate. Whether the causal mechanisms that are presupposed in the rules that are to be complied with are correct or not is a question that researchers from disciplines other than the social sciences have to find answers to. A point of departure here is, therefore, that the existing rules within the areas of the environment and the work environment are based on 'correct' knowledge, even if this may not be the case.[1] However legal frameworks and the regulations within it are always open to interpretation. As other public officials at the street level, inspectors must constantly use their discretion to interpret the regulations that are to be complied with.

> Praxis is always developing. We have a discussion here, there is a lot of talk about 'how high to set the bar', and you can feel unsure as an inspector. 'What do you think? What should we require? What did you require - and why did you do that? And it can go on like this between us, and when the 'buzzing' has gone on for a while, it will even itself out. And then a new paragraph in some regulation will turn up, and then the buzzing will start all over again and we reach some common ground. (Interviewed inspector)

In order to establish if and to what extent the chemical and labour safety inspectors have discretion in interpreting regulations, a comparison between the two most central regulations of the inspectorates will be made. The incidence of vague words and sentences is used as an indicator of vagueness in legislation and thereby of the leeway for various interpretations of the regulations by inspectors. A high proportion of vague sentences indicates that the leeway for discretion on the part of inspectors is, in principle, great. If the there is great leeway for discretion, the rule of law can be questioned, since it can be difficult, from the perspective of the inspection objects, to anticipate when regulations are being followed or not. Also, efficiency can be questioned, since discretion inevitable leads to regulations being applied differently.

From Table 18.2 the conclusion can be drawn that the differences between the two most key regulations of the two authorities are quit large. While almost 80% of the sentences in the the Swedish Chemicals Inspectorate regulation on 'classification and labeling of chemical products' can be classified as clear, the corresponding figure for the Swedish Work Environment Authority regulation on the 'design of the workplace' is 20%. A general difference between the two regulations is that the word 'must' is often, though not always, combined with exact instructions in the

[1]However, the research conducted on the effects of inspections, mainly in the USA, shows that inspections positively affect compliance with regulations both within the areas of the environment and the work environment and that they have environmental impacts (Weil 1996; Gray and Deily 1996). For example, results indicate that inspections have the greatest impact on average companies, that the size of the penalties is inconsequential, that injuries among employees and pollution decline and that permitted levels are maintained to a higher degree if inspections are carried out (Gray and Scholz 1993; Magat and Viscusi 1990).

Table 18.2 Proportion vague, difficult to determine and clear sentences in the two most key regulations from the Swedish Work Environment Authority, and the Swedish Chemicals Inspectorate

Sentences in the paragraphs of the regulation	The National Board of Occupational Safety and Health regulation on the design of the workplace (AFS 2000:42)	The Swedish Chemicals Inspectorate regulation on classification and labeling of chemical products (KIFS 1994:12)
Vague	58	11
Difficult to determine	23	10
Clear	19	79
Total (per cent)	100	100
Total (number)	268	120

Chemical Inspectorate's regulation, while 'must' in the Work Environment Authority's regulation is combined with vague words and with unclear instructions, as for example 'must if possible' and 'must if appropriate'. The chemical regulation, as compared to the labour safety regulation, is less vague, which indicates that the leeway for discretion among chemical inspectors is, in principle, lower than among their colleagues in labour safety.

One basic principle in the rule of law is that it should be possible to anticipate what requirements one, as a person or an organisation, is obliged to live up to. The vagueness that has been observed in the regulations is, therefore, problematic. The empirical analysis shows that the inspectors must interpret regulations before they conduct or while they are conducting an inspection, even if the leeway for such interpretations is greater within work environment regulation than within chemical regulation. Interpretations vary more between labour safety inspectors than between chemical inspectors, which can be explained partly by the fact that the chemical regulations are less open to interpretation but also by the fact that there are few chemical inspectors (15 in all compared to 500 labour safety inspectors), that they have similar educational backgrounds and that they are located at one office.

18.5 Priorities of Inspection Themes and Objects

Inspectors, as other street-level bureaucrats, have leeway for discretion, which makes it both possible and necessary for them to prioritise between different tasks and objectives. Since time is a scarce resource, they often develop routines that they use when making these priorities. How these routines are designed will affect both the efficiency of inspections and the policy outcomes. It is, therefore, important to identify the routines for prioritising that are used in practice.

One of the most important tasks that governmental authorities and agencies in Sweden have is, based on their specialist and expert knowledge, to weigh different alternatives against each other and to give priority to the most important ones. Within the environmental area, for example, the supervisory authorities are commissioned by the government to investigate the need for inspections in different areas, while SWEA is commissioned to follow developments in work environment and to take initiatives that the authority find necessary (Förordning (2000:1211) med instruktion för Arbetsmiljöverket). Even if it were possible to identify the most appropriate selection criteria, these might not be identical with the criteria that would engender the greatest legitimacy for the policy among the citizenry. There can even quite possibly be an opposite relationship between goal achievement and trust for the policy among the citizenry. Trust for public authorities and public policy might increase if the authorities focus on objects that the citizens regard as hazardous, even if experts claim that the area is one of low risk (Luhmann 1993).

Before inspectors can carry out inspections, two choices must, thus, be made. First, the authority or the inspector must choose themes and, second, objects for inspection within the framework of the selected themes. The question of which themes and objects that ought to be chosen is, on a theoretical level, related to notions of (a) what measures are most important to take in order to improve the environment? and (b) which human behaviours affect the environment most negatively?

In previous research three risk selection criteria have been identified. Priority can be given to: (1) activities that are most dangerous to humans, even if the number exposed to the risk are few; (2) activities that affect many people, even if the risk involved is quite low; (3) activities that affect many people, regardless of the calculated risk level (Hansson and Lindblom 2003).

Studies of labour safety and chemical inspectors suggests that the priority given to themes and objects can be attributed to several different factors and that theoretical risk assessment is not salient as a central selection and decision criterion. The decision regarding which inspection themes are chosen for inspection does, however, affect the possibility of achieving the goals of 'a non-toxic environment' and of 'a working environment that is sound and conducive to personal development for all'. Ideally, knowledge of how measures, effects and outcomes are linked and which measures or combination of measures yield the highest degree of goal achievement guides the inspectorates and the inspectors in their choice of relevant themes. In reality, however, parallel and conflicting opinions exist on which elements within a policy area ought to be given priority, due to the fact that actors understand the causal logic differently.

Nine factors that influence the choice of themes can be identified in the interviews: experience, the media, unfortunate developments, special commissions from the Government, budget documents, production of statistics, tips from other authorities, tradition and annual activity plans. The choice of inspection objects is influenced by eight corresponding factors: random sampling, what you can find, geographical site, hearsay, long time ago, piecework, self-prioritisation and previous problem kids.[2]

If we take a closer look at the nine selection criteria for themes and objects, we can conclude that few of them indicate that systematic comprehensive judgements are

[2]From the answers given in the interview study, it is possible to identify selection criteria but not to rank them in relation to how often they are used or how common they are.

Table 18.3 Identified selection criteria for themes and objects to inspect (Johansson 2006: 142)

Selection of themes and objects (criteria)	
Themes	Objects
Experience from previous inspections	OSU
The media	What you find
Unfortunate developments	Geographical site
Commissions from the Government	Hearsay
Budget documents	Long time ago
Production of statistics	Piecework(individual and/or collective)
Tips from other authorities	Self-prioritising
Tradition	Previous problem kids
Annual activity plans	

made, in which different risks are balanced and weighed against each other. The selection criteria found in Table 18.3 indicate rather that other methods are used. The criteria identified have one thing in common: all of them restrict the number of potential decision alternatives. The selection process is simplified and facilitated for both the supervision authority and the inspector and restricts the possibilities for authorities to choose other inspection themes – if you do one thing, you cannot do another.

Budget documents and special commissions from the Government involve selection being done on the political level, while annual activity plans involve the management of the central agency selecting a few themes that are considered to be of particular importance. In order to answer the question of whether priorities set by political and administrative management are based on knowledge of how different measures influence goal achievement, it would be necessary to conduct studies of the policy making process. In addition to expert knowledge, other factors can be used to explain why agencies and inspectors set the priorities they do. A decision design is a function of, among other things, political compromises, actions of interest groups, crises, the strength of professional groups and reports in and by the media (Hill 2005).

Characteristic of several of the selection criteria is that they are reactive. The authority or the inspectors choose themes on the basis of information received from others. They do not analyse information they have collected independently. Tips from other authorities and reports in the media are examples of such criteria. Events and environmental problems in the media are often related to specific serious accidents. It is not unusual for the media itself to act reactively under pressure from stake-holders, interest groups, public authorities and private companies (Asp 1986). From the perspective of environmental efficiency, this kind of selection can be problematic, specifically when environmental problems of lesser importance are chosen for inspection.

Even if the most important themes from the perspective of environmental efficiency are selected, the degree of goal achievement in the next step will depend on which inspection objects are selected within the prioritised themes. In the interviews, eight selection criteria can be identified: random selection, what you can find, geographical site, hearsay, long time ago, piecework, self-prioritisation and previous problem kids. Each criterion can be described as a way of routinising the selection process. The priorities are set in relation to decision patterns that both simplify decision making and minimise the time spent on risk assessment in relation to goal achievement. As with the prioritisation of themes, several of the selection criteria can be characterised as reactive.

Self-prioritising, such as in emergency cases, involves issues that the authority or its inspectors must give priority to, while hearsay is something that the inspectors themselves use to set priorities. In both cases, inspectors react to information provided by others and use this information to make a judgement. A central characteristic of street-level bureaucrats is that they usually have to deal with a lot of emergency cases. Street-level bureaucrats can, according to Lipsky, use emergency cases as a power resource when they negotiate over workloads and tasks with the administrative management. Since it is difficult to anticipate the number of future emergencies, is it not unusual for street-level bureaucrats to use emergencies as an argument for reducing their own normal workloads (Lipsky 1980). The knowledge of street-level bureaucrats of their organisation and its activities is normally greater than that of the management, which is why information asymmetry – a characteristic trait in the principal–agent dilemma – is present in the negotiations. Since the knowledge of agents exceeds the principals, the agents, according to the theory, can decide what information she wants to provide to the principal i.e. management. The principal is, on the other hand, dependent on the information she receives in order to make decisions (Pratt och Zeckhauser 1985). Emergency cases clearly exemplify the steering problem in a principal–agent relation.

One common method used by management in order to limit, direct and control the discretionary power and autonomy of street-level bureaucrats is work by contract (Lipsky 1980). When this method is used, employees must complete a certain specified number of matters within a specific time. A counter–strategy among street-level bureaucrats when subjected to this kind of method is to shift their focus from complicated and time-consuming matters to matters that can be completed quickly and easily. Within both SWEA and SCI work by contract is practiced on an individual and a collective level – 'piecework' is the phrase the interviewees use when referring to work by contract. Inspectors can, for example, receive one 'piece' for an inspection or a requirement they make on an inspection object. One way to live up to the contract is to accomplish many easy and simplified inspections, especially if there is a risk that the stipulated measures will not be reached. By doing so, the inspectors can create more leeway for their discretion and autonomy. However, at the same time the risk that the policy outcomes will deviate from expected and politically decided on outcomes increases.

We can, thus, conclude that the routines for prioritising used to choose both inspection themes and inspection objects can be characterised as reactive, while each of them can also be understood as a reaction and adjustment to command and control systems as well as to the working conditions of the inspectors.

Since the empirical data is qualitative, it is not possible to draw any definite conclusions as to which selection criteria are most commonly used by SCI and SWEA, respectively, and their inspectors. However, it is possible to draw the conclusion that both groups of inspectors use simplified decision models when making their prioritisations and that these models can be understood as reactive rather than proactive. Furthermore, chemical inspectors seem to have more influence over the selection of themes, while labour safety inspectors seem to have more influence over the selection of objects. These variations can be explained by differences in organisational structure, educational background and the number of inspectors employed.

18.6 Compliance Strategies Used by Inspectors

The compliance strategies used by inspectors when carrying out inspections are developed in relation to the bureaucratic control systems and the organisational conditions within their own authorities (May and Wood 2003). In order to foster compliance by the inspection objects, the inspectors employ different methods of control and persuasion to achieve compliance. From their position as street-level bureaucrats, is important to achieve compliance with as little 'work as possible', thus increasing their autonomy.

To both chemical and labour safety inspectors, the most important result of an inspection is that the inspection in itself generate an interest among those inspected to continuously improve their work within the inspection area and not only to fulfil the requirements made during the actual inspection or to follow existing regulations. The most desirable result is that the inspection objects take steps and measures that go far beyond existing regulations, that they become more dedicated to the policy goal than the law prescribes. In order to accomplish this kind of result, it is necessary to use strategies that promote confidence and respect for the authority and its inspectors among those inspected. In order to reach a positive result, i.e. to promote observance of existing regulations and measures that go beyond, the inspection process must generate both trust and respect. Acting as 'police' is something that should only be considered when no other methods yield results. The actual inspection process consists of three phases: the contact phase, the visiting phase and the assessment phase. The inspectors constantly consider the pros and cons of different combinations of soft and hard methods in each of these phases, but the actual choice of strategies varies depending on exiting regulations, inspection theme and overall inspection methods. Inspectors adjust strategies to how they understand the trust on the part of the inspection objects in the authority and their willingness to follow regulations. The inspectors also make a kind of 'mental diagnosis' of the inspection objects that guides them in their treatment of them. The inspection process and the different phases are similar in the two authorities studied. The same is true regarding the view of the inspectors of how compliance ought to be achieved. Both groups prefer soft methods, such as education, guidance and persuasion. The differences between how the inspectors actually go about choosing between compliance strategies rests above all in the regulatory framework and the consequences this has for the inspectors in the assessment phase. The chemical regulations are more detailed and precise and regulate, for example, situations in which inspectors are required to file police reports. Injunctions represent another course of action that is to be used when inspection objects fail to observe the law. Together, these circumstances make it more difficult for chemical inspectors to use soft methods, even if they would prefer to do so.

The legal framework for labour safety is different. First, it is to a greater degree open to interpretation. Second, injunctions are not the first course of action that inspectors are to use if inspection objects fail to observe the law. The first course of action to take in such a situation is rather to write an 'inspection memorandum' in which the inspector describes in what respects a workplace does not live up to existing regulations. The workplace is requested to present in writing on no later than a given date a description of the measures it intends to take in order to conform to the regulations. The contents of this memorandum cannot be appealed. If the workplace provides satisfactory answers within the time limit, the inspection is terminated. Injunctions are only issued if the answers are unsatisfactory or if there is no answer at all. Workplaces are usually given more than one chance to respond to the requirements in an 'inspection memorandum'. In addition to differences, there is another factor that makes it easier for labour safety inspectors to adopt situation-specific soft compliance strategies. Labour safety inspectors are often regarded by employees as their representatives, and it is common for employees to contact and request an inspectorate to make an inspection of their workplace. This is not the case with organisations working with chemicals. In these cases it is more common for enterprises to report competitors instead of welcoming inspectors to their own enterprises.

In sum, the leeway for discretion among labour safety inspectors in choosing compliance strategies is broader than that for chemical inspectors. However, both groups of inspectors use methods that aim at promoting respect for and trust in the inspection authority, and they try to adopt situation-specific compliance strategies that are based on soft persuasion and negotiating methods.

18.7 Is Chemical Inspection an Effective Policy Instrument?

Inspections can be an effective policy instrument. However, there are several factors that affect their degree of efficiency. This is largely due to the fact that inspectors and inspection authorities: (a) interpret legal rules; (b) set priorities between different types of rules and inspection objectives; (c) make use of situation-specific legal enforcement measures; and (d) have contacts with inspection objects that involve negotiations and persuasion, which are in turn influenced by the approach (style) adopted by the inspectors and by the monitoring system of management. The leeway for discretion on the part of inspectors can, therefore, both promote and counteract efficiency and legitimacy in theory as well as in practice.

Vagueness in regulations creates leeway for various interpretations of rules, and since different interpretations increase the possibility that rules are applied differently by different inspectors and by different inspection units within an authority, this can be problematic. The problem is double-edged. If the implementation of regulations is assessed differently in various parts of the country, the legitimacy of the regulations and of the authorities monitoring them can be questioned both by inspection objects and by the public. Many regulations and very detailed regulations can, on the other hand, cause similar problems. In a situation of limited resources, inspection authorities and inspectors must choose which rules to monitor, something that can yield the same result as vagueness in legislation – differences in the implementation of regulations.

The prioritisations of tasks and objectives made by inspectorates and inspectors affect policy outcomes and goal achievement. Several of the selection criteria used in practice in inspection activities can be characterised as reactive and routinised instances of decision making. From the perspective of goal efficiency, the use of such selection criteria can be problematic, since it becomes more likely that the most urgent inspections from an environmental perspective will not be carried out. However, priorities that are problematic from the perspective of environmental efficiency may not be problematic in relation to political efficiency. Political efficiency prevails when citizens consider the priorities made by authorities and their activities as being in line with their own opinions on what they believe to be important and necessary measures to take in order to improve the environment. Measures that might not be the most optimal in terms of reaching environmental goals may, thus, be quite adequate in relation to political efficiency (Lipset 1959). From a different point of view, however, citizens might come to regard the use of reactive strategies on the part of authorities as a lack competence – if authorities are always lagging one step behind. The ability of an authority to determine which environmental problems are most important to solve can, thus, be questioned. Such notions could in the long run weaken the legitimacy of authorities and their inspection activities.

References

Asp Kent (1986) *Mäktiga massmedier*. Stockholm: Akademilitteratur.

Birgersson, Bengt Owe (1975) *Kommunen som serviceproducent: Kommunal service och serviceattityder i 36 svenska kommuner*. Stockholm Studies in Politics 7. Stockholm Unviersity.

Braithwaite John (2006) The regulatory state? In: R.A.W Rhodes, Saraha Binder and Bert Rockman, eds., *Political Institutions*. Oxford: Oxford University Press.

De Bruijn, Hans och Hans Hufen (1998) The traditional approach to policy instruments. In: Peters, Guy and Frans van Nispen, eds., *Public Policy Instruments – Evaluating the Tools of Public Administration.* Cheltenham: Edward Elgar.

Fineman, Stephen (1998) Street–Level Bureaucrats and the Social Construction of Environmental Control, *Organization Studies* 19(6): 954–974.

Förordning (1998:525) med instruktion för Kemikalieinspektionen. Swedish Parliament.

Förordning (1998:900) om tillsyn enligt miljöbalken. Swedish Parliament.

Förordning (2000:1211) med instruktion för Arbetsmiljöverket. Swedish Parliament.

Gray, Wayne and Mary Deily (1996) Compliance and enforcement: Air Pollution regulation in the US Steel Industry, *Journal of Environmental economics and management* 31(1): 97–111.

Gray, Wayne and John Scholz (1993) Does Regulatory Enforcement Work – A Panel Analysis of OSHA Enforcement, *Law & Society Review* 27: 177–213.

Hansson, Sven Ove och Lars Lindblom (2003) *The Evaluation of inspections*. Rapport 2003:3: Arbetsmiljöverket.

Hill Michael (2005) *The Public Policy Process* Harlow: Pearson Longman.

Hood Christopher (2002) Control Bargains, and Cheating: The Politics of Public Service reform. *Journal of Public Administration Research and Theory* 12(3): 309–332.

Hutter, Bridget (2001) *Regulation and Risk. Occupational Health and Safety on the Railways.* Oxford: Oxford University Press.

Hydén Håkan (2002*) Rättssociologi som rättsvetenskap*. Lund: Studentlitteratur.

Johansson Vicki (2006) *Tillsyn och effektivitet. Statliga inspektörers yrkesroller och strategival.* Umeå: Borea.

Jewell Christopher J and Bonnie E. Glaser (2006) 'Toward a General Analytic Framework. Organisational settings, Policy Goals and Street-level Behavior' Administration & Society. 38(3): 335–364.

Kleinman Mark (2002) *A European Welfare State? European Union Social Policy in context.* Hampshire. Palgrave.

Lipset Seymour (1959) 'Some Social Requisites of Democracy: Economic Development and Political Legtimacy' The American Political Science Review 53(1): 69–105.

Lipsky, Michael (1980) *Street-level Bureaucracy. Dilemmas of the individual in public services.* New York: Rusell Sage Foundation.

Luhmann (1993) Risk: A Sociological Theory. New York: de Gruyter

May, Peter. J and Sören Winter (1999) 'Regulatory Enforcement and Compliance: Examining Danish Agro-Environmental Policy' Journal of Policy Analysis and Management 18(4): 625–651.

May Peter, J and Sören Winter (2007/2009) 'Politicians, managers, and Street-level Bureaucrats: Influence on Policy Implementation' *Journal of Public Administration Research and Theory*, 19(3): 453–476.

May Peter J. and Robert S. Wood (2003) 'At the Regulatory Front lines: Inspectors´ Enforcement Styles and Regulatory Compliance' Journal of Public Administration Research and Theory 13(2): 117–139.

MacDonald, Keith (1999) *The Sociology of Professions*. London: Sage Publications.

Magat, Wesly and Kip Viscusi (1990) 'Effectiveness of the EPA's regulatory enforcement: The case of industrial effluenct standards' Journal of Law and Economics 33(2): 331–361.

Majone Giandomenico (1996) *Regulating Europe*. London: Rutledge.

Mac Neil Michael, Neil Sargent and Peter Swan eds. (2002) *Law, Regulation, and Governance*. Oxford: Oxford University Press.

Meier Kenneth J. and Laurence J. Jr O'Tool (2007) *Bureaucracy in a Democratic State: A governance perspective*. John Hopkins University Press

Pierre Jon and Guy Peters (2000) *Governance, Politics and the State*. London: MacMillan Press ltd.

Pratt John and Richard Zeckhauser (1985) 'Principals and Agents: An Overview' in Pratt and Zeckhauser (eds.) Principal and Agents: The Structure of Business. Boston: Harvard business School Press.

Sannerstedt Anders (2001) 'Implementering – hur politiska beslut genomförs i praktiken' in Rothstein Bo (ed.) *Politik som organisation*. Förvaltningspolitikens grundproblem. Stockholm: SNS Förlag.

Schneider Anne and Helen Ingram (1990) 'Behavioural Assumptions of Policy Tools', The Journal of Politics 52(2):510–529.

Setterlind, Rikard (2000) *Offentlig tillsyn enligt miljöbalken*. Stockholm: Norstedts Juridik.

SOU (2002:14) *Statlig tillsyn – Granskning på medborgarnas uppdrag*. Stockholm: Frizes.

SOU (2004:100) Tillsyn. *Förslag om en tydligare och effektivare offentlig tillsyn.* Stockholm: Frizes.

van Meter, Donald and Carl van Horn (1975) The Policy Implementation Process. Administration and Society, Vol. 6.

Vedung, Evert (1998) *Utvärdering i politik och förvaltning*. Lund: Studentlitteratur.

Weil, David (1996) 'Assessing OSHA Performance: New evidence from the Construction Industry' The Rand Journal of Economics 27(3): 618–641.

Winter Sören, Peter Dinesen and Peter May (2007) 'Implementation Regimes and Street-Level Bureaucrats: Employment Service Delivery in Denmark' Paper Prepared for the Public Management Conference Tucson, Arizona 25-27 Octorber

Woodside Kenneth (1998) 'The acceptability and visibility of policy instruments' in Peters Guy and Frans Nispen (1998) Public Policy Instruments – Evaluating the Tools of Public administration. Edward Elgar: Cheltenham.

Part III
Conclusion

Chapter 19
Regulatory Futures in Retrospect

Willem Halffman and Roland Bal

Abstract In our 1998 volume *The Politics of Chemical Risk: Scenarios for a Regulatory Future* we envisioned four ideal-typical scenarios for the future of European chemicals policies. The scenarios focused on the nature of expertise (seen either as a universal or a localised phenomenon) and the organisation of the boundary between science and policy (as either diverging or converging). The four scenarios were titled International Experts, European Risk Consultation, European Coordination of Assessment, and Europe as a Translator. For all four scenarios, we hypothesised internal dynamics and articulated dilemmas related to the development of the sciences contributing to chemical assessment, the relation between the EU and member states and the role of the public. In this contribution, we look back on our four scenarios 15 years later, to see which ones have materialised and to explore whether the dilemmas we saw have indeed surfaced. We conclude that the International Experts scenario by and large has materialised and explore some of the underlying tensions and dynamics in this development.

Keywords Complexity • Controversy • European Union • REACH • Risk

19.1 Regulatory Futures

In the winter of 1995 we gathered thirty experts in an Amsterdam hotel to get to the bottom of the fundamental misunderstandings over chemical risk regulation. Half of the experts were natural scientists working on chemical risk issues: toxicologists,

---

W. Halffman (✉)
Department of Philosophy and Science Studies, Radboud University Nijmegen,
P.O. Box 9010, 6500 GL Nijmegen, The Netherlands
e-mail: w.halffman@gmail.com

R. Bal
Institute of Health Policy and Management, Erasmus University Rotterdam,
P.O. Box 1738, 3000 DR Rotterdam, The Netherlands
e-mail: r.bal@bmg.eur.nl

J. Eriksson et al. (eds.), *Regulating Chemical Risks: European and Global Challenges*,
DOI 10.1007/978-90-481-9428-5_19,

regulatory risk assessors, experts from regulatory agencies. The other half were social scientists who had been studying regulatory decision making, trying to analyse the frames and assumptions underlying regulation, and in some cases even challenging those assumptions, academically or through activism.

We were young and our plan was ambitious. Both of us were doing PhD research on the organisation of the science-policy boundary in chemical licensing and standard setting, either in occupational health (Bal 1999), or environmental hazards of chemicals (Halffman 2003). We imagined that, in the absence of concrete economic stakes or disagreement over specific substances, we could at least have a meaningful debate about the paradigm of risk assessment, the boundary between science and policy in regulatory regimes, and the world views underlying them. We had also hoped to consider some alternative ideas to organise regulatory regimes. To focus the debate, we aimed at the construction of four scenarios for the future organisation of regulatory chemical hazard assessment. The scenarios focused on the nature of expertise (seen either as universal or imbued with local context, such a national regulatory traditions) and the organisation of the boundary between science and policy (as either sharp and diverging or flexible and converging). The results were published in an edited volume, which included both the presentations and an account of the debates at the event (Bal and Halffman 1998).

We now write almost a decade and a half later. New issues have appeared on the regulatory agenda since: toxicity of nano-materials, the unravelling of BSE, chemical persistence, or endocrine disruption. Meanwhile also, the organisation of regulatory expertise has witnessed some important shifts, such as with the introduction of the REACH scheme and the creation of the European Chemicals Agency, as extensively documented in this book (see Chapters 2, 5, 6, 13, 14 and 17). When the invitation came from the editors of this volume to look back and reflect on how far the regulatory debate has come since, we were keen to make use of the occasion, although our research interests had moved on since.

In this chapter, we will give an account of the issues at stake around regulatory expertise, as we encountered them in our Amsterdam discussions and how we saw possible futures at the time. Second, we want to compare our expectations to regulatory developments since, as reflected in this volume. Last, we will once again try to look forward and explore regulatory futures.

19.2 The Future, As It Was

At the Amsterdam workshop, we presented four scenarios, entitled International Experts, European Risk Consultation, European Coordination of Assessment, and Europe as a Translator. For all four scenarios, we hypothesised internal dynamics and articulated dilemmas related to the development of the sciences contributing to chemical assessment, the relation between the EU and member states and the role of the public (Bal 1998; Halffman and Bal 1998).

The scenarios were based on what we saw as important tensions in regulatory regimes. First, we saw a tension between universalising risk assessment and a growing attention for local conditions of regulation. Risk assessment was clearly moving towards harmonisation, in Europe, but also beyond, as stated emphatically by the representatives of international initiatives (see Chapters 2, 9, 12, 13 and 16). Regulatory experts insisted that they were making progress with shared standards for how to assess chemical hazards, so that companies would not have to repeat permit applications with slightly different testing requirements.

At the same time, there were indications of a counter-tendency, pointing out the need for political choice in chemical hazard assessment, particularly with respect to the definition of what needs to be protected (e.g. all biodiversity of a stream or merely interesting species and their ecological support; see Chapter 6). As the political assumptions at the basis of what seems technical risk assessments were being made explicit, it seemed like they would require rich political institutions to address. Similarly, different economic conditions could be conducive to a continued importance of national governments, leaving at best an intergovernmental logic for limited regulatory cooperation (see Chapter 15).

The second tension we saw was between those who defended a clear-cut demarcation of the work of experts and regulatory policy makers on the one hand, and those who questioned the wisdom of this insistence. This divide ran roughly along the split between natural scientists and social scientists, but the two did not coincide completely.

The argument for a strict maintenance of a science-policy boundary was complex. The regulatory scientists generally insisted on the importance of universalist principles of science and on standardised methodologies as guarantees for regulatory independence. Essentially, they argued, as many still do, that experts should be kept apart from politics to be able to do their job, taking understandable and admirable professional pride in their work. With such a separation, a rational process of risk assessment would be possible, while undesired intervention of superstition, emotion, political passion, or private interest could be kept at bay (see also Chapters 10 and 17).

The counter-argument from the social scientists (and science and technology studies in particular) were also varied. One was that private interest cannot be kept at bay by well intentioned principles, but that industry (to call it by its name) always manages to be present on the science-side of the divide more than other interests. Another argument was that the insistence on a strict separation based on standardised methods leads to the exclusion of un-standardised knowledge of adverse effects. (At the time endocrine disruption was becoming a major issue that was not covered in standard methods). In addition, a strategy to insist on a strict boundary would always face a challenge of hypocrisy, as risk assessment is unable to purify itself completely from all politics, all value-laden assumptions. Rather than to see a universal scientific method, social scientists pointed to the co-construction of methods by both scientists and regulatory agencies together, in fact at the very construction of entire fields of scientific expertise in the shadow of policy agencies such as the US Environmental Protection Agency or the OECD (see Chapters 12, 14 and 16).

The answer of regulatory experts to such issues is that they can show that nothing untoward is going on by being transparent about their work. Rather than to hide behind secretive deliberations that relied on personalised expertise, regulatory experts were beginning to learn to document their evaluations. At the same time, however, governments noticed that publishing an overwhelming amount of technical information can just as much create inaccessibility to regulatory assessments as a total lack of public information. An increasing transparency creates its own intransparancies. In addition, transparency only works if the premises of regulatory action are shared by all involved, which was exactly the problem. To use the example of endocrine disruption again: no amount of transparency could solve the problem that critics wanted this included as a relevant endpoint for testing, while regulatory experts could only point at the lack of standard methodology to assess such effects. Similarly, critics pointed out the chemical-by-chemical assessment, ignoring the accumulated effect of chemicals already in circulation – a fundamental problem that continues to pose a serious challenge to all chemical regulation that considers exposure rates in assessments.

As an alternative, we proposed to look at the relation between experts and non-experts more symmetrically, for example by arguing that the principles by which experts (dis)trust each other are really not that different from the principles by which lay people (dis)trust experts. For example, experts rely on the reputation of other experts to assess whether a claim requires additional questioning, are more critical of a discrepant finding, or a finding that leads to higher costs, just as lay people do. With such arguments, we argued that more dialogical relations between processes of expertise and of value consideration were at least worth considering.

In the scenario International Experts, we extrapolated the development of standards harmonised on a European level, organised around a strict boundary between science and politics. We saw this to imply the creation of a European chemical risk assessment agency, effectively internationalising expertise, fixing lingering differences in style and approach between EU countries. This is the scenario that most resembled the development of EU chemical regulation since (see Chapters 10 and 17). Even though, with hindsight, this may seem like a necessary development, at the time it was not. It seemed at least as likely that European chemical assessment would remain under national control, the European Coordination of Assessment scenarios, so that countries maintained national resources to perform trusted assessments. Harmonisation would then have gradually proceeded through increased coordination of procedures, further protocolisation, and increased expert cooperation.

Our alternatives that suggested a more flexible attitude about the science-policy boundary seem further away from actual developments. In 'Europe as a translator' we saw a regulatory role for Europe as a mediator between chemical regulation and debate as they were differently framed in various countries. In the last scenario, 'European risk consultation', expertise shifts to a European level, but provisions are made for controversy through consultation with national experts. Assessment is routinised where possible, but evaluation details are publicly available, allowing for a shift of assessment to more consultative procedures.

On paper, it seems as if the 'International Experts' scenario has become the reality. However, the tensions we identified in this scenario remain present. The key element of trying to maintain a strict boundary between risk assessment and risk management is that reason is mobilised to tame the beast of politics. The assumption is that we can define the rules of the reason game beforehand (test standards, assessment protocols, expert decision rules) and then hold all parties involved to these rules once the game is played. However, no game is played like that. Players contest the interpretation of the rules, the application of the rule to a particular case, whether a particular case is subject to the rules, will try to question the rules, change them, and – if all else fails – undermine them, sabotage implementation, or just refuse to play. For most players, the foundational rule that risk assessment must be separated from risk management can become just as questionable as the other rules. Even though chemical regulation driven by expert risk assessment may 'fix' risk regulation *in principle*, the question remains whether they also do in practice.

19.3 New Key Tensions in Chemical Regulations: Controversy

The key test for REACH is not whether it will be able to assess thousands of chemicals. The real question is whether it will withstand escalated conflict over a handful of lucrative or strategically important chemicals. Risk assessment procedures in the past have been notoriously bad at anticipating and accommodating contestation. When we brought toxicologists and social scientists together at our meeting, the fiercest debates (and the most fundamental misunderstandings) were about the framing of risk research problems, rather than excessive influence or malpractice in regulatory assessments.

Let us pay some extra attention to this point, for it is an extremely important one. Basic regulatory contestation accuses regulators of inappropriate sympathies, intentional bias, or even outright corruption. The large majority of regulatory experts are of good faith and would equally condemn such practices. Most contestation of regulatory expertise is not of this base level. Rather, citizens ask questions such as: Are you sure that your findings also hold for children and pregnant women? Have you actually measured that in the soil, or is this a model? What will happen to this substance if it should escape from your laboratory (even if you claim it never will)? Such questions ask whether the risk assessments have been appropriately framed, whether an assessment protocol (by necessity a simplified representation of the world) covers all the relevant processes in the world, who has decided which processes should be considered relevant, and what kind of certainty is appropriate for action to be taken. The GMO debate offers a nice example: for the opponents the question is not just what GMOs will do in the soil or in human bodies, but also what they will do to the power relations between farmers and seed producers, an issue well beyond the expertise of your hard-working, honest, and sincere toxicologist.

Chemical risk assessment is a means to contain such conflicts. It defines a series of relevant endpoints and exposure processes in the world that we will accept as relevant, while others (such as changing property relations) are not. The boundary between risk assessment and risk regulation is not just a boundary that keeps dirty politics out of disinterested science, but also a boundary that prevents new concerns from making the assessment process unpredictable, for applicants, policy makers, as well as environmentalists. Such new concerns can be new health or environmental concerns, but also new arguments for mitigating factors or alleged over-estimation of risk. The alternatives to this strict boundary are basically propositions that try to find conditions for a reasonable debate about what should be considered relevant consequences. Chemical controversies are not just about whether the facts are right, but more interestingly also about whether these are the right facts.

The question for REACH is whether it will be able to deal with these framing issues (on framing, see Chapter 4). In the next section we discuss some underlying tensions that question the ability of the REACH framework to do so in a sustainable way.

19.4 New Key Tensions in Chemical Regulations: Complexity

Looking at the development of chemical regulation with the benefit of distance, some key developments can be identified. Most striking is the growing complexity of the regulatory regime. REACH involves more tests, testing more endpoints, and has more complex decision rules (see Chapters 2, 5 and 13). For example, the extension of environmental effects from acute and chronic effects to persistence as a cause for concern shows how regulatory assessment is trying to include more of the complexity around chemicals interacting with the world.

This complexity is not just the result of a regulatory system that tries to mirror the complexity of the world out there, however. From a more political perspective, it also reflects the complexity of the negotiations around chemical regulation. Environmental concerns have insisted on new endpoints in the micro-politics of technical meetings and grass-root campaigning, such as with endocrine disruptors. Industrial organisations have argued for flexibility, regulatory restraint, or even de-regulation, such as with low volume chemicals (see Chapter 16). Different EU member states have argued the case for 'their' industries (see Chapter 15). The complexity of the regulatory regime therefore also shows the complexity of the compromises that seemed necessary to prevent conflict escalation.

At the base of this approach lies a particular strategy for dealing with chemical hazards. This strategy is utilitarian at heart, as has been argued poignantly by philosopher and activist Anne Chapman (Chapman 2007; Halffman 2009). The utilitarian logic states that the free enterprise of actors will lead to a collective good, provided that unwanted consequences ('externalities') are contained, typically through rules. In other words: the basic premise for chemical regulatory regimes is that chemical producers are free to pursue their commercial activity, provided these activities do not cause unacceptable harm, as far as specified in general rules that are policed, ultimately, by the state. The rules have to be general, as they have to create a 'level playing field' for all players, that is: equal competitive conditions in an international chemical market (see also Chapter 15 and 16). The question of which harms are unacceptable and should be caught in general rules, has dominated the EU debate over chemical regulation, for example on the issue of whether mere persistence (rather than toxicity) constitutes harm.

The growing complexity of regulatory regimes is a problem in itself (see Chapter 13). First, the complexity of regulation can create a distorted playing field, where only companies capable of maintaining a large staff of regulatory specialists can participate. Rather than creating a free market, complexity can become a means to corner the market and avoid competition. For these reasons, the largest players in a market with complex regulation may even support increasing regulation, as it allows them to take control. The objection is a recurrent one, as smaller firms are concerned about the ability to compete, or small innovators, such as in the ecological pesticides sector, fear they will not be able to pass the regulatory post. Regulations that claim to create a level playing field can thus become self-defeating.

Growing complexity also implies a threat of growing regulatory cost, especially as complex regulation may make it hard to assess regulatory overheads beforehand. The objection has recently also been raised against REACH, not only in terms of financial cost, but also test animal lives (Hartung and Rovida 2009), but is one that has been made since the birth of regulatory regimes, usually from the side of industry. In the markets for pesticides, but also pharmaceuticals, it is often claimed that high regulatory costs stifle innovation or reduce profitability. From a public perspective also, growing complexity raises collective costs, through the need to maintain regulatory agencies, monitoring, and enforcement (see Chapters 16 and 18).

To a certain extent, such objections have been met: with exceptions for small volume chemicals to allow innovation, public regulatory costs being reclaimed through registration fees, with support for innovation through public research funds, or with attempts to harmonise chemical regulation internationally, such as through the creation of a European regulatory agency, rather than to rely on many national ones. However, such solutions have also created additional complexities, exacerbating rather than solving some of these problems.

The growing complexity of chemical regulation also constitutes a problem for the democratic order. The claim that the fairness of regulatory policies is guaranteed by a public, transparent process becomes more and more unreal as this process itself is so complex that only a small group of experts can really understand the intricacies of the regulatory filigree. Even 15 years ago, experts in regulatory agencies were already worried that they were outmatched by the intricate knowledge of regulatory procedure of industry experts. Citizens are therefore increasingly dependent on third parties and their experts in order to assure them that just decisions are being made. At the same time, inexplicable regulatory issues become less and less appealing to the public spaces where citizens discuss the collective good: how many journalists would dare tackle a regulatory decision under REACH for the evening news? (On the news media and chemical regulations, see Chapters 3 and 4.)

Regulatory complexity also reduces the possibility for effective policy intervention. Complex rules create new loopholes, new possibilities for delay, exception, exemption, objection, appeal. In this respect, chemical regulatory policy has an appalling track record. Effective regulatory action on a chemical of major significance typically takes several decades between early warning and effective policy action; see DDT, PCBs, dioxins, asbestos, phtalates, or anti-fouling paints (see Chapters 8 and 11). It is quite disconcerting that it does not seem to matter much whether a refined regulatory apparatus is in place or not. Most major cases eventually require ad-hoc policy intervention anyway, with specific compromises and measures, specific regulations, and large amounts of non-standard research into unanticipated chemical effects through unforeseen pathways (see Chapter 10 for an example). A cold-hearted look at the history of chemical regulation has to ask why such a complex regulatory apparatus has lead to so few effective regulatory actions, especially with respect to chemicals already in use.

Lastly, complexity itself can become a source of hazards, as Charles Perrow has showed for complex technological systems (Perrow 1999). Where complexity undermines overview and an ability to see how rules affect the concrete life of the regulated subject, regulation can become its own worst enemy, creating a false sense of security. There are also political risks in complexity. Contradicting rules may undermine effective regulatory action, while the resulting absurd regulatory effects are easy pickings for populist anti-regulatory rhetoric that may delegitimise risk policy as a whole. (Health and safety regulations are notorious for unintended contradictions in regulatory requirements; e.g. the floor that has to be both smooth for hygiene and rough so as not to be slippery). In the US, the populist use of regulatory costs and unintended regulatory consequences has contributed to undermine effective regulatory innovation for the last decade (see Chapter 14 for a comparison of the US Toxic Substance Control Act and the REACH programme). Thus regulatory complexity might create conditions for a regulatory backlash.

Regulatory complexity is likely to increase further (see Chapters 2, 12 and 13). One of the major gaping holes in chemical risk assessment is substance interaction. We still largely assess hazards of chemicals one by one, ignoring the fact that, for example, toxic chemicals are added to a world where a lot of toxic chemicals circulate already (see Chapter 11). We thereby ignore additive effects, let alone mutually reinforcing effects. There are a few exceptions, such as greenhouse gases, such as carbon dioxide, methane, or nitrous oxide. We have learnt to express greenhouse gas effects in terms of their warming effect. That is: we assess risk from the perspective of the receptor, in this case the atmosphere. Environmentalists have long argued that something similar should be done for chemicals and concerns have been raised about cumulative neurotoxic effects of organophosphates. Experts have already recommended to look at toxic equivalents in food (Committee on toxicity of chemicals in food 2002; Gezondheidsraad 2002; Health Council of the Netherlands 2004). Similarly, in the past, we have had to group organochlorine pesticides because of similar effects, dioxins, or man-made fibres. Even if we were to identify a handful of key receptors in the human body, a handful in the physical environment, and a handful in the biological environment, the re-assessment of all chemicals from the perspective of these receptors would present an almost impossible task.

Complexity in chemical regulation is also likely to increase because the forces that drive it are still there. Experts continue to research new pathways, continue to develop new tests, or further refine exposure models. The main purpose of maintaining a competitive internal chemical market requires general rules, rather than case-by-case judgement. Every compromise over assessment procedures is embedded in new protocols. Furthermore, societal, including industry interests (see Chapters 15 and 16), are bound to further drive the complexity of regulation as each new controversy tends to add new rules to existing ones.

It would be wrong to see our argument as a pro-industry plea for deregulation. First, because regulatory simplification is not necessarily pro-industry. In fact, the most radical de-regulation proposal on chemicals was made in circles of Greenpeace, suggesting that perhaps we have enough chemicals already and we should consider a ban on new chemicals. Second, de-regulation would be meaningless as long as there is no significant alternative to deal with the hazards created by the growing human circulation of chemicals through the world, man-made or not.

The growing complexity of regulation may be presented as an increasing correspondence between the complexity of the world and the complexity of assessment protocols, but we would argue that there is a limit to this development.

19.5 Conclusions

In this reflection on the developments of and in chemicals regulation in Europe since we wrote our scenarios more than 10 years ago, we have argued that the course of the developments has by and large followed the International Experts scenario that we then sketched. The separation between risk assessment and risk management has hardened in this period (see Chapters 9, 10 and 17) and the regulation of chemicals has increasingly become an international, European, rather than a local affair (Chapter 12 and 13). We have also argued that some of the key tensions that we then saw resulting from this scenario are still present. Framing issues remain contentious and the increasing complexity of regulation proves to be transparent only for a few experts, thus threatening the democratic character of chemicals regulation. Whether this will result in a backlash in the future remains to be seen, but the conditions for this are most certainly present.

References

Bal, R. (1998) Boundary Dynamics in Dutch Standard Setting for Occupational Chemicals. In R. Bal and W. Halffman (Eds.), *The politics of chemical risk: scenarios for a regulatory future* (pp. 75–103). Dordrecht: Kluwer Academic Publishers.

Bal, R. (1999) *Grenzenwerk: Over het organiseren van normstelling voor de arbeidsplek* (Vol. 6). Twente: Twente University Press.

Bal, R., and Halffman, W. (Eds.). (1998) *The Politics of Chemical Risk: Scenarios for a Regulatory Future*. Dordrecht: Kluwer.

Chapman, A. (2007) *Democratizing Technology: Risk, Responsibility and the Regulation of Chemicals*. London: Earthscan.

Committee on toxicity of chemicals in food, C.P.A.T.E. (2002) *Risk assessment of mixtures of pesticides and similar substances*. London: Food Standards Agency.

Gezondheidsraad. (2002) *Blootstelling aan combinaties van stoffen*. The Hague: Health Council of the Netherlands.

Halffman, W. (2003) *Boundaries of Regulatory Science: Eco/toxicology and aquatic hazards of chemicals in the US, England, and the Netherlands, 1970-1995*. Unpublished doctoral thesis, University of Amsterdam, Amsterdam.

Halffman, W. (2009). Republican Chemicals (Review of Anne Chapman, Democratizing Technology, London: Earthscan 2007). *Journal of Risk Research, 12*(7–8), 1035–1037.

Halffman, W., and Bal, R. (1998) The politics of chemical risk: Scenarios for regulatory policy. In R. Bal and W. Halffman (Eds.), *The Politics of Chemical Risk* (pp. 303-322). Dordrecht: Kluwer Academic Publishers.

Hartung, T., and Rovida, C. (2009) Chemical regulators have overreached. *Nature, 460*(27 August), 1080–1081.

Health Council of the Netherlands. (2004) *Pesticides in food: assessing the risk to children.* The Hague: Health Council of the Netherlands.

Perrow, C. (1999) *Normal Accidents: living with high-risk technologies.* Princeton: Princeton University Press.

Index

MIX
Papier aus verantwortungsvollen Quellen
Paper from responsible sources
FSC® C105338

If you have any concerns about our products,
you can contact us on
ProductSafety@springernature.com

In case Publisher is established outside the EU,
the EU authorized representative is:
Springer Nature Customer Service Center GmbH
Europaplatz 3, 69115 Heidelberg, Germany

Printed by Libri Plureos GmbH
in Hamburg, Germany